WELL-BEING AND QUALITY OF LIFE - MEDICAL PERSPECTIVE

Edited by **Mukadder Mollaoğlu**

Well-being and Quality of Life - Medical Perspective
http://dx.doi.org/10.5772/66728
Edited by Mukadder Mollaoğlu

Contributors

Susana Castaños-Cervantes, Ann-Marie Svensson, Ulla Hellström Muhli, Cristiana Lucretia Pop, Murat Darçın, Mahmut Surmeli, Özlem Çinar Özdemir, Andrea Vranic, Aborlo Nkporbu, Eva Žiaková, Karolína Barinková, Sezen Tezcan, Carmen Cristescu, Maria Suciu, Shawn Y. Wu

Published by InTech
Janeza Trdine 9, 51000 Rijeka, Croatia

Publishing Process Manager Iva Simcic
Technical Editor SPi Global
Cover InTech Design team

First published September, 2017
Printed in Croatia
Legal deposit, Croatia: National and University Library in Zagreb

Additional hard copies can be obtained from orders@intechopen.com

Well-being and Quality of Life - Medical Perspective, Edited by Mukadder Mollaoğlu
p. cm.
Print ISBN 978-953-51-3513-5
Online ISBN 978-953-51-3514-2

For EU product safety concerns:
IN TECH d.o.o., Prolaz Marije Krucifikse Kozulić 3, 51000 Rijeka, Croatia,
info@intechopen.com or visit our website at intechopen.com.

Meet the editor

Professor Mukadder Mollaoğlu graduated from Cumhuriyet University, Faculty of Health Sciences in 1991. She completed her doctorate at Istanbul University Health Sciences Institute. Mollaoğlu especially examined the effects on the quality of life of people with chronic diseases. She also lectures and publishes in her field, focusing on chronic diseases on the quality of life, particularly chronic neurological diseases, chronic renal failure, and endocrine and cardiac diseases. Her doctoral thesis is "Quality of life in people with epilepsy." By doing the validity-reliability study of many international quality of life scales, she has enabled the use of these scales in her country. She is an editorial board member and a reviewer in many medical journals and has several publications in eminent journals as well as books in the field of chronic diseases and quality of life. Professor Mollaoğlu has participated in a number of national and international panels, conferences, and congresses related to quality of life, where she has held various duties such as speaker and session chairperson.

Contents

Preface

QOL has a wide range of contexts, including the fields of international development, healthcare, politics, and employment. When quality of life is considered in the context of health and disease, it's commonly referred to as health-related quality of life. Health-related quality of life is a multidimensional concept that includes domains related to physical, mental, emotional, and social functioning. It goes beyond direct measures of population health, life expectancy, and causes of death and focuses on the impact health status has on quality of life. A related concept of health-related quality of life is well-being, which assesses the positive aspects of a person's life, such as positive emotions and life satisfaction.

Well-being is a relative state where one maximizes his or her physical, mental, and social functioning in the context of supportive environments to live a full, satisfying, and productive life. Health-related quality of life and well-being also reflect individuals' assessment of the impact of their health and functional status on their participation in society.

Most of the chapters in this book are concerned with the quality of life of people with chronic diseases. The quality of life of individuals with chronic diseases is affected in many ways. Improvement on the quality of life is determined in one of the main objectives of chronic disease treatment. To give training and counseling by health professionals is an important process to empower individual patients and their families cope with the challenges of the disease and therefore raise the quality of life. I hope this book can contribute to improving the quality of life.

In this book, besides quality of life of individuals with chronic diseases, the relationship between air pollution, leisure activities, and quality of life is also examined. At the same time, social studies that study the quality of life of homeless girls are also included in the book. Through different perspectives, quality of life has been handled in a very wide range.

Thus, the reader will be provided with up-to-date information and practices about quality of life, well-being, and the influencing factors in this book. I appreciate tremendous effort of the authors to complete this book.

Finally, I would like to thank the writers, Iva Simcic, and the publisher for the valuable cooperation in the emergence of the book.

Professor Mukadder Mollaoğlu, PhD
Cumhuriyet University
Health Science Faculty
Sivas, Turkey

Chapter 1

Physical Activity, Body Image, and Subjective Well-Being

Cristiana Lucretia Pop

Additional information is available at the end of the chapter

http://dx.doi.org/10.5772/intechopen.68333

Abstract

The notion of well-being is in tight relation and sometimes used intercheangebly, with quality of life, physical and mental health promotion, good living, or happyness. Physical appearance is one of the first individual characteristics noticed by others and has an important impact on social interactions and therefore has become very important construct in contemporary societies. The aim of this chapter is to discuss the influence of physical exercise on the subjective well-being dimension related to better health and eventually happiness. In contrast, the physical inactivity determined by the increasing amount of time spent in sedentary activities is becoming an issue with serious consequences, being the cause of more than 5 million death/year globally. Physical activity favorably influences mental health, improves the emotional state and reduces the incidence and severity of diseases and pathological conditions, such as cardiovascular disease, type II diabetes, osteoarthritis, osteoporosis, and obesity. Promoting physical activity, physical and health education and sport as healthy lifestyle components in schools, universities and lifelong learning targets the enhacement of vigour, resilience, employement, and social outcomes for graduates and comunities. Motives of physical inactivity were discussed aiming to underpin possible remedial solution for better health, quality of life and well-being.

Keywords: physical inactivity, weight status, self-esteem, self-image, healthy lifestyle, co-morbidity

1. Introduction

Well-being is a topic in trend in social sciences, even the definition of this concept is not yet established and unitary. The notion of well-being is in tight relation and sometimes used interchangebly, with quality of life, physical and mental health promotion, good living, or happyness.

Quality of life is also the subject of academic debate in economics aiming to measure and compare changes in quality of life within and between communities, cities, regions, and countries [1]. Over the past 40 years grow the idea that economic indicators alone could not reflect accurate the quality of life of populations. New indicators and datasets were created to capture social and environmental aspects that Gross National Product failed to incorporate. Therefore, social and psychological indicators have been developed to assess various facets and dimension of subjective well-being. This included indicators measuring education achievements, health outcomes, and environmental degradation [2]. Major studies on well-being and quality of life are now undertaken by the most important global organizations such as UNO, the OECD, or WHO.

Subjective well-being is related to individual perceptions, opinions, beliefs, cultural patterns, and feelings about own life. The range of variables showing significant associations with subjective well-being includes health, employment status, income and material wealth, education, marital status, social relationships, migrant status, trust in others, volunteering, governance, confidence in institutions, freedom, water and air quality, personal safety, and crime—among others [3]. Still the most important drivers of well-being seem to be considered the income, social connection, and health.

Diener et al. [4] proposed a structure of four groups of concepts that summarize subjective well-being:

1. Positive emotions: glad, strong, proud, determined, interested, etc.
2. Negative emotions: worried, sad, guilty, insecure, angry, etc.
3. Satisfaction of life: satisfaction, fulfillment, sense, achievement, aims, etc.
4. Domain satisfaction: family, career, health, financial status, living conditions, etc. [4].

Well-being develops individually and depends on the attitude individuals evaluate their lives. The psychological aspect of well-being has important repercussions on self-esteem and self-confidence and has a wide range of consequences for how a person deals with society and life. In the last 20 years, psychologists have had a constant regard of self-esteem as a significant psychological predictor for health and quality of life. An important number of studies has linked the self-esteem concept with a wide range of topics from violence and aggression [5] to life satisfaction [6], moderated by age, gender or ethnicity.

Regarding the relationship between self-esteem and body image perception, studies have revealed a preference for Caucasian female samples confronted with Western cultural patterns.

http://dx.doi.org/10.5772/intechopen.68333

A possible explanation is the unconscious association between feminine beauty and tall, slender figures with Caucasian skin promoted on catwalks. In 2007, a very inspired New York University Ph.D. student counted every single model on the runway, and of the 677 models that were hired, only 27, or less than 4%, were non-white [7].

2. Body image: a constant stressor

Physical appearance is one of the first individual characteristics noticed by others and has an important impact on social interactions. Appearance in general and body image in particular have become very important constructs in contemporary Western societies [8]. Body image is not just a cognitive construct, but also a reflection of attitudes and interactions with others. The tendency to link physical attractiveness with positive personal qualities has become a cultural stereotype, not only in western culture, but also globally. The image is powerful, but also superficial, and we are all over surrounded by images. The avalanche of perfect bodies in mass media, advertising, and social media is burdensome to the subconscious, causing people to accept that "what is beautiful is good" with physical attractiveness often being linked with success.

Physical appearance was far less important in earlier times. The ancient Greek ideal was "kalokagathia"—a man refined in mind and body: Physical beauty (kalos) was acquired by exercising in the palestra, practicing agonistic disciplines, while the intellectual and spiritual goodness (agathos) were refined by practicing music, song, dance, rhetoric, and philosophy [9]. In a Hymn to Hygieia (V or VI century BC), it was said that "To have good health is the best for a mortal./Second is to be born handsome in appearance." The link between health and beauty trespasses the centuries through the medieval age. Despite the tones of pink flash painted by Rubens (1577-1640), depicting the presumptive beauty ideal of the epoch, some studies have suggested that a small waist was actually a symbol of feminine beauty, health and fertility. Singh et al, after examining over 7000 documents containing prose, poetry, and drama references to women's physical appearance in XV-XVII century, concluded that "the marker of health and fertility—a small waist—has always been an invariant symbol of feminine beauty" [10], not only in the European countries, but also in the Indian and Chinese cultural spaces [17].

In the middle of the nineteenth century, society admired overweight people. Generally, those in the lower socioeconomic classes would not be overweight due to physical work, walking as their primary form of transportation, and lack of quantity or quality of food. On a contrary, the upper socioeconomic classes viewed excess weight as a sign of success and prosperity. Men of robust proportions were often thought to have correspondingly large bank accounts. As a woman became older, she was almost expected to have a larger figure and extra weight was connected to successful motherhood [11]. Yet, in a very short amount of time, a full figure changed from ideal to unattractive. At the turn of a century, the small waist was again appreciated, and roundness was out of fashion. The twentieth century debuted with new beauty standards for women: a slender, graceful, and healthful silhouette was promoted in motion pictures, on stages, post cards, and illustration.

In present times, the personal physical image is a mean of gaining a distinct place in the real or virtual social environment. In order to achieve this status, investments in body appearance (cosmetics products and procedures, piercing and tattoos, plastic surgeries, sportive material and equipment, etc.) have notably increased for women and men as well.

As they grow up, children are building a picture or image of themselves. This image develops through the things that they can or cannot do and by how other people see them. Poor opinion of our body can cause low self-esteem and self-confidence. An important contribution in constructing the youth's body image has the media. Constantly, watching "perfect" bodies can feed their insecurities over attractiveness and weight. Studies show that idealized body image contributes to eating disorders as anorexia nervosa or bulimia, steroid use, protein supplements [12] and even plastic surgery.

Before and in parallel with formal education, children acquire life habits in family, tending to adopt the example of their parents. Therefore, parents have an important responsibility in promoting a healthy lifestyle in family and thereby giving a good example to their children. The effect of family life style is tracking more then one generation. In a similar way, overweight and obesity tend to run in families. There are 50% chances for a child of being overweight if one parent is overweight or obese, and if both of parents have weight problems the chances will increase to 80%.

During adolescence girls, more than boys, have particulary concerns about weight, body shape, and self-image. There is scientific evidence that body image is experienced negatively by the majority of women and girls [13]. Many are dissatisfied with their body size and weight because slimness is seen as the desirable standard or the beauty pattern especially for young women. In adulthood, also the underweight is much more prevalent among women compared with men [14].

Disturbed self-perception is usually associated with preoccupation, insecure attitude or seeking reassurance in peer's opinion. An explanation for this insecurity is the media and fashion industry's controversial promotion of underweight models and unrealistic imagery, which in time created the Western cultural pattern. The question which arises is how could this post-modern society to overcome the twenty-first century frustration, depressions, anxieties, and psychoses when natural beauty has been annulated by botox's dictatorship [15], the cult of anorexia and emaciated super models?

Media prizes way ahead the relationship between health and beauty than the relationship between health and well-being. Health standards have changed with culture's definition of the attractive figure [16]. The manipulative use of technology in advertising creates unrealistic images of ultra-thin women bodies and muscular, fit males. In the same time, the contemporary society creates a perfect paradox promoting beside those idealized bodies personifying everlasting youth and beauty, a physical effortless life style in a culture of abundance.

3. Body image and subjective well-being

In my research, I have found a significant correlation between self-body image perception and self-esteem, mediated by weight and subsequent by fat deposits. Results indicates a consistent statistically significant correlation between body mass index (BMI) and body dissatisfaction ($r(158) = 0.56$, $p < 0.0005$), with a prevalence of 79% of body dissatisfaction on young women [17]. Even 87.7% of the subjects had a BMI that place them in normal and underweight category, most of them, 66% wanted to lose weight, for an ideal, slimmer silhouette. I found as well a significant negative correlation between body image dissatisfaction and self-esteem; the higher the dissatisfaction, the lower the self-esteem: $r(158) = 0.56$, $p < 0.0005$ and also a significant correlation between health perception and self-esteem ($r = 0.36$; $p < 0.005$). Health perception and self-esteem are variables incorporated in subjective well-being concept.

These data confirm the thesis that landmarks that society promotes are very severe for most girls and young women and put them in a position of inferiority, repercussions on self-esteem and self-confidence. These conclusions are confirmed by others studies' results investigating women samples, generally students, from divers cultural and ethnic samples. Jaworowska and Bazylak have found in a sample of Polish female students almost the same percentage (65.6%) of body dissatisfaction [18].

In Brazil, women seem to be more indulgent with their appearance, whereas the prevalence of dissatisfaction reported was 47.3% [19]. Among a Pakistani students, sample body dissatisfaction was complete: 45.6% of the respondents perceived their self-body as too heavy, while 54.4% perceived it as too thin [20]. In a different cultural environment, in Saudi Arabia, among a group of young female students aged 21.02 (±1.48 SD) with a BMI mean value of 22.79 (±4.71 SD) 26.4% reported to be satisfied, 18.6% perceived self-body as too thin, while 55% perceived as being too heavy and wanted to lose weight [21]. These results confirmed the percentage obtained in my study: 21% satisfied, 13% wanted to gain weight, and 66% wanted to lose weight.

There are scientific studies, which are linking women's body dissatisfaction and low self-esteem for physical appearance [22]. This perception can have repercussions over lower self-esteem especially in teenagers and emergent adults, who are still on the education period, but right before important life choices as a working place or a life partner. A systematic review concluded that control weight and behavioral interventions could be successful by boosting self-esteem and increasing satisfaction with body areas too [23].

Speaking about the subjective young women's body image perception, we conclude that it surpass the normal anthropometrical references regarding average body mass. Under the social and cultural patterns pressure, a majority of women are not satisfied with their body shape regardless the BMI values and identifies their ideal body with a thinner version. Even underweight persons are willing to lose weight in order to "over adapt" to an expected physical attractiveness standard. The normality for women under 25 years of age seems to be a constant preoccupation for losing weight and an ultrathin body is their beauty ideal.

By prizing women's physical attractiveness, western society encourages them to evaluate their social value in terms of image and also perpetuates this societal objectification through continuous cultural scrutiny, the strengthening of negative stereotypes and prejudices against overweight people. Among women, social and cultural context shape a self-critical orientation toward their physical appearance that is manifested in certain comparison tendencies associated with negative self-body perception [24], anxiety, eating disorders, social reluctance, and depression.

Integrating physical and health education in overweight preventative strategies would have effect in reducing the occurrence of physical and emotional disorders and co-morbidities associated with these later, over the lifetime and could offer an alternative to this eternal and quasi-globalized body dissatisfaction.

4. Physical inactivity

If 5 years ago we were worried because physical activity among European children tends to drop significantly between the ages of 11 and 15 years and only 20% of them exercise regularly, nowadays it is certain that physical inactivity accounts for more than 5 million deaths each year globally [25]. The estimated proportion of mortality due to physical inactivity ranges from a high of 19% in Malta to 1% in Bangladesh. Self-reported physical activity levels vary substantially around the world as well, with six countries reporting 90% or more of the adult population reaching the 150 min/week recommended by the World Health Organization and 16 countries reporting 40% or less of the population meeting the physical activity recommendation [26].

The average percentage of mortality due to physical inactivity is estimated to 9% globally. The European average is equal with the calculated rage of 9% worldwide, but there are important differences between the lowest and the highest values of European countries in estimated proportions of mortality due to lack of physical activity, as could be seen in the following graphs (**Figures 1** and **2**). Observing data distribution is hard to discern a pattern; countries

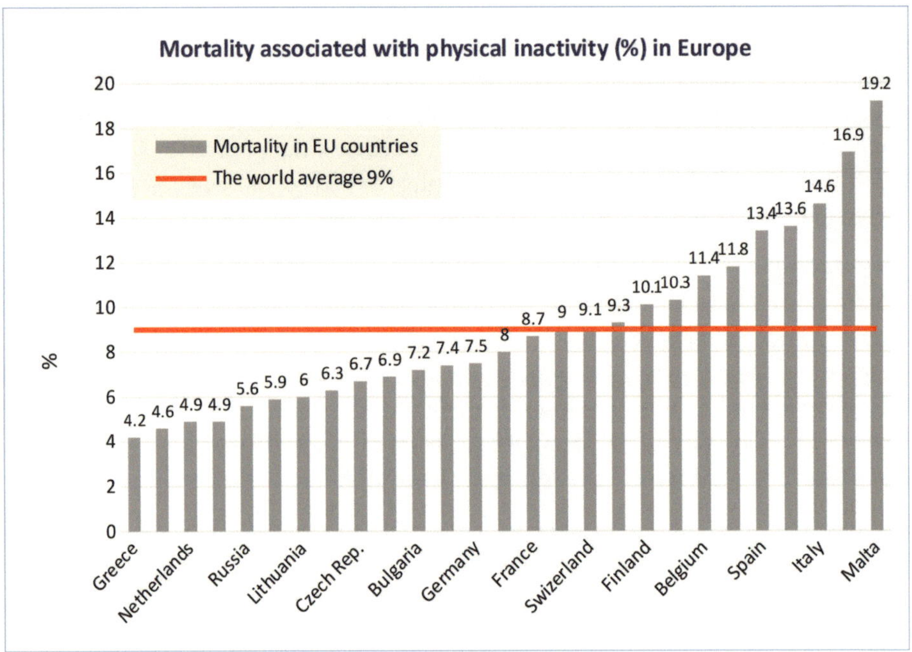

Figure 1. Mortality associated with physical inactivity (%) in Europe.

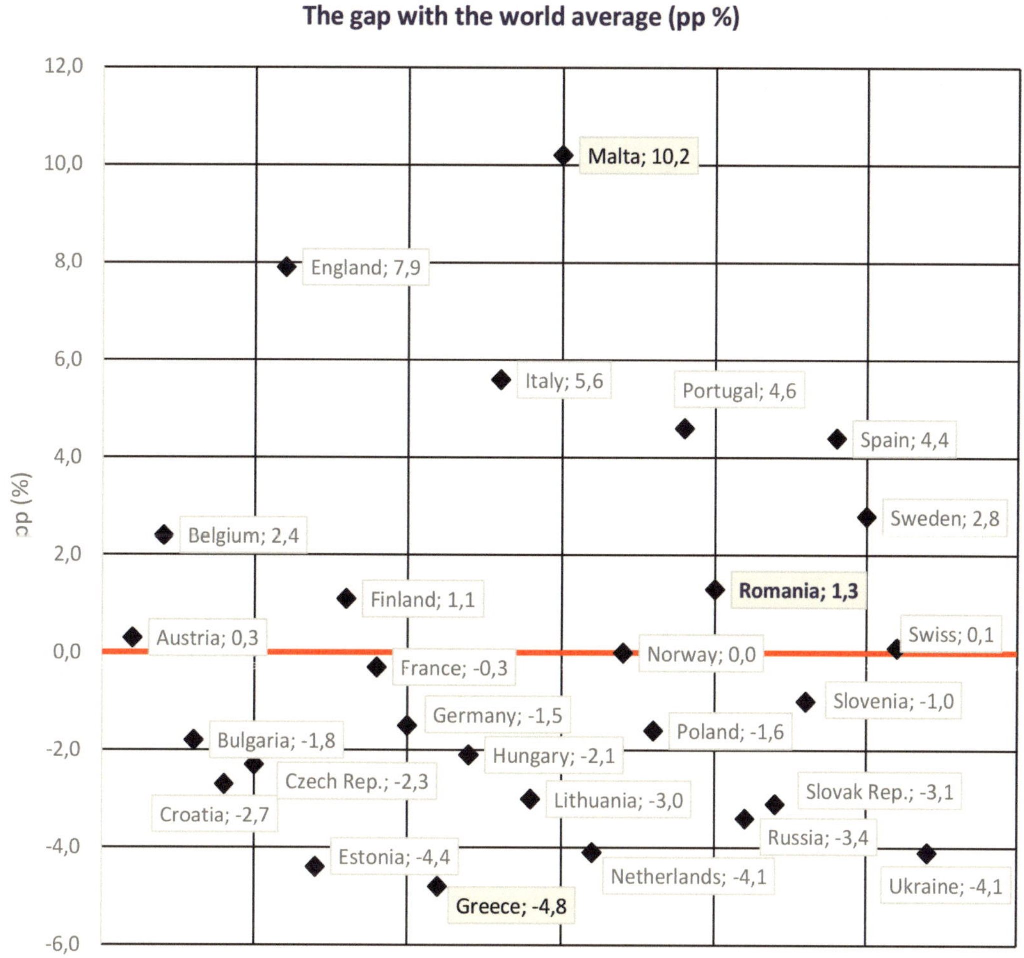

Figure 2. Mortality associated with physical inactivity (%) in Europe comparing with world average.

from same geographical region, with same economical status have different ranking. Greece and Estonia have the best ranking, while Malta and England are the European countries with the highest percentage of death due physical inactivity.

According to a report from the Organization for Economic Cooperation and Development (OECD) entitled "Health at a Glance: Europe 2012," obesity in Europe has more than doubled over the past 20 years in most EU countries for which data were available. In addition, the report explains that physical activity among European children tends to drop significantly between the ages of 11 and 15 years. Only 20% of children have exercised regularly in 2012 [27], and if the tendency was correctly estimated, in 2017, less European children are involved in regularly formal and informal physical activities and more have a precarious weight status. Physical activity includes all activities which involve bodily movement and are done as part of playing, working, active transportation, house chores, exercise, and recreational activities. Formal physical activities are planned, structured, repetitive, and aiming the improvement or maintenance of one or more components of physical fitness. Whether is about physical education, exercise or training, the main point is the balance between calories intake and energy expenditure.

The globalization of overweight and obesity, the acute perception of being stressed most or all the time, the sedentary life style and the co-morbidities of those behaviors might be the most comun consequences for the new millenium generation and their families. Education and healthcare systems are attempting a weak counterattack to this aggressive epidemic and against the consumption culture which pushes the younger generation toward a greedy lethargy. It is striking that in the United Kingdom between about 1942 and 1947, when very strict rationing was imposed as a result of the Second World War, the British people were probably better nourished than ever before or after. Severe restrictions were put on each family, particularly regarding the amount of meat, butter, eggs, edible fat and other foods of animal origin in their diets. Fruits and vegetables were not rationed. The population benefited nutritionally and mortality rates from diabetes and heart disease were significantly lowered [44].

The Internet is already taking a big share of our time, keeping us seated and staring many long hours (days if we are talking about gamers) at a screen. Everyone fathoms what an important change the Internet had made in their communication, learning, and behaving patterns. We are talking already about a young generation surfing attitude related to learning and working and also about internet addiction. These constant exposures to the Internet have shaped how they search for and acquire information, how they learn and how they socially interact. In a short time, they will not need to memorize any more information, because everything will be accessible in one click. In the same time, it will be a challenge to act without technological backup. Handwriting will be as rare as a handmade lace and calligraphy will became an art like sculpture or ballet.

Making a small imagination, exercise we can figure out how the new technology will affect our daily lives. People already work, shop, pay taxes and entertain themselves online, spending less time for transportation and having less interpersonal interactions. For holydays, virtual travels in a personal paradise is already a project and feeling the breeze or a friend's handshake without leaving the room is an achievable dream. Domestic robots for housekeeping and easy conversations will be available on the Internet (where else?).

Meanwhile, because of Internet addiction, people could suffer a retrogression of imagination, memory, and discernment. The sedentary behavior, the indoor living in absence of sunlight and fresh air, in addition to unhealthy food will transform most adolescents into flaccid, wick adults with narrow shoulders, underdeveloped lungs, limited physical effort possibilities, but really quick in typing. This is not the most optimistic scenario, but in some points is a possible one [42].

The obsessive informatic and communication technology use results in a progresive physical skills decay and less social interaction. Today, when the inteligent phones are shaping our daily life, our bodies are also shaped by bending the neck and back in a "tapping position." Among the technology use consequences could be mentioned the posterior musculature atrophy because the prolonged sitting position and the postural deficiencies associated with "text neck." Bending the head forward and down in a hunched position over a device for typing or gaming causes a higher pressure in the spine. This pressure increases whit every degree of head flexing; at 45°, the head exerts 22.5 kg comparing with 5.5 kg in normal position [28]. Tipping and computer work could cause pain and stiffness in the wrist and fingers of preferred hand. A prolonged mouse using for example could cause carpal tunnel syndrome (CTS). The swollen flexor tendons of the hand compress the median nerve in the wrist and may result in tingling, numbness, weakness, or pain in the fingers or hand. Women are three times more likely to experience CTS than men. This may be because women and obvious children generally have smaller wrists, creating a tighter space for nerves and muscles.

Poor posture can have wide-ranging detrimental effects on our body, the most common being: shoulder, neck, and back pain, degenerative disc dieses, kyphosis, tension headache, restricted breathing, depression, increased stress, and diminished levels of energy [29]. A hunch posture compresses the internal organs restricting their function and making the body appear heavier.

A bad posture has not only physical consequences, but psychological also: an upright, open, expansive posture is associated with power, self-confidence, and good mood. When sitting in a collapsed position and looking downward to a smart phone or other screen device, participants in a study found it much easier to recall hopeless, helpless, powerless, and negative memories, than empowering, positive memories [30].

Constant playing of console and computer games has an influence in reducing the inhibition level of children. They are more self-confident and eager to experience risky situation, similar with virtual characters. This opinion is supported by a study about potential signals for addictive behavior, which reveals that inhibitory control deficit (similar to impulsivity) was significantly related to high intercept levels of both video gaming and high calories and low nutrients intake. Boys exhibited higher levels of both video gaming and HCLN intake than girls [31].

This lower level of self-protection instinct or maybe a rush for excitement generates a new approach in physical activities, especially in urban locations. Leisure time activities have their own place in this contemporaneous trend and are in tight relationship with web cultural elements, exploring new spaces and possibilities. The architectural structures and elements are used by rollers, skate-boarders, bikers, and free runners who use their own body for creating acrobatic, original but also risky movements, and tricks.

5. The benefits of physical activities on subjective well-being

Physical activity is fundamentally important for the maintenance of life functions, and it is an essential part of having a healthy lifestyle, as it has been proven to have a protective role against the development of cardiovascular disease, metabolic disorders, skeletal disorders, and even mental illness.

Researchers and physicians, and even non-specialists in physical education and sport domain recommend regular physical exercises for their substantial and sustainable health benefits.

Practical interventions and scientific studies demonstrate without any doubt the potential positive effect for exercise to improve both physical and psychological well-being. In the psychological well-being category, we can frame the perceptions, opinions and feelings related to body image, health condition, self-esteem, etc., most of them improvable through physical activity.

Scientific evidence suggests that exercise enjoyment is positively associated with body image change. A pleasant and supportive work climate will help people to take part enthusiastically in fitness, aerobic, dance or climbing classes. Working in a friendly group, watching other people exercising, receiving constructive feedback and assistance could motivate people to join a physical activity.

Investigating the potential of music in managing health and well-being is a new research approach, considering human beings holistically, as bio-cultural persons. Music can provide a resource for enhancing well-being, understood as the positive flourishing of identity, relationship and community, regardless of "objective" health status [32]. Among the most popular physical activities, using music support is aerobic gymnastics with its variants (zumba, tae-bo, step aerobics, etc.) dance therapy, Pilates or yoga. Non-competitive aerobic gymnastics' goal is to produce the optimal function of the human body, redefining body image, individual and social identities and facilitate social interaction. It offers a large variety of moves and multiple possibilities for structuring the motion sequences by combining various elements and positions. The variety of technical elements, which are permanently combined in a multitude of dynamic structures, combinations, compositions, rhythms and various choreographies, usually develop in conjunction with a dynamic musical background.

An inspired choice of musical pieces adds an esthetic component to the physical benefit. A controversy regarding the influence of music over body and the overall well-being is provided by Dr. Masaru Emoto's water experiments. His experiments revealed the responsive nature of water to human emotions and music. Since water in human body takes 70% is more likely that music has an effect on us, therefore a good music selection could increase the physical activity effect and restore body well-being. Beneficial effects of music listening on subjective well-being and physical health outside clinical contexts have been reported by a number of reputed researchers also [33–35].

Dance therapy is an alternative way to cure people emotionally and mentally through movement, appealing the sensorial motor system. Using different dance movement dynamics, practitioners become aware themselves of their emotional, mental, and physical immediacy. The philosophy of this therapy is to consider that we are in constant change and adaptation, and the movement is the very thing that reflects this process. The practitioners can observe the body shape or some movement patterns, but the dancer should become aware about his own physical, mental, or emotional malaise. It is this self-awareness that brings acceptance, change and healing and helps them improve their psychic and somatic consciousness of their body image.

In South America, the dance ritually use for healing is a heritage from the pre Columbian ancestors, having important spiritual connotations. In Europe, dance therapy is a practice, with scientific reports, used in improving the mental representations linked to body image among obese patients. The authors have drawn encouraging conclusions: "Obese patients enrolled in the dance therapy workshop displayed a significant improvement in health-related quality of life ($p < 0.03$), body consciousness ($p < 0.001$), and mental representations linked to self-body image ($p < 0.001$)" [36]. Dance therapy is turned out to be a useful means for Parkinson patients to alleviate pain and help them maintain control over increasingly uncooperative bodies.

Dance and movement have encouraging results in emotional disorders caused by domestic violence. In those cases, the reconstruction of own body image is crucial for victims emotional healing. The poor self-image results from the constant critical and negative reflection in their partner abusive behavior. Someone constantly worried about their safety or depressed, instinctively adopts a collapsed posture, tends to look smaller and powerless, and having a defensive attitude. Dance and movement therapy enable the domestic abuse victims to shape a better self-image perception and to experience stronger poses. Our body perception seems to influence our mind and our mind further change our attitude.

A possibility to mediate the I-Generation fascination for technology and the beneficial effects of physical activity are the exergames. The energy expenditure from exergaming is similar to skipping, walking, or jogging on a treadmill [37] being preferred by children who are already overweight or obese. The main critic against using these means in replacing traditional physical activities is that turns over more energy than sedentary gaming, but not as much as authentic sports, are mainly indoor activities and over time children lose interest in exergaming due to the repetitive and predictable nature of some of the games [38]. Another use of intelligent phones, which stimulates the people interest for physical activities, can be recording their heart rate, step counts and energy expenditure, among other things, demonstrating increases in physical activity and fitness level.

Urban sports have the potential to appeal mainly to teenagers in the 14–18 age group, who often have little interest in more structured team sports activities. They can contribute to skill building, self-confidence, social inclusion, and healthy active lifestyles. Urban sports and leisure are parts of city's young generation culture. They are a kind of adaptation to the build environment in which most of us are living. First of all inline skates, skateboards and bikes are vehicles and are used for transportation in the urban space. At least from this point of view, they differ fundamentally from other sports like football, rugby or cross country which suppose the field idea, a natural surface, or an unmodified soil, with a natural texture.

The spectacular character of tricks and series of acrobatic elements and the speed of performing requires an impressive volume of practice. The training develops all types of physical skills and especially balance, spatial awareness in unusual conditions, a good reaction and execution speed. In skate parks, the beginners practice with advanced people and they learn together and one from another. Instead of explanations, they use the mobile phone to video recorder the drills and the internet connection for instant sharing the hits. The young people communities formed by common interests, sharing experiences and knowledge in a genuine active learning process, have a certain social and cultural value.

The high speed, the use of IT applications, and most of the time a high level of adrenaline are, in my opinion, the three main factors that define the spirit of this new generation physical activities. Urban sports and leisure are mainly outdoors activities, during which air, sunlight and (sometimes) water can act to harden and strengthen the immunity system and body vigor. Beside these positive aspects, the risk of injuries especially in a lower level of self-protection instinct conditions is higher than in traditional physical activities.

6. Motives and possible solution

6.1. Man has to be encourage to live

"*Man is the only animal who has to be encouraged to live*" as Nietzche said in the nineteenth century. The extensive promotion and use of pharmaceutical products permit the avoidance of any physical pain or stress, inducing anxiety and reluctance against effort or suffering in physical challanges. People lose the capacity to invest hard work for a postpone reward and to enjoy the succes. The balance between effort and joy perishes in a insesisable, bouring oscilation [39]. As a consequence people might sink into a greedy lethargy, fiting in the large frame of the consumerism culture. Due the effort and sometimes the pain suffered while exercising, motivation for physical effort is an important issue in attracting and maintaining young people involved in physical activities. On the interpersonal level, the teacher should maintain a dialogue with performers; this ensures awareness, cognitive achievement, and commitment [17]. The means of pedagogical communication applied in physical education classes like giving feedback related to tasks; recognition of accomplishments; encouragement and support in difficult moments could have good results with students of all ages. These means of communication and motivation have also a beneficial effect on youngster's confidence in their own skills and strengths and self-esteem for physical appearance.

6.2. Exercise, stress management and diet can prevent 90% of chronic illness

In the past years of twenty-first century, mounting research has shown how lifestyle changes, including exercise, stress management, and diet can prevent almost 90% of chronic illnesses in our society and improve the quality of life and well-being. Promoting physical activity as a well-being component in schools and universities targets the enhacement of vigor, resilience, employement and social outcomes for graduates and comunities. Further, health is conditioned by our own habits and behavior and the accumulation of positive and negative effects on health and well-being is for over the life-course. Therefore, cognitive acquisition related to a healthy and active lifestyle would be a useful support for physical activity.

Improved well-being in youth should contribute to reducing school and college/university drop-out on short term, strengthening personal confidence and cognitive function, improving educational efforts and enhancing employability on long term. The education level is correlated with health; educated individuals report higher sense of control which conducts to a better health.

Integrating physical and health education in preventative strategies would have a real effect in reducing the occurrence of physical and mental disorders and co-morbidities associated with these later, over the life time. Lasting acquisition of behavior is through social learning and a team sport or a walking, yoga or jogging group could be a proper social context for learning and eventually behavior and attitude changes.

6.3. "Adds are propaganda too" (Konrad Lorincz—Nobel Prize laureate)

Often youth and children are the targets of advertising for high-calorie, high-fat snacks, and sugary drinks. The goal of these ads is to sway people to buy these high-calorie foods, and often they do. Children are easily tempted by instantaneous pleasures, for example, sweets, candies or chocolate bars, and they are not necessarily in a position to balance their short-term satisfaction versus the long-term consequences. Companies are exploiting this luck of self-control and discernment by the way they provide information to consumers, and for children are not easy to understand the permeable boundaries between education, advertising, and entertainment. Research shows that exposure to food advertisements produces significant increases in calorie intake in all children and the increase is largest in obese children [40].

As a biological entity, the body has a functional role, but as a social entity it conveys important messages about social status, personality or cultural group membership. For an overweight person, the awerness of his or her body size and volumen will determine a social reluctance, timidity, and low self-confidence reflected in her/his posture, attitudes and non-verbal communication (gesture, body language, tone of voice). Beside, the sociocultural patterns associate fatness with laziness and overweight persons are easily labeled as indolent, careless or greedy. The isolation, emotional insecurity or being bullied sometimes can induce anxiety or even depression symptoms.

Children and young people are often afraid they will be blamed or ridiculed because of their weight and size. Therefore, a safe, non-threatening approach to tackling overweight and obesity in a productive pedagogical climate is the recommended inclusive method. Studies proved that in a positive emotional situation, information is easier memorized, and on the contrary, a stressful situation (fear, excessive effort, tension) gets along inhibition and forgetfulness. Tension, anxiety or fatigue can cause a breakdown of physical skills and increase the risk of injury [45]. The students' involvement in physical tasks is gained by showing concern for their executions; giving them positive or negative feedback as appropriate, making suggestions to solve difficulties, listening actively and making sure that recommendations have been understood.

6.4. Acceptance and cooperation in teams and groups

Being part of a supportive team is a good motivator for an overweight or just sedentary person. Building a team is a process that starts at the individual level, and the first step is accepting the idea that everyone has a different set of values, skills, and needs. While individual qualities and skills will be used for team goals, each member will receive from the team honest feedback on which he or she can assess his/her strengths and weaknesses. Team and its spirit are built on interpersonal relationships and through a communication based on respect and trust among teammates. The team leader, through his attitude and style, will determine the climate in which these relations will be favored or hampered [41].

Experiences shared by team members resonate differently in each of their consciences. If the observations, thoughts, and feelings that these experiences determine will remain to an individual level, the team will remain a sum of individuals [43]. Sharing with others the effects of what common experience causes at intimate level, helps creating interpersonal relationships, mutual understanding, and finally embracing the idea of acceptance and cooperation. Supportive relationships diminish the exposure to stress. For individuals, a physical activity in a group or team is an entertaining way to achieve knowledge and social skills as an active form of learning, especially for a generation which is set on "fast-forward" pattern.

An increased number of adequate physical activities, spending time outdoors, healthy lifestyle programs, or cutting out unhelthy foods on children meals are a few examples of remedial measures taken in schools and families. Other critical changes could be part of the solution for a better active life:

- Take care of children to spend a maximum of 2 h of screen activities especialy if they are in primary school.
- Parents are responsible for adopting a healthier lifestyle, being aware of personal example they give their sons and daughters.
- Encourage the youngsters to set and assume realistic goals and motivate them to manage their weight by combining diet and physical activities.
- Develop critical thinking which enable young people to choose exercises and practice methods suited to their age and personal goals.
- Enhance individuals understanding of their own and others corporeality.
- Recommend ergonomically suitable physical activities for overweight and obese persons.
- Establish a good relationship between effort and recovery and between stress and leisure.
- Make clear that individuals are responsible for their own task success or failure [45].

7. Conclusions

Changes in objective indices of physical fitness play a minor role in body image change, whereas improvements in perceived fitness and self-efficacy appear to be important mechanisms by which exercise improves body image. There is no direct correlation between objective progress in fitness level and the subjective perception of being fit or more functionally efficient. Paradoxically shifting the focus away from appearance and emphasizing the physical and mental benefits of exercise, we can inspire our students with self-body acceptance, self-confidence and a better self-esteem. Body image and health perception are intermediate variables between physical activity and subjective well-being improvement [41] as next model shows:

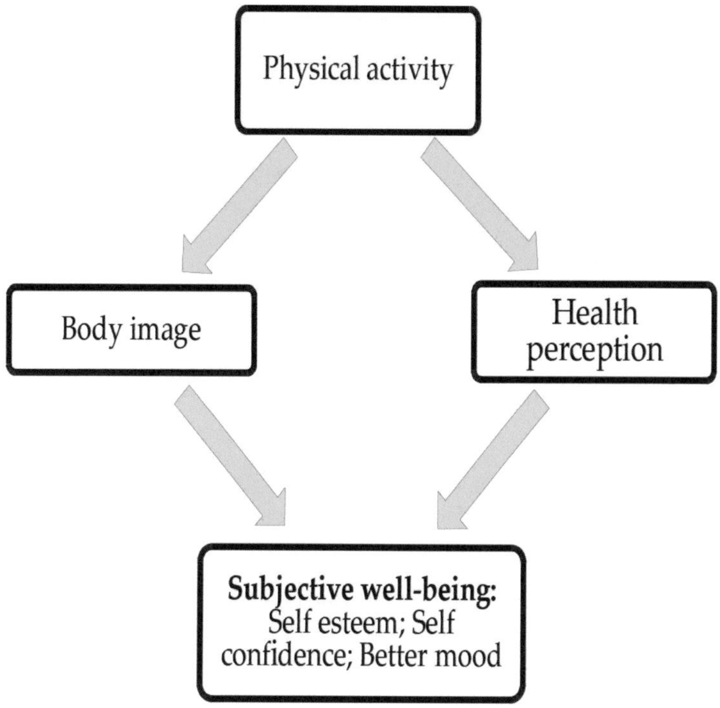

Physical education may contribute more to young people's self-image acceptance if lessons are planned and delivered with this specific goal in mind. Having this intention and confident in the positive effect of physical activity on body image, I'll make a few recommendations with practical and managerial use for physical and health education, exercising, and other physical activities related to quality of life and well-being.

- Emphisize that the accumulation of positive and negative effects on health and well-being is for over the life-course and investing in prevention reduces personal and social health costs.

- Provide all young people, of all sizes, with meaningful, relevant and positive physical education and physical experiences and eventually deliver a healthy, valuable working force for society.

- Provide a safe exercising and emotionally environment for all practitioners.

- In improving body image, the perception of how fit or functionally efficient someone becomes is more important than the objective level of fitness. Body image is a subjective parameter and depends on feelings, beliefs, and the psychological context. By shifting the focus away from appearance and emphasizing the physical, mental, and social benefits of exercise, we can inspire practitioners with self-body acceptance, self-confidence, and improved self-esteem.

- There are positive changes in perceptions of body image when exercise is performed on more days per week and at moderate intensity. The exercise chosen depends on personality, preferences, and personal goals but the effect on improvements in body image is the same.
- Enjoyment of exercise is positively associated with changes in perceptions of body image. Teachers and trainers can enhance enjoyment by creating a good working environment, by adding variety to workouts, and by ensuring that fitness programs are physically challenging and respect the subject's preferences and personal goals.
- Physical activities program leaders should have a supporting and encouraging attitude and motivate practitioners for physical effort. The satisfaction gained from exercising can eventually become a motivation in itself, especially when the effort has positive effects on enhancing perceptions of body shape and self-image [41].

Many references I made in this chapter are about children and young people, having in mind that for an obese adult the chances to attain a normal weight are less than 1%. Also overweight children starting primary school have low odds in terms of having a normal BMI in future, in that 80% of them will become overweight and end up as obese teenagers and adults. In present days, when the daily pursuits of our young generation are mainly sedentary and the obesity is widely distributed, too, to spend time playing a sport could bring health and freshness in their life and the physical effort could provide enough satisfaction, in order to become an intrinsic motivation [45]. A good strategy against the sedentary and greedy life style, which tempts us nowadays, is promoting physical activities in preschools, schools, universities and among adults through active transportation (walking, cycling), spending time or exercising outdoors as leisure time, or joining sport competitions in a proper environment.

Acknowledgements

Parts of this chapter are reproduced from the authors' previous publications [17, 41, 42, 45].

Author details

Cristiana Lucretia Pop

Address all correspondence to: crispotir@yahoo.com

Physical Education and Sport Department, Bucharest Economic Studies University, Bucharest, Romania

References

[1] Galloway S. Quality of Life and Well-being: Measuring the Benefits of Culture and Sport: Literature Review and Tink piece.2006. p. 7; Scottish Executive Social Research, Glasgow, Scotland

[2] Conceição P, Romina B. Measuring Subjective Wellbeing: A Summary Review of the Literature. Working Papers, edited by UNDP New York: U N Development Program. 2008. Available from: http://web.undp.org/developmentstudies/docs/subjective_wellbeing_conceicao_bandura.pd

[3] Boarini R., M. Comola C. Smith, R. Manchin, and F. De Keulenaer. What makes for a better life? The determinants of subjective well-being in OECD countries: Evidence from the Gallup world poll. Technical report, OECD WP47, 2012

[4] Diener E, Scollon C, Lucas R. The evolving concept of subjective well-being: The multifaceted nature of happiness. Advances in Cell Aging and Gerontology. 2003;**15**:187-219

[5] Baumeister RF, Smart L, Boden JM. Relation of threatened egotism to violence and aggression: The dark side of high self-esteem. Psychological Review. 1996;**103**:5-33

[6] Zang L, Leung JP. Moderating effects of gender and age on the relationship between self-esteem and life satisfaction in mainland Chinese. International Journal of Psychology. 2002:**37**(2):83-91

[7] Russel C. Looks aren't everything. Believe me, I'm a model, TED Conference. 2013. Available from: https://www.ted.com/talks/cameron_russell_looks_aren_t_everything [Retrieved December 12, 2017]

[8] Tiggemann M. Sociocultural perspectives on human appearance and body image, In Body Image. New York: The Guilford Press; 2011. pp. 12-19

[9] Corral CD, Perez-Turpin JA, Vidal AM, Padorno CM, Patio JM, Molina AG. Principles of the olympic movement. Journal of Human Sport and Exercise. 2010;**5**(1):5

[10] Singh D, Peter Renn P, Adrian Singh A. Did the perils of abdominal obesity affect depiction of feminine beauty in the sixteenth to eighteenth century British literature? Exploring the health and beauty link. Proceedings of the Royal Society B. 2007;**274**(1611):891-894

[11] Patricia V. 'Weighs and means': Examining the surveillance of fat bodies through physical education practices in North America in the late nineteenth and early twentieth centuries. Journal of Sport History. 2008;**35**(3):449-468; 454; Segrave, Obesity in America

[12] Hogan MG, Strasburger VC. Body image, eating disorders, and the media. Adolescent Medicine. 2008;**19**:521-546

[13] Grogan S., "Body Image: Understanding Body Dissatisfaction in Men, Women, and Children", Routledge, London / New York, 2008. pp 45-46

[14] Ali SM, Lindstrom M. Socioeconomic, psychosocial, behavioural, and psychological determinants of BMI among young women: Differing patterns for underweight and overweight/obesity. European Journal of Public Health. 2006:**16**(3):324-330

[15] Delerm P. Telefonul mobil. In: Garcin J, editor. Noile Mitologii. Bucuresti: Art Pub; 2009. p. 45

[16] Owens NK. Imago: The Rhetoric of Women's Magazines. Charlottesville: University of Virginia; 1991

[17] Pop CL. Self esteem and body image perception in a sample of university students. Eurasian Journal of Educational Research. 2016;**64**:31-44. DOI: 10.14689/ejer.2016.64.2

[18] Jaworowska A, Bazylak B. An outbreak of body weight dissatisfaction associated with self-perceived BMI and dieting among female pharmacy students. Biomedicine & Pharmacotherapy. 2009;**63**(9):679-692

[19] Costa CL, de Vasconcelos A. Influence of socioeconomic, behavioral and nutritional factors on dissatisfaction with body image among female university students in Florianopolis, SC. Revista Brasileira de Epidemiologia. 2010;**13**(4):665-676

[20] Khan AN, Khalid S, Khan H, Jabeen M. Impact of today's media on university student's body image in Pakistan: a conservative, developing country's perspective. BMC Public Health. 2011;**11**(2011):379

[21] As-Sa'edi E, Sheerah S, Al-Ayoubi R, Al-Jehani A, Tajaddin W, Habeeb H. Body image dissatisfaction: Prevalence and relation to body mass index among female medical students in Taibah University. Journal of Taibah University Medical Sciences. 2011; **8**(2):126-133

[22] Daniali S, Azadbakht L, Mostafavi F. Relationship between body satisfaction with self esteem and unhealthy body weight management. Journal of Education and Health Promotion. 2013;**2**:29

[23] Poobalan AS, Aucott LS, Precious E, Crombie IK, Smith WC. Weight loss interventions in young people (18 to 25 year olds): A systematic review. Obesity Reviews. 2010;**11**:580-592

[24] Neagu A. Body image: A theoretical framework. Proceedings of the Romanian Academy. Series B. 2015;**17**(1):29-38

[25] Lee IM, Shiroma EJ, Lobelo F, Puska P, Blair SN, Katzmarzyk PT. Effect of physical inactivity on major non-communicable diseases worldwide: An analysis of burden of disease and life expectancy. Lancet. 2012;**219**:219-229

[26] Pratt M, Ramirez A, Martins R, Bauman A, Heath G, Harold Kohl III H, I-Min Lee Powell K, Hallal P. 127 Steps toward a more active world. Journal of Physical Activity and Health, 2015;**12**:1193-1194

[27] OECD, Health at a Glance: Europe 2012, OECD Publishing, pp 62-65, http://dx.doi.org/10.1787/9789264183896-en, www.oecd.org/publishing

[28] Hansraj K. Assessment of stresses in the cervical spine caused by posture and position of the head. Surgical Technology International. 2014;**25**(2014):277-279. Available from: http://www.massageandbodyworkdigital

[29] Peper E, Lin I-M. Increase or decrease depression: How body postures influence your energy level. Biofeedbeack. 2012;**40**(3):125-130

[30] Scutti S. Change Your Posture To Improve Your Mood, Memory, and 5 Other Aspects of Your Life. 2014. Available from: http://www.medicaldaily.com/change-your-posture-improve-your-mood-memory-and-5-other-aspects-your-life-289724 [Retrieved November 21, 2016]

[31] Pentz MA, Spruijt-Metz D, Chou CP, Riggs NR. High calorie, low nutrient food/beverage intake and video gaming in children as potential signals for addictive behavior. International Journal of Environmental Research and Public Health. 2011;**8**(12):4406-4424

[32] Ansdell, G. & DeNora, T. Musical Flourishing: Community Music Therapy, controversy, and the cultivation of wellbeing. In R. Mac Donald, G. Kreutz and L. Mitchell (eds), Music, Health & Wellbeing. 2012. Oxford University Press.

[33] DeNora T. Health and music in everyday life—A theory of practice. Psyke & Logos. 2007;**28**(1):271-287

[34] Pelletier CL. The effect of music on decreasing arousal due to stress: A meta-analysis. Journal of Music Therapy. 2004;**41**:192-214

[35] MacDonald R.A.R. Music, health, and well-being: A review Int J Qual Stud Health Well-being. 2013; 8: Published online 2013 Aug 7. doi: 10.3402/qhw.v8i0.20635

[36] Muller-Pinget S, Carrard I, Ybarra J, Golay A. Dance therapy improves self-body image among obese patients. Patient Education and Counseling. 2012;**89**(3):525-528

[37] O'Louglin EK, Dugas EN, Sbiston CM, O'Louglin JL. Prevalence and Correlates of Exergaming in Youth. Pediatrics Nov 2012, 130 (5) 806-814; DOI: 10.1542/peds.2012-0391

[38] Bailey BW, McInnis K. Energy cost of exergaming: A comparison of the energy cost of 6 forms of exergaming. Archives of Pediatrics and Adolescent Medicine. 2011 Jul;165(7):597-602. doi: 10.1001/archpediatrics.2011.15. [Retrieved November 20, 2015]

[39] Konrad L. Die acht Todsunden der zivilisierten Menschheit. Munchen; Piper Verlag GmbH: 1973

[40] Montgomery KC, Chester J. Interactive food and beverage marketing: Targeting adolescents in the digital age. Journal of Adolescent Health. 2009;**45**(3 Suppl):18-29

[41] Pop C, Ciomag V. The relationship between subjective parameters of well-being in a sample of young Romanian women. Procedia – Social and Behavioral Sciences. 2014; **149**;737-740

[42] Pop CL. Physical and health education facing the technology challenge. Physical Education of Students. 2016;**20**(2):45-49

[43] Pop CL. Teamwork effect on student's involvement in physical activities., Revista Marathon. 2013;**5**(1):90-96

[44] Pop CL. Overweight – Cultural and educational aspects. Revista Romaneasca Pentru Educatie Multidimensionala. 2013;**5**(2):57-65

[45] Pop CL. Physical activities for overweight and obese children – An inclusive approach. Procedia – Social and Behavioral Sciences. 2014;**163**:142-147

Chapter 2

From Relaxation Response, Building Power for Health to an Advanced Self-Cultivation Practice: Genuine Well-Being

Shawn Wu, Lin Jiang and J. Wang

Additional information is available at the end of the chapter

http://dx.doi.org/10.5772/intechopen.68678

Abstract

Recently, much more mind-body/mindfulness interventions have been used in different people including frail elders, patients with major depressive order, pain suffering, chronic fatigue, infection, and patients in intensive care unit, etc. In the field of Complementary and Alternative Medicine, many Chinese mind-body exercises are under very active investigation for evidence-based decision making. At the same time, there is a fact that Yoga and Health *Qigong*/Tai Chi, especially meditative practice based on ancient oriental civilizations, have been practiced for thousands of years and are getting more attention. The use of Traditional Chinese music as therapy was documented around 94 BC and that was foundational to traditional Chinese medicine, building power for health as well. Meditation, as a mind-body practice, originally introduced by cultivators in ancient Eastern religious and/or spiritual traditions has raised increasing awareness recently in the Western world. Medical research results suggest that the benefits of Eastern meditative practices not only include the promotion of emotional and mental health, resilience, stress and anxiety reductions, but also cure other health-related conditions. It was suggested that improvement of moral level would be of the same importance with those achieving mind-body health. Through experiential learning, and what delivering big sample size survey results and numerous case reports, the authors are illustrating the mindful practice of Falun Gong (also called "Falun Dafa") and have real health-wellness effects to practitioners.

Keywords: meditation, Falun Dafa, mind-body, cultivation, well-being

1. Introduction

Different types of mind-body exercises have been criticized by researchers and practitioners to be lacking in scientific support in two aspects, though these interventions have been practiced by people from different cultures and clinical effects have been observed for centuries. To build a body of empiric evidence in this related field is a kind of challenge. It was found on decade ago that, 19.2% of U.S. adults (more than 55 million people) had used at least one form of mind-body therapy during the previous 12 months, according to data from the 2007 National Health Interview Survey, the National Centers for Complementary and Alternative Medicine (CAM), and the National Center for Health Statistics [1].

The mind-body interaction and related methods originates from ancient Eastern tradition [2, 3]. CAM has not been viewed globally as distinct entities for decades, and there was an example showing that, for many conditions, people tend to integrate Traditional Chinese Medicine (TCM) and Western medicine, with the treatment principles based upon recognition of the pattern of disharmonies [2, 4], and upon diagnosis, respectively. And there are much more evidence to support the powerful effects of practices that are directed toward the mind or spirituality besides items of human body when talking about health, well-being concepts. While the health benefits of yoga, meditation, and prayer have been aware of and recognized by the general public [5–9], many of the practices that are called *Qigong* (pronounced chi-kung) were getting recognized as transitional health practices and healing techniques.

When the mind-body interaction had been got interested in, and the mechanisms underlying them need to be further understood and systematically studied, also, mindfulness meditation is specifically an unique approach to get health in terms of calling for a clearer conceptualization, and assessing the potential application. Although mindfulness meditation has been formalized for clinical interventions with Mindfulness Based Stress Reduction [10] and Mindfulness Based Cognitive Therapy [11], there are many fields needed to explore that have potential to make mindfulness meditation adopted by a larger population. Meanwhile, a group of high school students' recent outcome from a transcendental meditation (called "Quiet Time Program") conducted by the University of Chicago Crime Lab, has been reported and recommended by staff of Chicago Public Schools because of its goal to address the effects of toxic stress on young people that had also been implemented by students in San Francisco, Los Angeles and New York City from 2015 [12]. The immediate results yielded for students (such as, suspensions are down, a recent round of SAT prep scores showed improvement, teachers see improvement in students' behavior and ability to concentrate in class) and staff from meditation indicated this approach holds promise in a larger range of utilization scope in human society.

Moreover, people may acknowledge that, Chinese culture was composed of an abundant and very profound system of values, and in Chinese people's mind "man and nature must be in balance" and "respect the heavens to know one's destiny" were the dominant concepts during the whole life. Meanwhile, benevolence, righteousness, propriety, wisdom, and faithfulness, called five cardinal virtues (ren yi li zhi xin, or 仁義禮智信 in Chinese), are all products of Buddhism, Daoism, and Confucianism, the three religions' teachings over China's 5000-year-long history, to which belief in the divine is so central, nowadays presented by Shen Yun Performing Arts that was established in New York, USA in 2006 by elite Chinese artists [13]. Present authors are considering that, people need to know the nature of *Qigong*, as well as mindful practices, even need to find the genuine relationship between the practices, keeping moral level based on following traditional culture, and achieving body and mind health.

2. Relaxation response

Many items will be explored in the field of the mind-body therapies that elicit the relaxation response (RR), which was described almost half century ago, as a status associated with decreases in oxygen consumption, respiratory rate, and blood pressure, along with an increased sense of well-being [14, 15], after a three-stage model of the body's response to stress was introduced by endocrinologist Hans Selye based on the understandings more than 100 years ago of the connection between the mind and body [16], and given a term as stress response (SR) for the systemic adaptations by experts of the Integrative Health Research Jeffery Dusek and Herbert Benson [15].

Some platforms have introduced RR programs [17–19], and a lot of study results have also been reported for brain science, including for brain signals making participant's muscles and organs slow down and increase blood flow to the brain, and effectiveness for stress-related disorders.

After evidence-based research from different Institutes provided the results indicating the benefits of mind-body intervention for some health issues (cardiovascular and neuromuscular conditions) [20–26], or psychological conditions (including depression) [27–32], investigators in one of them demonstrate in pilot studies that the RR-based group intervention, and other modalities (i.e., tai chi, *Qigong*, mindfulness training) have effect on treating mild and moderate depression symptoms [33–35] and maybe regarded as adjunctive treatment for other conditions. Their efforts also have been added into the processing of number increasing of studies [36–38] from different continents of the world showing the promise of multimodal mind-body group interventions for patients with depression.

In 2009, Jeffery Dusek and Herbert Benson presented a model of the physiological and biochemical changes (with two main pathways activated, the sympatho-adrenomedullary (SAM) axis and the hypthothalamus-pituitary-adreno (HPA) axis) taking place during exposure to acute stressors or elicitation of the RR, and the relationship between these two responses [15]. The hypothalamus secreting corticotrophin-releasing hormone (CRH), which causes the pituitary gland to release adrenocorticotropic hormone (ACTH), can activate both axes while the later one is thought as one of the other stress hormones to be modulated by nitric oxide (NO) than those (such as cortisol) within the adrenal glands easily inhibited during the initial biosynthetic step in steroid production. NO was shown to play a central role in the development of atherosclerotic plaque, and the regulation of platelet function, vascular smooth muscle cell proliferation, and leukocyte interactions with vascular endothelial cells, as well as to mediate diverse physiological processes including neuronal function and neurotoxicity, immune and cardiovascular functions [39]. Studies from this group also reported that 8 weeks of RR training was an effective therapeutic intervention to counteract the adverse clinical effects of stress in individuals with systolic hypertension, with 8 more weeks training, elicitation of the RR made 32% of participants be able to eliminate one or more of their antihypertensive medications [40].

A most recent study on the links between cardiac health and psychological stress, with the use of ^{18}F-fluorodexoyglucose PET/CT showed that, high levels of activity in the amygdale, a region known to be involved in emotional processing at the start of the study were associated with an increased risk of experiencing a cardiac event, and the association was significant even after adjusting for other cardiovascular risk factors and atherosclerosis. This first study to link regional brain activity to subsequent cardiovascular disease illustrated findings providing novel insights into the mechanism that how emotional stressors lead to cardiovascular disease in human beings [41]. These procedures may perform as a novel tool for evidence-based medicine to evaluation the whole body effectiveness of the programs (in CAM) mentioned above.

Meanwhile, the number of experiments investigating on gene expression stimulated by the simple activity of interpersonal experience is growing now, and studies have examined gene expression cascades following the use of RR [42, 43] and Chinese *Qigong* [8]. Dusek et al. evaluated possible gene expression changes by RR with the test subjects to elicit RR, and their study included 19 long-term RR practitioners (Group M) versus 20 healthy controls who were tested at baseline (Group N1) and who participated in 8 weeks of training in guided relaxation techniques and were tested again (Group N2). Polymorphonuclear cells (PBMCs) were isolated for the blood sample, and global transcriptome profiles were determined using microarrays (interrogating approximately 47,000 genes and gene variants), and pairwise comparisons of the transcriptomes between the three groups then performed. A 2209 genes (1275 up- and 934 down-regulated) were found to be differentially expressed between Groups M and N1, 1504 genes (774 up- and 730 down-regulated) between Groups M and N2, and 1561 genes (874 up- and 687 down-regulated) between Groups N1 and N2. They suggested that the gene expression changes in the M and N2 (428 genes were shared between the short and long-term RR participants) groups might indicate a greater capacity to respond to oxidative stress and associated detrimental effects; basal gene expression changes in PBMCs can be caused by RR.

3. Building power for health

Where language is lost, music may help bring it back, researchers said at the AAAS (AMERICAN ASSOCIATION FOR THE ADVANCEMENT OF SCIENCE) Annual Meeting (2010) [44]. During the topics of "In a range of New Science, Researchers Find the Power of Music to Build the Brain," Aniruddh Patel challenged the conventional idea that that music and language are processed independently, saying that "evidence suggests at least a degree of overlap." It was expected that scientists utilize a range of assessment techniques drawn from psychology and medical studies to fully understand the relationship between language and music and to explore its full power.

Neuroplasticity, with its definition like "the natural tendency of the brain architecture to shift in negative or positive directions in response to intrinsic and extrinsic influences" [45], was regarded as a kind of index when assessing "Building Brain Power for Health" [46] during clinical intervention. Musicians are supposed to show stronger activation in an audiovisual incongruency response and to have response different from unisensory mismatch negativity (MMN) mainly depicted by functional MR imaging (fMRI), so that possess an enhanced responsiveness in a genuine audiovisual process, and all of these would indicate plasticity effects on multisensory processing in musicians [47], which belonged to some forms within the complex music enrichment to positively influence on neuroplasticity in a number of brain regions [45]. Bottiroli et al. [48] performed a study for non-musician healthy individuals aged 60–84 who listened to the background music of Mozart (as compared to silence and white noise) and the results showed that they improved declarative memory tasks.

Traditional Chinese music can be regarded as a kind of methodology with medical effectiveness, and this was reflected by the similar writing approach between the Chinese characters of "樂" ("music") and "藥" ("medicine") (**Figure 1**) [49]. The use of traditional Chinese music as therapy was documented around 94 BC. It is an aspect of the Chinese Theory of Five Elements, which is foundational to traditional Chinese medicine (TCM), building power for health as well.

In Traditional Chinese music, the music notes had their relationship with corresponding human body organs in Chinese medicine. In the "Treatise on Music" "樂書" (included within"史記," "The Scribe's Records," a monumental history of ancient China and the world finished by the Han dynasty official Sima Qian), it was introduced that Traditional Chinese music was composed from five notes or sounds (in Chinese) —"宫"(gong), "商" (shang), "角"(jue), "徵"(zhi), and "羽"(yu), having relationship with internal organs of human body in Chinese medicine, like spleen (脾), lung (肺), liver (肝), heart (心), and kidney (腎), respectively, to achieve different healing purposes (**Figure 2**) [50]. In terms of the Chinese Theory of Five Elements, these organs belong to metal, wood, water, fire and earth, respectively. Chinese medicine uses the relationship between internal organs and five-element correspondences, such as traditional Chinese musical notes and their comprehensive effect, to achieve different healing purposes, using musical instruments to relate their states of mind. Western music, meanwhile, focuses on the overall effect of the musical ensemble—and to achieve that, arrangement and harmony are of utmost importance. Both East and West have a long history of artists integrating spirituality into their work.

Traditional Chinese music focuses on expressing inner feelings, and the ancient people always used musical instruments to relate their states of mind. At the same time, music in Western world focuses on the overall effect of the musical ensemble, and in order to achieve that, arrangement and harmony are of utmost importance. Shen Yun Symphony Orchestra blends the spirit of Chinese music with the power of a Western orchestra. All-original compositions draw upon five millennia of culture and legends. Western strings, percussion, woodwinds, and brass accentuate the sound of ancient Chinese instruments—like the two-stringed erhu and the plucked pipa. Never before have the exquisite beauty of Chinese melodies and the grandeur of a Western symphony have been so seamlessly combined.

Music of Shen Yun features the perfect harmony of classical music of East and West. First, the Western orchestra serves as a foundation, accentuating the distinct sound of Chinese instruments. Second, the bedrock of soul-stirring melodies from the ancient Middle Kingdom is fully brought to life by a Western symphony. This is what makes Shen Yun's music unique and is a new frontier in classical music [50]. Bogdan Zvoristeanu, Concertmaster of the Orchestre de la Suisse Romande said about Shen Yun music that, "A different sensuality and a power of expression… It comes from the heart and it goes to the heart." Present authors propose that, audients who physically attend to Shen Yun Symphony Orchestra would get a kind of benefits even achieving well-being.

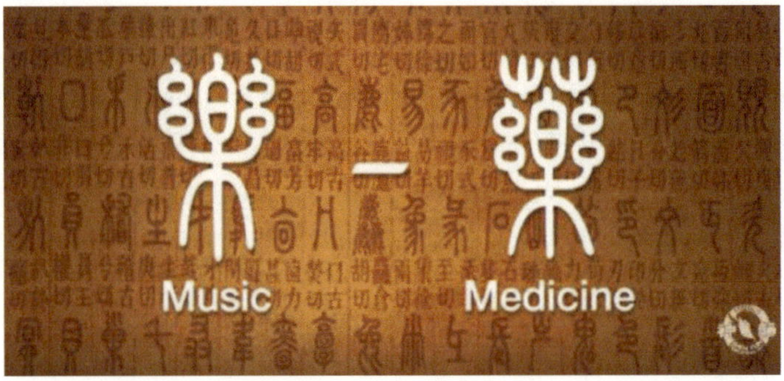

Figure 1. Screen shot from Shen Yun website shows the similarity between the Chinese characters of "樂" ("music") and " 藥." (Screen shot from Shen Yun website, refer [49].)

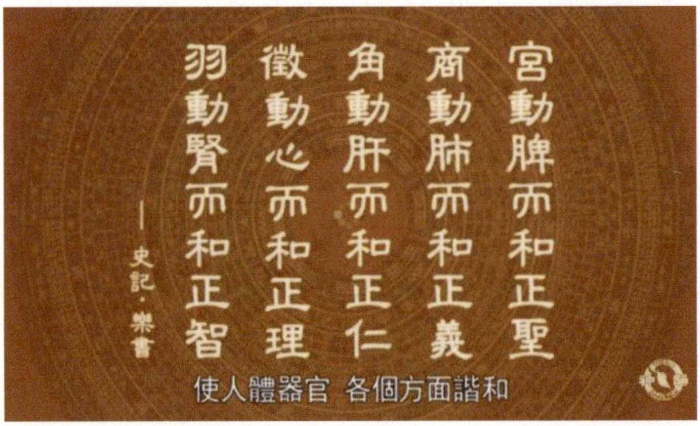

Figure 2. Traditional Chinese music notes and their relationship with corresponding human body organs in Chinese medicine. In the "Treatise on Music" "樂書" (included within"史記," "The Scribe's Records," a monumental history of ancient China and the world finished around 94 BC by the Han dynasty official Sima Qian), classical Chinese music was composed of five notes or sounds (in Chinese) —"宮"(gong), "商" (shang), "角"(jue), "徵"(zhi), and "羽"(yu), having relationship with internal organs in Chinese medicine, like spleen (脾), lung (肺), liver (肝), heart (心), kidney (腎), respectively. (Screen shot from Shen Yun website, refer [49].)

Meditation, as a mind-body practice originally introduced by cultivators in ancient Eastern religious and/or spiritual traditions thousands of years ago, has raised increasing awareness recently in the Western world. Medical research results suggest that the benefits of Eastern meditative practices not only include the promotion of emotional and mental health, resilience, coping skills, stress and anxiety reductions, but also cure other health-related conditions.

Meditation provided participators in an eight-week mindfulness meditation program [51] cognitive and psychological benefits that persist throughout the day, that was measurable changes in brain regions associated with memory, sense of self, empathy and stress. Within the study led by an investigating team at Massachusetts General Hospital, they looked at the brain MR scans of 16 people before and after taking Mindfulness-Based Stress Reduction (MBSR) Program, and it was found that reductions in gray matter volume in the right basolateral amygdala (structure for anxiety and stress occurrence) obtained from MRI scans had relationship with reductions in perceived stress following MBSR. It was suggested that plasticity-related alterations in brain regions implicated in stress can occur after short term of mindfulness meditation training, and they concluded that parts of the participants' brains associated with compassion and self-awareness grew, and parts associated with stress shrank. Earlier studies with the analysis of MR images, which focused on areas where meditation-associated differences were illustrated, showed that increased grey-matter density in the hippocampus, known to be important for learning and memory, and in structures associated with self-awareness, compassion and introspection. They also suggested that longer-term meditation practice might be needed to produce changes in a self-awareness-associated structure called the insula though no change was seen in their studies. And they were glad to know the brain's plasticity from imaging data, and know that practicing meditation can play an active role in changing the brain and can increase people's well-being and quality of life [52, 53]. Based on some scientists' observation, music has some reward value beyond the pleasurable sounds and direct feedback, it also plays an important role in social interactions, both in contexts of group listening and music making, which needed to be further investigated, at the same time, research can be performed in some important aspects in the context of music and learning, both including pupil-teacher interactions and imitation learning, social reward and influences on self-perception, and some negative influences (such as stress in professional situations and performance anxiety).

A group of US and UK scientists, by using fMRI has extended the observations in the field of empathy processes studies that can be modulated by the implicit context of the empathic experience, and their findings showed that brain regions associated with empathic processes were modulated by voluntary regulation of one's emotional responses with compassion meditation [54].

4. An advanced self-cultivation practice: Falun Dafa

Falun Dafa (also called Falun Gong) has a heritage dating back thousands of years when it existed as teachings passed down secretly from one master to one disciple for generations. It was introduced to the public in 1992 when its founder, Mr. Li Hongzhi, the practice's master, gave the first series of lectures on the practice in Changchun, Jilin province and in northeastern China [55]. Master Li had traveled across China (and later to France and Sweden) giving similar nine-day seminars 54 times during the following 2 years [56]. Over the next few years, interest in the practice skyrocketed, fueled by word of Falun Dafa's moral and health benefits; around 1998 State Sports Administration of China estimated over 70 million people practiced Falun Dafa. Master Li accepted a proclamation from the City of Houston, Texas, USA, which declared October 12, 1996 "Li Hongzhi Day." At the same time, the city also named Master Li an Honorary Citizen and Goodwill Ambassador [57]. In 2007, Master Li was ranked as one of the Top 100 Chinese Talents in the World, and he was awarded the Outstanding Spirit Leader Award in 2009. He has received over 3000 letters of citations, proclamations, and awards (**Figure 3**).

Figure 3. Photos of Falun Dafa practitioners in China (practicing) (left side), and pictures (right side) of proclamations, and awards etc. from worldwide to Master Li and Falunn Dafa. Upper left one showed Falun Dafa practitioners in Shuangcheng city, China were doing group practicing; while down left one showing five thousand people practicing in Wuhan city, China, forming the Chinese characters for Truthfulness, Compassion, Forbearance (1998). Refer [58].

4.1. Zhen, Shan, Ren (Truthfulness, Compassion, Forbearance)—the foundation of practice

Falun Dafa's basic tenets are based on ancient cultivation practices. In China, cultivation has a history that is much longer than that of Buddhism, Daoism, and Confucianism. The ancient wisdom of cultivation, including those ideas expressed in Falun Gong, may far precede all the religions we observe today, based on unearthed archaeological relics. Falun Dafa is an advanced practice of Buddha school self-cultivation, founded in China by Master Li. It is a discipline in which "assimilation to the highest qualities of the universe—Truthfulness (Zhen 真), Compassion (Shan 善), Forbearance (Ren 忍), is the foundation of practice. Practice is guided by these supreme qualities, and based on the very laws which underlie the development of the cosmos." Master Li's teachings are set forth in a number of texts, among which are included Falun Gong, Zhuan Falun, The Great Perfection Way of Falun Dafa, Essentials for Further Advancement, and Hong Yin (The Grand Verses). These and other works have been translated into 38 languages now, and are published and distributed worldwide. The focus of Falun Dafa practice is the mind, with the cultivation of one's mind and thoughts, or "*Xinxing*" (expressed as "Charater" in the text Falun Gong), being singled out as the key to increasing Gong energy. The height of a person's Gong is directly proportionate to that of his *Xinxing*. The concept of "*Xinxing*" encompasses the transformation of virtue (a white form of matter) and karma (a black form of matter). It also includes forbearance, discernment, and abandonment—that is, forsaking ordinary human desires and attachments, and managing to endure the most trying of ordeals. Much is encompassed by the concept. Falun Dafa also includes the cultivation of the body, which is accomplished by performing specific exercises, including five sets of exercises (and the fifth is the comprehensive sitting exercise, meditation (see in below)). One purpose of the exercises is to strengthen the practitioner's supernatural abilities and energy mechanisms by means of his or her powerful Gong force [58].

4.2. Its relation to well-being

First introduced to the public by Master Li in 1992, the practice nowadays has gained more than 100 million practitioners of all ages and backgrounds in over 100 countries, including the United States (**Figure 4**). In Taiwan, the number of Falun Dafa practitioners increased from 3000 in 1999 to over 300,000 in a dozen of years ago [59]. Taiwan's former Vice President went in 2002, to a Falun Dafa conference to give a congratulatory speech so as to acknowledge the positive contributions of Falun Gong has brought to the well-being of the people of Taiwan [60].

Practitioners around the globe are presenting how Falun Dafa has brought positive changes to those who practice this advanced self-cultivation. As early as 1996, a survey report titled "A report on the effect of Falun Gong in curing diseases and keeping fit based on a survey of 355 cultivators of Falun Gong at certain sites in Beijing, China" was completed and delivered [61]. On May 15, 1998, the then director of the State Sports Bureau of China went to Changchun City,

Figure 4. Falun Dafa practitioners in all over the world. Left, practitioners in Washington DC (upper), little disciples of Falun Dafa at an elementary school in Taiwan; Middle, in France, UK, Netherland, and Spain; Right, in Mexico City (Forum on Falun Gong Held in Chamber of Deputies). (Refer [58])

Jilin province, where Falun Dafa originated, to conduct an investigation into the practice. Then the Bureau surveyed 12,553 Falun Dafa practitioners and found that the illness healing rate was 77.5%. Adding 20.4% who reported experiencing improvement in their overall health, the overall effective rate was 97.9%. Each person on average saved more than RMB ¥1700 of medical bills each year. The yearly saving was RMB ¥21 million and even more. Following the teachings of Falun Dafa, genuine practitioners constantly raise their *Xinxing* (or mind-nature and morality), besides practicing exercises. As a result, they have gained both physical and mental health and much more. With an average annual savings of RMB ¥3270 (~US $654) per capita estimated, over 100 million practitioners worldwide have cut healthcare costs tremendously. People can find that six independent reports online summarized provide a more detailed picture of the healing efficacy of Falun Dafa. There were at least 10 surveys with definite sample size (**Table 1**) [62–65], as well as numerous individual and/or summary case reports.

The book "A Journey to Ultimate Health" edited by William McCoy, MD, Lijuan Zhang, MD, PhD many years ago [64] was composed of 40 vivid cultivation stories. Those were made revisions and adjustments so as to include only the parts relevant to health improvement. Since health improvement in all persons cannot be separated from changes in their situation of cultivation practicing, cases were purposely kept the content in some stories that described improvement in practitioners' moral character. All the stories depict the circumstances at the time they were written, which were specified after each story. Among 39 practitioners, 27 of them were residing in North America, five in Taiwan, three in Mainland China, two in Australia, and two in Europe. Actually they came from a wide scope of backgrounds, from a 10-year-old school student to retirees of over 70 ages, from housewives to highly educated professionals (including engineers and medical doctors). For the goal of better bridging the readers into the stories, editors of the book discussed some about the relationship between modern medicine with the disease-healing effects obtained from practicing Falun Dafa. The cases described were all so impressive, while two of them are attached in this text. One of them is: A 46-year-old male was very unhealthy for a long time before discovering Falun Dafa. He had multiple diseases and syndromes such as Petit mal seizure (from 10 years old, also causing fainting spells, migraines), depression, being unable to properly chew food, learning disabilities, autistic tendency, sleep apnea, and a foot injury that severely limited his range of motion. After he had started to do Falun Dafa practicing for 6 months, his ability to stand and support himself improved significantly. His medical support group has been surprised by his change. One physician who had been following his progress for some time concluded that Falun Dafa contributed to putting him back on his feet. A member of staff in the hospital was moved to tears by his progress.

Another one is about a 13-years-old seventh grade student in Toronto, Canada. She came from China with her parents in 1996. When she was 6 months old, it was discovered that her spleen and liver were severely enlarged and hardened. Doctors indicated that her case was rare. She could eat little, and was skinny and sick. She was much shorter than average for her age and was easily fatigued after just a short walk. Chinese and Western medicine became her daily meals. She was encouraged to do Falun Dafa practicing when she was 9 years old. Then she got rid of the "incurable" disease. She became a healthy middle-school student and a young, sincere Falun Dafa practitioner.

Survey name	Year	Sample size (and basic information)*	Methods	Results	Significances	Notes
A report on the effect of Falun Gong in curing diseases and keeping fit based on a survey of 355 cultivators of Falun Gong at certain sites in Beijing, China	1996	355	On a voluntary basis, the sample was asked to complete the questionnaires in person.	Complete or partial disappearance of diseases of Falun Gong practitioners after their practicing. The average rate of disappearance of diseases is 79.4%.	The medical costs of the cultivators reduced greatly; therefore, both social effect and economic benefits have been achieved.	By Zhang, Rongjia, The College of the Basic Courses, Beijing Medical University Xiao, Jun, the Institute of Environmental Sanitation and Sanitation Engineering, Chinese Academy of Preventive Medicine, China [61]
The State Sports Bureau Survey (China)	1998	12,553	Pending further investigation	The illness healing rate 77.5% was shown in the survey and the overall effective rate could be 97.9% when adding 20.4% from those practitioners who reported experiencing improvement in their overall health.	Each person on average saved more than RMB ¥1700 of medical bills annually, then the yearly saving would be RMB ¥21 million and even more.	Government organized. Refer [62]
The Beijing Survey Report (China)	1998	12,731	The participants voluntarily filled out the self-evaluated health status questionnaires according to instructions.	After practicing Falun Gong, 58.5%, or 6962 people recovered from their illnesses completely, and 24.9%, or 2956 people, had a general recovery. The overall efficacy rate of Falun Gong was thus 99.1%.	The overall efficacy rate of Falun Gong was thus 99.1%.	Researchers did a survey over five districts in Beijing (Xicheng District, Chongwen District, Dongcheng City, Xuanwu District and Chaoyang District). Refer [62]
Survey Report from Wuhan City (China)	1998	2005	A random sampling, at over 50 practice sites within three	75.15% had their health problems resolved, while 23.3% found that their	37 participants had spent more than RMB ¥10,000 annually on medical	Refer [62]

Survey name	Year	Sample size (and basic information)*	Methods	Results	Significances	Notes
			districts of the three main towns in Wuhan City	conditions had improved. (1899 or 94.7%, had different kinds of health problems prior to practicing cultivation.)	expenses before, but through practicing Falun DaFa, they have became illness-free and most of them have stopped visiting their physicians and/or taking medicine. Based on to the survey results, 95.51% of these people had no need to use their medical insurance benefits.	
Survey of over 6000 Cultivators in Dalian, Liaoning Province (China)	1998	6478	A health status survey	92% reported total disappearance of the symptoms, 7.74% observed moderate improvement; the disease recovery rate shows no major difference among practitioners with multiple illnesses or with a single illness, which are 89.73% and 88.83% respectively.	6192 participants (95.59%) of them suffered various diseases and medical conditions of the cardiovascular system, nervous system, digestive system, respiratory system, urogenital system, hemic and immune system or musculoskeletal system. It is very interesting to note that those who achieved the best results didn't use any medical treatment at all.	Refer [62]
North American Survey Report	1999	235	A small-scale survey (questionnaires)	224 practitioners, or 97%, had great health improvements after the practice	The first one in North America.	Inspired by results of Falun Gong health surveys conducted in China, several

Survey name	Year	Sample size (and basic information)*	Methods	Results	Significances	Notes
				(230 surveyed practitioners returned questionnaires with complete health records before their practice, and 226 practitioners turned in complete records after they began the practice); 103 practitioners consumed alcohol before the practice, 100 of them quit drinking after taking up the practice.		medical researchers conducted a small-scale survey of practitioners in USA and Canada. Refer [62]
Russian Survey Report	2001	12 practitioners out of 32 candidates（took random samples）; Two of them practiced Falun Gong for over 1 year and 11 for over 2 years.	Investigations covered the following aspects: Cultivators' medical records, the health index of practitioners before and after cultivation, the contents of Falun Gong, and the daily lives of those who practice Falun Gong.	After they began practicing Falun Gong practitioners reported no more complaints of ailments. Blood and urine tests were all normal. Physical examinations showed no abnormalities. All subjects showed normal mental states– positive, responsible, easy to get along with. (The subjects all suffered some illnesses before cultivation. Three had stomach ulcers or infections. One had an internal hormone imbalance. One	The subjects' subjective opinion: Practicing Falun DaFa improved their physical and mental health. The investigating team's conclusion: (1) The effective rate of Falun DaFa in healing illnesses and improving health is 75%; (2) The medical examination of Falun DaFa practitioners showed that cultivation practicing had made remarkable improvement in one's physical and mental condition; and (3) Falun DaFa	Senior forensic specialist, Professor Guluoqi of the Forensic Office of the Russian Internal Department, and Principal Forensic Doctor Simintani of the Judicial Forensic Bureau teamed up to conduct the survey. Refer [62]

Survey name	Year	Sample size (and basic information)*	Methods	Results	Significances	Notes
				had respiratory tract problems. Eight of them had relationship difficulties with their families or coworkers, and they were depressed and tired easily.)	has no negative physical or mental effects.	
Survey Report from Taiwan	2002	1210	By way of a stratified sampling, with questionnaires distribution （selecting 20% of towns and cities for the research）	81% of the respondents quit smoking, 77% quit drinking, 85% quit gambling, and 85% also completely stopped their habit of chewing betel nuts. The satisfaction rate regarding personal health increased from 24% prior to practicing to 78% after practicing, and the rate concerning carrying out daily activities increased from 36% to 81%.	Falun Dafa offered great psychological and mental benefits.	A survey completed by Dr. Hu Yuhui from the Department of Economics at National Taiwan University. Refer [62]
Australian Survey	2016	590 (360 FG & 230 non-FG respondents); Chinese: 47%, n = 170; Caucasians: 24%, n = 88; Australians: 7%, n = 26 (and totally from 29 countries); Males: 42%, n = 151; Females: 57%, n = 206	2 questionnaires	91%, improving *Xinxing* or moral character based on the principles of Falun Dafa practice; 44%, positive change of attitude towards life since practicing Falun DaFa.	The first one including participants in so many countries worldwide.	Refer [63]

Survey name	Year	Sample size (and basic information)*	Methods	Results	Significances	Notes
Others Other surveys Cases reports	Around 1999	Pending further investigation (at least on Changchu, Nanchang, Guangxi, Anhui, Tianjin data, and other surveys performed earlier by Beijing practitioners)** 39	Pending further investigation	Pending further investigation Mind and body improvement in all cultivators of the group.	Genuine practitioners constantly raise their *Xinxing*, besides practicing exercises of Falun Dafa.	"Pending further investigation" due to persecution starting from July 1999; Refer [64, 65, 89]; for individual case, refer dataset through the website (www.minghui.org)

* indicated that the total number being about 36, 208 cases (except ** that included cases number in China still being collected).

Table 1. Nine Survey results from China, North America, Russia Taiwan and Australia as well as many case reports about Falun Dafa practitioners.

The health benefits commonly experienced by Falun Dafa practitioners are very scientific and revealed the intimate link between mind-body-spirit manifested at the molecular level. Scientists have performed some investigations in this field, which provide evidence-based information. A team in University of Texas Southwestern Medical Center and Baylor College of Medicine, Houston initiated a set of studies in 2005 [8] with gene expression differences investigated in the neutrophils of six Falun Dafa practitioners (who had practiced the program for at least 1 year doing daily book reading [66] and daily Falun Dafa exercises lasting 1–2 hours each time), comparing them with a control group of six healthy Asian candidates. Among the approximately 12,000 genes interrogated by microarrays (when the neutrophils isolated from fresh blood with detecting gene expression profiling), 250 genes consistently showed difference in expression between the Falun Dafa practitioners and the control group, with 132 down regulated and 118 up regulated genes. Cellular stress response genes were generally down regulated in Falun Dafa practitioners compared to the control group, but the expression of two of the heat shock proteins was increased. Expressions of some genes that are related to immunity were also increased in the Falun Dafa practitioners' group, such as interferon gamma (IFN-g) and IFN-related and IFN-regulated genes. Results showed that the cells of Falun Dafa practitioners exhibited drastic reduction of metabolism, a key feature of longevity, and enhanced anti-bacteria function at the cellular level; at the molecular level, an advanced biological technique was applied and yielded clean-cut results that demonstrated drastic down-regulation of stress response genes as well as proteins involved in protein synthesis and protein degradation. The studies showed that the ubiquitin (characterized as a stress-inducible protein) [67, 68] related pathways and apoptosis can be affected, suggesting that Falun Dafa practices give rise to gene expression changes consistent with improved response to environmental stress, and improve the survival of immune cells (**Figure 5**).

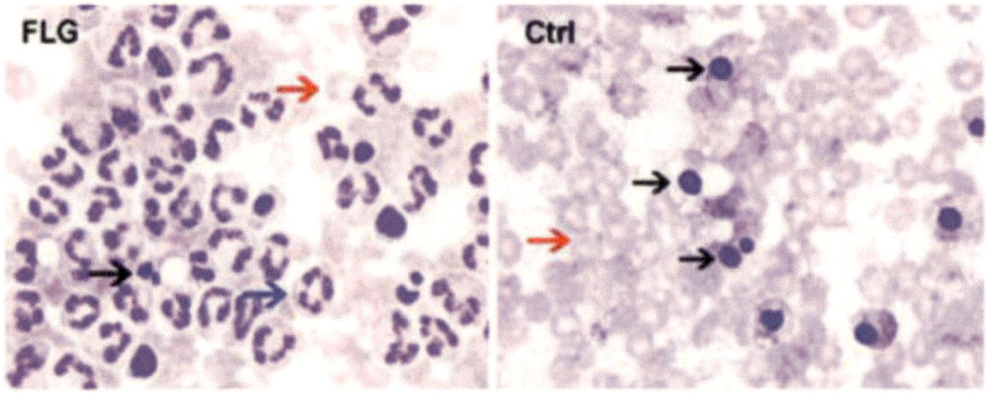

Figure 5. Micrographic photos of altered neutrophil apoptosis in the absence of lipopolysaccharide (LPS) after 16-hour culture.

Because delayed neutrophil apoptosis has been linked to a number of chronic inflammatory injuries resulting from enhanced immunity [69], the authors interpreted those data above as beneficial, in terms of the effects on immunity, metabolic rate, and apoptosis through modulating gene. Their finding that Falun Dafa practices lead to resending gene expression changes is consistent with the finding in the aforementioned surveys conducted among practitioners. Their studies also indicate that modern technology may be used as a scientific tool to study the molecular mechanism of health benefits seen in people practicing spirituality or employing complementary and alternative medicine (**Figure 5**) [8, 70].

Most of the neutrophils from a Falun Gong practitioner (left) were alive without apoptosis in the absence of LPS. In contrast, neutrophils from a normal control (right) were apoptotic. (Blue arrow, normal neutrophils; black arrow, apoptotic cell indicated as shrunken neutrophils with chromatin condensation, rounded nuclear profiles, and presence of cytoplasmic vacuolization; red arrow, red blood cell.) Wright-Giemsa stain. Original magnification: ×400. (Refer [70])

(Most of the neutrophils, when stimulated with LPS (25 ng/mL), were apoptotic in the Falun Gong practitioner but alive in the normal control. Refer [8])

A research outcome entitled "Study at Cellular Level on the Psychological and Physical Healing Effects of Falun Dafa Meditation" (by Jason Liu and Gwendalle Cooper) was presented at the 65th Annual International Council of Psychologists (ICP) Conference, held at San Diego in August 2007 [71]. Psychologists from around the world shared their results probing the relationship between mental health and environment, and the scope of research is to promote multi-cultural relations, peace between ethnic groups, as well as health. Dr. Liu talked about his research on how Falun Dafa improved practitioners' situation at the cellular, psychological and energetic levels. His data showed practicing Falun Dafa helped significantly in reducing practitioners' mental stress, improving their mental and physical health, healing diseases, enhancing mental and moral levels, and developing human potential and intelligence.

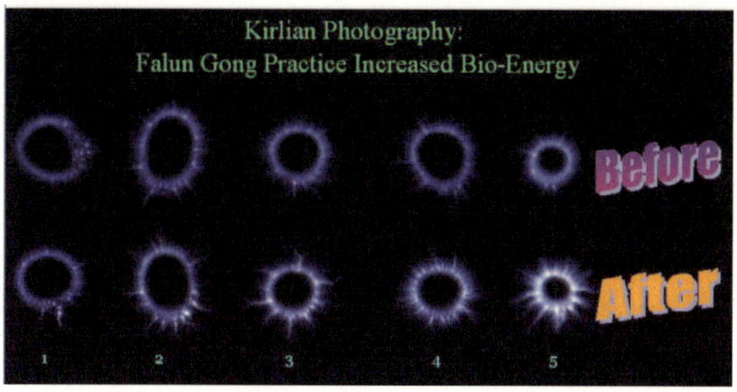

Figure 6. Bio-energy photography showed that practicing Falun Dafa was able to intensify people's energy field and to improve energy circulation in the human body. The numbers 1–5 were bio-energy photographs taken of a practitioner's thumb, index finger, middle finger, ring finger, and small finger (left hand). The bio-energy on the right hand was the same and was not shown in the picture.

Dr. Liu carefully analyzed his results from different scientific point of views. One kind of results (**Figure 6**) showed bio-energy photography taken of a practitioner's fingers before and after an hour-long meditation. The photography was taken with a commonly used high-voltage high-frequency Kirlian CV6000 photoelectric technology with a Sony DCR-VX2000 NTSC camera. It indicated that bio-energy in a practitioner's finger increased significantly after meditating for an hour, means better peripheral circulation. The 10 fingers connected peripheral meridians and acupuncture points. This experiment proved the practice opens the body's meridians and increases life energy. The results provided scientific evidence of health benefits of Falun Dafa from the perspectives of energy medicine and Chinese medicine.

During the psychological experiment, the Heart Math Monitor developed by the Heart Math Institute was utilized by Dr. Liu to record and calculates the Heart-Brain Entrainment ratio of a practitioner in the process of meditation. The ratio was used as an index of mental health and showed whether a person's mind was pure and in harmony with the person's body. The result indicated that practitioners' Heart-Brain Entrainment Ratio went from 40% before practicing to 94% after practicing. It demonstrated that practitioners, after improving their personality and mind in cultivation, experience enhanced mental clarity.

The research on the healing power of Falun Dafa added an alternative approach to modern science and medicine. It created opportunities in the future to introduce people to a purer insight of the real meaning of life within the greater context of the universe–and the science behind cultivating a human life. Liu's research also focuses on combining modern technology and traditional natural healing methods from China such as meditation, hypnosis, music and energy to heal illnesses. This research got a warm welcome from the psychologists in the conference. The study and other works by Dr. Liu have extended to a book published [72].

It has been reported that Falun Dafa exhibits very dramatic and powerful effects on practitioners. Falun Dafa practitioners have shared amply about the kinds of benefits they have experienced through doing the practice. One recent case [73, 74], for example, was about how a 67-year-old female practitioner recovered from suffering complex fractures at her right leg and foot—knee joint fracture (comminuted fracture of proximal end of tibia) with articular surface involved, and ankle joint facture (trimellolar fracture, comminuted fracture) with multiple fragments, as well as cuboid bone fracture due to a traffic accident. This lady's fractures appeared a great degree of recovery after she has diligent practiced Falun Dafa excises in home for 46 days (showing visualized fracture healing), without receiving any formal orthopedic intervention treatment. She then has had entire recovery, and afterwards participated in Falun Dafa related memorial events in New York City, USA in May 2016, and marched in the 17th World Falun Dafa Day parade after travelling internationally from her home all by herself (a little more than half year after the traffic accident) (**Figure 7**).

During the 2016 ASCO Annual Meeting held from June 3 to June 7, a group of investigators presented their results of an observational cohort study on terminal cancer survivors practicing Falun Dafa in China [75]. After 152 terminal cases (predicted survival, PS $\leq$ 12 months, using the NIH SEER data if the treating physician's Clinical Prediction of Survival (CPS) was unavailable) of Chinese cancer patients between 2000 and 2015 in China were collected through a web platform's (www.minghui.org) database searching function, and the participating candidates deemed eligible for the quality of life (QoL) evaluation, Falun Dafa practitioners' data who have had different types of primary tumor, such as cancer in lung (n = 38), liver (n = 29), stomach (n = 17), leukemia (n = 12), esophagus (n = 10), gynecological (n = 9), pancreas/bile duct (n = 8), colorectal (n = 7), and other organs (n = 22), were included. They found, as of the report date, 149 patients were still alive, among this 152 Chinese cohort (the onset age was 53.3 $\pm$ 15.6 years; the Falun Dafa practice duration was 53.1 $\pm$ 58.9 months), with excited values in several parameters like, time to effect was 1.3 $\pm$ 1.7 months, time to symptom recovery was 3.6 $\pm$ 3.3 months, and symptom free survival (SFS) was 52.7 $\pm$ 61.1 months. A total of 147 patients (96.7%) reported complete symptom recovery with 60 patients confirmed by treating physicians, and QoL after Falun Dafa practice was significantly improved. Investigators drew the interim conclusion that terminal cancer patients practicing the advanced self-cultivation practice survive significantly longer, in addition to seeing notable improvement in cancer symptoms, and they continue to make more profound observations in related investigations on a broader range.

Modern medicine tended to believe that human illnesses were caused by social, environmental, and biological factors. These three factors, to a large extent, affected the body's system through mental impact, which eventually led to illnesses or a generally unhealthy physical condition. This showed that psychology was an important factor regarding one's health.

4.3. Achievement of well-being

Just like what was mentioned in the survey report drawn up by The Chinese National Sports Bureau in 1998, "The unusual phenomena exhibiting among Falun Gong practitioners indicated that Falun Dafa has extraordinary supernormal power. In summary, the unusual phenomena observed among Falun Dafa practitioners far exceeds what can be explained by modern medicine. These phenomena deserve thorough discussions and researches by the medical and scientific communities. It has a very positive impact toward improving the physical and mental well-being of all people, and suggests an all new possibility for the further advancement of science" [62].

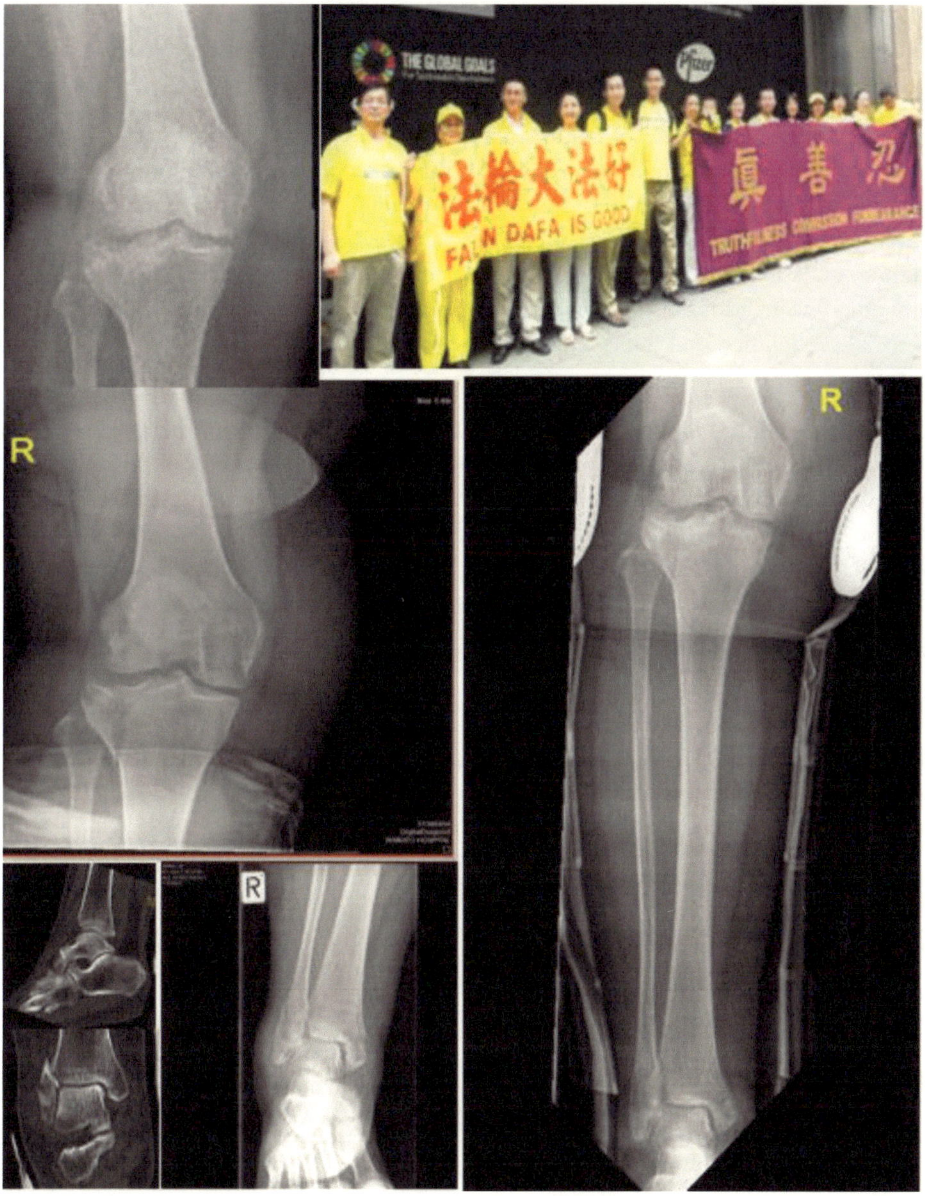

Figure 7. A report (with X-ray and CT images) on the effect of Falun Dafa practicing in curing multiple-site comminuted fractures (at 67 years old female's body due to traffic accident). Lower left, complex fractures (on November 7, 2015) shown at the senior lady's right leg and foot—knee joint fracture (comminuted fracture of proximal end of tibia) with articular surface involved, and ankle joint facture (trimellolar fracture, comminuted fracture) with multiple fragments, as well as cuboid bone fracture (not shown), lower right, showing visualized fracture healing on her knee and ankle joints X-ray images (on December 4, 2015); upper left, showing (on December 23, 2015) almost disappearance of fracture line at her right knee joint (ankle joint not shown), upper right, presenting a picture on which her healthy status was obvious when she (left two) has marched in the 17th World Falun Dafa Day parade at New York city (May 13, 2016) with her old friends. (Refer [73, 74].)

From our point of view, practicing Falun Dafa is about achieving a state of holistic well-being. Curing illnesses is not the goal of practicing Falun Dafa, since the practice is a spiritual cultivation of both the mind and body.

One of the distinctive features of Falun Dafa cultivation is that it puts elevating one's mental state as the top priority in one's cultivation. It requires practitioners to follow the principles of Truthfulness-Compassion-Forbearance, in order to truly improve one's mental state by elevating moral character. This will in turn allow one to reach an ideal physical state. By cultivating one's inner state of mind as well as doing physical exercises, achieving physical health is a natural outcome. In addition, cultivating in Falun Dafa helps end harmful addictions, improve one's social capabilities, and thus generate an overall positive social impact.

Aside from the power of healing illness and improving health as discussed above, Falun Dafa emphasizes improving one's moral character, and guides practitioners to become more honest, kind, tolerant, and peaceful. Its discourse lays emphasis on looking inward when faced with conflicts, be considerate of others, put others first with kindness, and help those in need. The teachings of Falun Dafa can guide a dedicated practitioner to an advanced, transcendent state.

4.4. Practice is free of charge and easy to begin

4.4.1. General information

Falun Dafa practice begins at a high plane right from the outset, thus providing the most expedient, fast, ideal, and precious means of practice for those with a predestined connection or who have been practicing for years using other means but failed to develop Gong.

Falun Dafa practice is an advanced self-cultivation practice. The Chinese word for cultivation is "修炼"（"Xiulian"） and it was used to describe those ancient practices which had adopted religious-sounding names. In the aftermath of the "Cultural Revolution" in China, however, people saw that using the word "Xiulian" could lead to trouble from the authorities, they therefore formulated a new word, "qigong," a term made up of two words that were more acceptable: "Qi" (pronounced "chee"), meaning universal life energy, and "Gong," meaning cultivation energy, which is a higher and more refined substance than Qi.

In the text FALUN GONG [76], Master Li said:

> "Falun Gong is Buddhist qigong, but it far exceeds the scope of the Buddhist system: what we cultivate is the entire universe. In the past, Buddhist cultivation only taught Buddhist principles, while Daoist cultivation only taught Daoist principles. Neither one gave a full explanation of the universe from its fundamental level. The universe is similar to human beings in that it has its own nature, along with its material composition. Its nature can be summarized in three words: Zhen, Shan, Ren (('juhn, shahn, ren'). Zhen means 'true, truth, real, truthfulness'; Shan, 'compassion, benevolence, kindness, goodness'; Ren, 'fortitude, forbearance, endurance, tolerance, patience, self-restraint, self-control'). Daoist cultivation focuses its understanding on the Zhen part: telling the truth, doing truthful deeds, returning to one's original, true self, and finally becoming a true

> being. Buddhist cultivation focuses on the Shan part: developing great compassion, and offering salvation to all beings. With our discipline, we work on all three of Zhen, Shan, and Ren, and we directly adhere to the fundamental nature of the universe in our cultivation, ultimately assimilating with the universe."

Master Li's teachings are set forth in a number of texts, among which are included Falun Gong, Zhuan Falun, The Great Perfection Way of Falun Dafa, Essentials for Further Advancement, and Hong Yin (The Grand Verses). These and other works have been translated into 38 languages and are published and distributed worldwide [58, 66].

The focus of Falun Dafa practice is the mind, with the cultivation of one's mind and thoughts, or "*Xinxing*," being singled out as the key to increasing Gong energy. The height of a person's Gong is directly proportionate to that of his (or her) *Xinxing*. The concept of "*Xinxing*" encompasses the transformation of virtue (a white form of matter) and karma (a black form of matter). It also includes forbearance, discernment, and abandonment—that is, forsaking ordinary human desires and attachments, and managing to endure the most trying of ordeals. Much is encompassed by the concept.

Falun Dafa also includes the cultivation of the body, which is accomplished by performing specific exercises. One purpose of the exercises is to strengthen the practitioner's supernatural abilities and energy mechanisms by means of his or her powerful Gong force. Another purpose is to develop many living entities in the practitioner's body. In advanced practice, the Immortal Infant will come into being and many abilities will be developed. The exercises of Falun Dafa are necessary for the transformation and cultivation of such things. A comprehensive mind-body cultivation system such as this requires both self-cultivation and physical exercises, with cultivation taking priority over exercises. A person's Gong simply will not increase if he or she merely does exercises while failing to cultivate *Xinxing*. The exercises are thus a supplemental means to achieving spiritual perfection.

Falun Dafa "brings a person to a state of wisdom and harmonious existence. The movements of the practice are concise, as a great way is extremely simple and easy" [58]. Falun Dafa is unique in eight ways:

1. A Falun is cultivated, rather than an energy elixir;
2. The Falun refines the person even when he or she is not doing the practice's exercises;
3. One's primary consciousness is cultivated, such that it is the person him or herself who obtains Gong energy;
4. Both mind and body are cultivated;
5. The practice consists of five exercises, which are simple and easy to learn;
6. The mind is not used to direct anything, there are no associated risks, and Gong energy increases quickly;
7. Location, time, and direction are not of concern when exercising, nor is how one concludes one's exercise session;

8. Protection is provided by the master's Fashen, so one needn't fear harm from malevolent entities.

The five sets of exercises include: The First Exercise, Buddha Stretching a Thousand Arms; The Second Exercise, Falun Standing Stance; The Third Exercise, Penetrating the Cosmic Extremes; The Fourth Exercise, Falun Cosmic Orbit; The Fifth Exercise, Reinforcing Supernatural Powers (refer to The Great Perfection Way of Falun Dafa) [66].

Especially after the main text Zhuan Falun was published in 1995, under most circumstances, people in China learned of Falun Dafa by the sharing of information among individuals, meaning word-of-mouth sharing out of goodwill. Reading books is always the first step. Practitioners who began the practice earlier than others voluntarily set up exercise sites at locations such as neighborhood parks, formed home-based Zhuan Falun studying groups, and practitioners shared studying experiences. Practitioners spontaneously came together to do exercises in groups of varying sizes in parks or other neighborhood venues in the morning and during weekend. Coworkers from the same working place often did exercises together during break time or lunch time. Master Li travelled and gave lectures in Europe, other parts of Asia, and Australia. Since 1996, he has only made public appearances to speak at annual Experience-Sharing Conferences at the invitation of practitioners around the world, including North America. Aside from those occasions, Master Li prefers to live a private life so as not to interfere with practitioners' cultivation practice [55].

With a heart of compassion, practitioners of Falun Dafa around the world are spreading the word about this advanced self-cultivation practice to the public through various means. For instance, the Falun Dafa Club at Columbia University (in USA), established in August 1999, hosts an annual information exhibition event on campus from October 17 to 20 to inform the truth about Falun Dafa to staff, students, and visitors alike. And recently, the student-run Falun Dafa Club at the University of the West of England (in UK) participated in the annual Fresher Fair on September 16, 2016 for the second time, and nearly 140 people registered to join the club [77, 78].

Regardless of how one learns of the practice, it's easy to find more information by going online to Falun Dafa's official website, and access books, and audio and video files on the teachings of Falun Dafa free of charge. This is a good approach for individuals to be introduced to the practice. Sometimes, friends or acquaintances who have already begun the practice often offer newcomers with suggestions or recommendations on how to go about with the practice.

The exercises have characteristics of being natural, slow and smooth without any strenuous movements. They are very good for people of all ages (including young children). A 12-year-old boy in Toronto, when asked recently what Falun Dafa had done for him, said, "I used to like to boss around the other kids at school, but I don't want to do that anymore." Many practitioners enjoy practicing with a group because of the strong energy field that is generated, which radiates peace and serenity. Chances are, you might have seen practitioners doing exercises early in the morning in a park close to your home.

It was estimated that since years ago, there had already been several thousand people in Canada practicing Falun Dafa. And in most major cities, one can find Falun Dafa information centers and practice sites. Through learning and practicing Falun Dafa, practitioners have improved their health, relieved the stress in their lives, uplifted their moral and ethical standards and are achieving gradual spiritual enlightenment.

In India, after Falun Dafa was introduced to the Department of Public Instruction, Karnataka, the Commissioner expressed a keen interest in introducing Falun Dafa to all primary school teachers in the state. On January 17–18, 2012, the director of the Department of State Education and Research Training and the principal and deputy director of Public Instruction in Bangalore organized a pilot program to introduce Falun Dafa to the coordinators of the government schoolteachers. The program was conducted by six practitioners at the District Institute of Education and Training, Bangalore. It was attended by 70 teachers from the rural districts of Karnataka. Through a presentation, practitioners shared their understandings. The five sets of exercises were then taught to the entire group. The next day, additional teachings of the exercises were conducted in smaller groups to ensure accuracy of the movements, followed by group reading of *Lunyu* (*On Dafa*, Master Li's recent article) [79] and a portion of Lecture One of the main book Zhuan Falun. The participants were happy to learn and experience Falun Dafa. Over the 2 days, they experienced peace and harmony through the practice. They said that they felt energy was being activated in their bodies and expressed their happiness at being introduced to this wonderful cultivation practice. Most of them said in their written feedback that Falun Dafa seemed very good for them and that they will spread the practice to the schools under their jurisdiction [80].

The above was an example showing that, local practitioners sharing what they have achieved through the practice to the person-in-charge of an institute (or an organization), so that more people could get the chance of being introduced to the principles of Falun Dafa. It should also be noted that in the context of this scenario, the term "program" only refers to the various means with which practitioners introduced the practice of Falun Dafa to others over a 2-day duration, rather than a fixed curriculum practitioners follow.

The practice is completely voluntary, where people choose to take part in practice-related activities (or not) of their own free will. This is adhered to by Falun Dafa practitioners around the world. Therefore, there is no real sense of "joining a program" other than the fact that anyone who thinks of himself or herself as a dedicated practitioner studies Falun Dafa texts and does the exercises regularly. Additionally, practitioners can set up teaching program for children and teenagers for free to learn more about traditional Chinese culture, give aid in their studying of the main book Zhuan Falun.

In view of the above, we have chosen the term "scenario" to describe in below some of the ways with which practitioners around the world study and advance themselves in Falun Dafa.

4.4.2. Several scenarios

After the year 1999, in countries and regions outside of China, the most commonly seen scenario is as follows: practitioners having the available time and a strong sense of community often volunteer to set up exercise sites so that other practitioners from the neighborhood could do the exercises in group. Such exercise sites, since they are mostly public venues, also serve the function of introducing the practice to interested individuals whose first encounter of Falun Dafa are usually seeing practitioners doing group exercises at public venues. For those who are new to the practice and the exercises, more experienced practitioners could assist the newcomers in studying the texts and making sure they are doing the exercises correctly. Such group activities also make it easy for practitioners to share their cultivation experience in a group setting.

In the scenario where an interested individual has some kind of disability or is suffering from illness and cannot do the exercises due to his/her physical condition, other practitioners could instead introduce Master Li's main book Zhuan Falun to such individuals for them to get started in the practice.

When an individual studies the texts of Falun Dafa regularly and has made up his or her mind to follow the principles of Truthfulness-Benevolence-Forbearance, the individual becomes a practitioner and is enlightened to the essence of this advanced self-cultivation practice as long as he/she remains a genuine practitioner.

Master Li said, in the book of "The Great Way of Spiritual Perfection":

> "Falun Dafa cultivators must cultivate their character, along with performing the movements. Those who focus solely on the exercises but neglect character cultivation will not be acknowledged as Falun Dafa disciples. Dafa students thus need to make studying the Law and reading the books the essential part of their daily cultivation" [81].

Falun Dafa practitioners strive to become better people through improving "*Xinxing*" ("心性," in Chinese), a term generally translated as "character" (perhaps initially understood as "heart-nature"). By placing others before themselves and letting go of material and emotional attachments, practitioners of Falun Dafa continue the Daoist tradition of non-intention or the Buddhist creed of giving up worldly desires. Like what has been discussed before, the cultivation of one's mind and thoughts will be regarded as the top priority in one's practice.

Many raise this question: if a student who studies Falun Dafa mainly focuses on doing the exercises, could he/she advance in the practice without putting effort into cultivating one's *Xinxing*? We should think about this: those who only focus on doing the exercises will eventually observe a disparity between their overall status and the achievement of genuine well-being.

It is very interesting to note that in the survey results listed above, those who achieved the best results didn't use any medical treatment. They did not administer alternative medicine, take vitamin or mineral supplements, or other natural health products. The application of such products may be considered as being inconsistent with the advanced self-cultivation and may even disrupt practitioners' health improvement process.

When a Falun Dafa disciple's *Xinxing* and the strength of his or her Gong reach a certain height, he or she can attain an imperishable, adamantine body while still in the secular world. A person can also achieve the "unlocking of Gong," enlightenment, and ascension of the whole person to higher planes. Those with great determination should study this upright teaching, strive to achieve their ultimate rank, elevate their *Xinxing*, and forsake their attachments. Only then is spiritual perfection possible.

In the text FALUN GONG, Master Li said:

> "It is easy to be a good person, but it's not easy to cultivate character—a cultivator must ready his mind. Sincerity is a prerequisite if you are to rectify your mind. People live in a world where society has become complicated. Though you might want to do good things, there might be some people who don't want you to; you might not want to harm others, but others might harm you for various reasons. Some of these things happen for unapparent reasons. Will you be able to enlighten to the reasons? Then what should you do? The challenges in this world test your character at every moment. When experiencing indescribable humiliation, when losing out, when tempted by money and lust, when in a power struggle, when rage and jealousy emerge in conflicts, when discord in society and in the family take place, and when experiencing all kinds of hardships, can you always follow closely the character criteria? Of course, if you could handle everything then you would already be an enlightened being. Most practitioners start as everyday people after all, and the cultivation of their character occurs gradually; it moves upward little by little. Determined cultivators will eventually gain a Noble Attainment (zheng-guo) if they are prepared to endure great hardships and face ordeals with an unwavering mind." [82]

In China today, the scenario is notably different due to the nationwide and state sanctioned persecution against Falun Dafa, which began in full swing since July 20, 1999 [83]. Prior to the onset of the persecution, practitioners in China studied the texts of Falun Dafa and did the exercises often in groups and in public venues much in the same way as those in the rest of the world, as described in the scenarios above. The persecution could be partially explained by the Chinese Communist Regime's (the CCP's) inherent tendency to control and suppress all individuals and groups that adhere to a different set of ideologies. Naturally, all spiritual practices in mainland China, including various expressions of religion, have been the targets and victims of persecutions by the CCP. The persecution against Falun Gong, in terms of its scale and scope, and the extent of egregious physical and psychological abuses imposed on practitioners, however, make this possibly the worst spiritual persecution we have seen in this day and age. Against this backdrop, practitioners in mainland China, in adherence to the principles of Truthfulness-Compassion-Forbearance, began an exceptionally arduous peaceful resistance effort against the persecution. On top of which, they have remained faithful to the practice and diligent in studying the texts and doing the exercises of Falun Dafa, mostly in secrecy and away from the prying eyes of state agents, so as to avoid risking their freedom and sometimes even their lives.

One other scenario worth noting are people who became practitioners after the persecution had begun in 1999. For those in China, doing so meant exposing themselves to great personal risks. Despite the smear campaign launched by the CCP to misinform the public about Falun

Dafa, many individuals, both within and outside of China, have learned about the truth of this invaluable cultivation practice. Through their own rational judgment, these people have chosen to become practitioners because they have come to understand that the wisdom and overall well-being in both the spiritual and physical sense one receives from Falun Dafa cultivation are of inestimable worth.

It is also indicated that, the current technology and the persecution in China limit scientific research on Falun Dafa, means the whole picture of benefits that can be delivered by Falun Dafa cannot be shown so far due to the persecution, and this point was reflected during above mentioned 65th Annual International Council of Psychologists (ICP) Conference in 2007 [71]. Dr. Jason Liu said the result he presented during those days was only a small part of what the practice can achieve. The current technology and the persecution in China limit scientific research on much more topics in Falun Dafa. But his research received a warm welcome from conference participants. When the psychologists learned that Falun Dafa has been under the Chinese Communist regime's persecution for so many years (since 1999), they showed their support and sympathy for Falun Dafa. They hoped to introduce this profound practice to others and wished to personally learn more about Falun Dafa to assist with their research on psychology and medicine. Many psychologists agreed that future psychological research and validation should not be confined only to western medicine that focuses on and treats only the symptoms of illness. Psychology should seriously begin researching the theory of eastern traditional health care such as Falun Dafa, a universal view of body and mind practice, and energy medicine. From achievement of health and well-being perspective, we believe, with support from those scientists like above psychologists showing righteous thoughts to Falun Dafa, investigators like in the study team [75] in the 2016 ASCO Annual Meeting will present much more data, if without persecution, from scientific research on terminal cancer survivors practicing Falun Dafa and other beneficial cohorts with more larger sample size, for people in all over the world.

5. Conclusion

Results from NIH-funded studies on CAM mind-body therapies include a lot [84]. Reports on various mind-body therapies to help at least treat certain neurological diseases involving pain providing some evidence for positive effects from some therapies. A spiritual perspective in medical practice and research had been adopted by some mainstream medical journals around the year of 1999 [85–88]. The origin of Chinese music can be dated back to distant antiquity, with ancient Chinese instruments sharing a deep connection with Heaven and Earth. Music in Shen Yun Performing Arts features the perfect harmony of classical music of East and West, and leading the melody amidst a full Western orchestra, so that they together create a profound musical experience that resonates deep in the heart, delivering pure energy. To revive traditional culture has benefits including letting people keep in moral level in daily life, and there is an example from ancient China, that people respect the Heaven to possess the five cardinal virtues of benevolence, righteousness, propriety, wisdom, and faithfulness.

It has been reported that Falun Dafa exhibits very dramatic and powerful effects on practitioners [89], and Falun Dafa is a mind-and-body practice that incorporates meditation with spiritual improvement, known traditionally as cultivation. Guided by the supreme principle of Truthfulness-Benevolence-Forbearance to always cultivate their hearts and minds, Falun Dafa practitioners strive to have increasingly better behaviors in all environments and conditions, and always try best to give up various bad habits and attachments, like those of being anger, anxiety, jealousy and having desire for personal fame and gain, etc. Falun Dafa practitioners also conduct the five sets of exercises to achieve bodies purified and energy increased. Based on all of these, the spreading of Falun Dafa is helping to improve family harmony, perform co-operation at work and deliver safety in neighborhoods. Simultaneously, this has also encouraged care for the environment, stabilization in societies, and goodness in human's hearts. Falun Dafa teaches practitioners to live by basic principles of the self-practice system combined with meditation exercises, to become very good people in the society, having genuine well-being as well.

Acknowledgements

The authors appreciate all the efforts by the staff collecting the materials, documents and data cited in the article, having contribution to dataset of the websites of "Minghui" (www.minghui.org), "EPOCH TIMES" (http://www.theepochtimes.com), Shen Yun Performing Arts (www.shenyunperforming arts.org), and others (such as, www.falundafa.org, NTDTV, etc.), and the research teams in different institutes for sharing all the related data. We also appreciate the chance of us to do discussion for the editing and sharing with each other based on our understanding and enlightment, which are not representing the entire contents of the advanced self-cultivation practice (herein Falun Dafa). S. W. reports no conflict; while both L. J. doing as an independent translator and J. W. report no conflict.

Author details

Shawn Wu*, Lin Jiang and J. Wang

*Address all correspondence to: wushawn@unseen.is

Falun Dafa, Association of New England, USA

References

[1] Barnes PM, Bloom B, Nahin RL. Complementary and alternative medicine use among adults and children: United States, 2007. National Health Statistics Report. 2008;**10**(12):1-23

[2] Hanser SB. From ancient to integrative medicine: Models for music therapy. Music and Medicine. 2009;**1**:87-96

[3] Tang YY, Ma Y, Wang J, Fan Y, Feng S, Lu Q, Yu Q, Sui D, Rothbart MK, Fan M, Posner MI. Short-term meditation training improves attention and self-regulation. Proceedings of the National Academy of Sciences. 2007;**104**(43):17152-17156

[4] Aung SK, Fay H, Hobbs RF III. Traditional Chinese medicine as a basis for treating psychiatric disorders: A review of theory with illustrative cases. Acupuncture in Medicine. 2013;**25**(6):398-406

[5] Gimbel MA. Yoga, meditation, and imagery: Clinical applications. Nurse Practitioner Forum. 1998;**9**:243-255

[6] Sovik R. The science of breathing: The yogic view. Progress in Brain Research. 2000;**122**: 491-505

[7] Krucoff MW, Crater SW, Green CL, Maas AC, Seskevich JE, Lane JD, Loeffler KA, Morris K, Bashore TM, Koenig HG. Integrative noetic therapies as adjuncts to percutaneous intervention during unstable coronary syndromes: Monitoring and actualization of noetic training (MANTRA) feasibility pilot. American Heart Journal. 2001;**142**:760-769

[8] Li QZ, Li P, Garcia GE, Johnson RJ, Feng L. Genomic profiling of neutrophil transcripts in asian qigong practitioners: A pilot study in gene regulation by mind-body interaction. Journal of Alternative and Complementary Medicine. 2005;**11**(1):29-39

[9] Cha KY, Wirth DP, Lobo RA. Does prayer influence the success of in vitro fertilization embryo transfer? Report of a masked, randomized trial. The Journal of Reproductive Medicine. 2001;**46**:781-787

[10] Kabat-Zinn J. An outpatient program in behavioral medicine for chronic pain patients based on the practice of mindfulness meditation: Theoretical considerations and preliminary results. General Hospital Psychiatry. 1982;**4**(1):33-47

[11] Teasdale JD, Segal ZV, Williams JM, Ridgeway VA, Soulsby JM, Lau MA. Prevention of relapse/recurrence in major depression by mindfulness-based cognitive therapy. Journal of Consulting and Clinical Psychology. 2000;**68**(4):615-623

[12] O'Connell PM. Can in-school meditation help curb youth violence? Chicago Tribune [Internet]. January 2, 2017. Available from: http://www.chicagotribune.com/news/ct-classroom-meditation-disadvantaged-students-met-20161231-story.html [Accessed: 10-January-2017]

[13] Shen Yun Performing Arts. An Introduction to Traditional Chinese Culture [Internet]. Available from: http://www.shenyunperformingarts.org/learn/article/read/item/uXrrqgyDZPo/level-one/ItNLnxh8WBE [Accessed: 10-January-2017]

[14] Wallace RK, Benson H, Wilson AF. A wakeful hypometabolic physiologic state. American Journal of Physiology. 1971;**221**(3):795-799

[15] Dusek JA, Benson H. Mind-body medicine: a model of the comparative clinical impact of the acute stress and relaxation responses. Minnesota Medical Solutions. 2009;**92**(5):47-50

[16] Cannon W. Emergency function of the adrenal medulla in pain and the major emotions. American Journal of Physiology. 1914;**33**:356

[17] Benson H. Steps to Elicit the Relaxation Response [Internet]. Available from: http://www.relaxationresponse.org/steps/ [Accessed: 10-January-2017]

[18] Mindful Family. Welcome to Mindful Family [Internet]. Available from: http://www.mindfulfamily.co.uk/about-1/brain-science/ [Accessed: 10-January-2017]

[19] Relaxation Techniques - UC Davis Health System [Internet]. Available from: https://www.ucdmc.ucdavis.edu/hr/hrdepts/asap/Documents/Relaxation_Techniques.pdf [Accessed: 10-January-2017]

[20] Kung WW. Cultural and practical barriers to seeking mental health treatment for Chinese Americans. Journal of Community Psychology. 2004;**32**(1):27-43

[21] Abe-Kim J, Takeuchi DT, Hong S, Zane N, Sue S, Spencer MS, Appel H, Nicdao E, Alegría M. Use of mental health-related services among immigrant and US-born Asian Americans: Results from the National Latino and Asian American Study. American Journal of Public Health. 2007;**97**(1):91-98

[22] Freedenthal S, Stiffman AR. "They Might Think I Was Crazy": young American Indians' reasons for not seeking help when suicidal. Journal of Adolescent Research. 2007;**22**(1):58-77

[23] Loya F, Reddy R, Hinshaw SP. Mental illness stigma as a mediator of differences in Caucasian and South Asian college students' attitudes toward psychological counseling. Journal of Counseling Psychology. 2010;**57**(4):484-490

[24] Eisenberg D, Downs MF, Golberstein E, Zivin K. Stigma and help seeking for mental health among college students. Medical Care Research and Review. 2009;**66**(5):522-541

[25] Chu JP, Sue S. Asian American mental health: What we know and what we don't know. Ocean Renewable Power Company. 2011;**3**(1):1-18

[26] Gee CB. Assessment of anxiety and depression in Asian American youth. Journal of Clinical Child & Adolescent Psychology. 2004;**33**(2):269-271

[27] Kung WW. Chinese Americans' help seeking for emotional distress. Social Service Review. 2003;**77**(1):110-134

[28] Jenkins GR, Hale R, Papanastassiou M, Crawford MJ, Tyrer P. Suicide rate 22 years after parasuicide: a cohort study. The British Medical Journal. 2002;**325**(7373):1155

[29] Runeson B, Tidemalm D, Dahlin M, Lichtenstein P, Langstrom N. Method of attempted suicide as predictor of subsequent successful suicide: National long term cohort study. The British Medical Journal. 2010;**341**

[30] Biglan A, Metzler CW, Wirt R, Ary D, Noell J, Ochs L, French C, Hood D. Social and behavioral factors associated with high-risk sexual behavior among adolescents. The Journal of Behavioral Medicine. 1990;**13**(3):245-261

[31] Cheng JKY, Fancher TL, Ratanasen M, Conner KR, Duberstein PR, Sue S, Takeuchi D. Lifetime suicidal ideation and suicide attempts in Asian Americans. Asian American Journal of Psychology. 2010;**1**(1):18-30

[32] Hahm HC, Lee CH, Choe J-YU, Ward A, Lundgren L. Sexual attitudes, reasons for forgoing condom use, and the influence of gender power among Asian-American women: A qualitative study. Journal of AIDS & Clinical Research. 2011;**S1**:1–8

[33] Yeung A, Slipp LE, Jacquart J, Fava M, Denninger JW, Benson H, Fricchione GL. The treatment of depressed Chinese Americans using Qigong in a health care setting: A pilot study. Journal of Evidence-Based Complementary and Alternative Medicine. 2013;2013:1-5

[34] Yeung A, Lepoutre V, Wayne P, Yeh G, Slipp LE, Fava M, Denninger JW, Benson H, Fricchione GL. Tai Chi treatment for depression in Chinese Americans: A pilot study. American Journal of Physical Medicine and Rehabilitation. 2012;**91**(10):863-870

[35] Yeung A, Slipp LE, Niles H, Jacquart J, Chow CL, Fava M, Denninger JW, Benson H, Fricchione GL. Effectiveness of the relaxation response-based group intervention for treating depressed Chinese American immigrants: A pilot study. International Journal of Environmental Research and Public Health. 2014;**11**(9):9186-9201

[36] Chan AS, Wong QY, Sze SL, Kwong PP, Han YM, Cheung MC: A Chinese Chan-based mind-body intervention for patients with depression. Journal of Affective Disorders. 2012;**142**:283-289

[37] Sreevani R, Reddemma K, Chan CL, Leung PP, Wong V, Chan CH. Effectiveness of integrated body-mind-spirit group intervention on the well-being of Indian patients with depression: A pilot study. Journal of Nursing Research. 2013;**21**:179-186

[38] Wurtzen H, Dalton SO, Elsass P, Sumbundu AD, Steding-Jensen M, Karlsen RV, Andersen KK, Flyger HL, Pedersen AE, Johansen C. Mindfulness significantly reduces self-reported levels of anxiety and depression: Results of a randomised controlled trial among 336 Danish women treated for stage I-III breast cancer. European Journal of Cancer. 2013;**49**:1365-1373

[39] Bredt DS, Snyder SH. Nitric oxide: A physiologic messenger molecule. Annual Review of Biochemistry. 1994;**63**:175-195

[40] Dusek JA, Hibberd PL, Buczynski B, Chang BH, Dusek KC, Johnston JM, Wohlhueter AL, Benson H, Zusman RM. Stress management versus lifestyle modification on systolic hypertension and medication elimination: a randomized trial. Journal of Alternative and Complementary Medicine 2008;**14**(2):129-138

[41] Tawakol A, Ishai A, Takx RA, Figueroa AL, Ali A, Kaiser Y, Truong QA, Solomon CJ, Calcagno C, Mani V, Tang CY, Mulder WJ, Murrough JW, Hoffmann U, Nahrendorf M,

Shin LM, Fayad ZA, Pitman RK. Relation between resting amygdalar activity and cardiovascular events: A longitudinal and cohort study. The Lancet. 2017;**389**(10071): 834-845

[42] Dusek JA, Otu HH, Wohlhueter AL, Bhasin M, Zerbini LF, Joseph MG, Benson H, Libermann TA. Genomic counter-stress changes induced by the relaxation response. PLoS ONE. 2008;**3**(7):e2576

[43] Saatcioglu F. Regulation of gene expression by yoga, meditation and related practices: A review of recent studies. Asian Journal of Psychiatry. 2013;**6**:74-77

[44] Lempinen EW. In a Range of New Science, Researchers Find the Power of Music to Build the Brain. In: AAAS (American Association for the Advancement of Science) Annual Meeting; 26 February 2010; Available from: https://www.aaas.org/news/range-new-science-researchers-find-power-music-build-brain

[45] Shaffer J. Neuroplasticity and positive psychology in clinical practice: A review for combined benefits. Psychology. 2012;**3**(12A):1110-1115

[46] Shaffer J. Neuroplasticity and clinical practice: Building brain power for health. Frontiers in Psychology. 2016;**7**:1118

[47] Paraskevopoulos E, Kuchenbuch A, Herholz SC, Pantev C. Musical expertise induces audiovisual integration of abstract congruency rules. Journal of Neuroscience. 2012;32: 18196-18203. DOI: 10. 1523/JNEUROSCI.1947-12.2012

[48] Bottiroli S, Rossi A, Russo R, Vecchi T, Cavallini E. The cognitive effects of listening to background music on older adults: Processing speed improves with upbeat music, while memory seems to benefit from both upbeat and downbeat music. Frontiers in Aging Neuroscience. 2014;**6**:284

[49] Shen Yun Performing Arts. Music of Shen Yun [Internet]. Available from: http://www.shenyunperformingarts.org/symphony/index?lang=en-us#music-of-shen-yun [Accessed: 11-January-2017]

[50] Shen Yun Performing Arts. Music of Shen Yun [Internet]. Available from: http://zh-tw.shenyunperformingarts.org/symphony/index (in Chinese) (http://www.shenyunperformingarts.org/spirituality/news-and-resources) [Accessed: 11-January-2017]

[51] Hölzel BK, Carmody J, Vangel M, Congleton C, Yerramsetti SM, Gard T, Lazar SW. Mindfulness practice leads to increases in regional brain gray matter density. Psychiatry Research: Neuroimaging. 2011;**191**:36-43

[52] Davidson RJ, McEwen BS. Social influences on neuroplasticity: Stress and interventions to promote well-being. Nature Neuroscience. 2012;**15**:689-695

[53] Hölzel BK, Lazar SW, Gard T, Schuman-Olivier Z, Vago DR, Ott U. How does mindfulness meditation work? Proposing mechanisms of action from a conceptual and neural perspective. Perspectives on Psychological Science. 2011;**6**:537-559

[54] Lutz A, Brefczynski-Lewis J, Johnstone T, Davidson RJ. Regulation of the neural circuitry of emotion by compassion meditation: Effects of meditative expertise. PLoS ONE. 2008;**3** (3):e1897

[55] Minghui. Some Background Information of Falun DaFa [Internet]. Available from: http://en.minghui.org/emh/download/infopack/bkgrnd_info.html [Accessed: 11-January-2017]

[56] The Falun Dafa Information Center. 1992- 1994, Mr. Li travels throughout China giving 54 talks and class series on Falun Gong. Classes typically last 8-10 days, two hours per day. Seminars are often arranged by local government-run qigong organizations. Attendees range from a few hundred to upwards of 6,000 per event [Internet]. Available from: http://faluninfo.net/topic/24/ [Accessed: 11-January-2017]

[57] Clearwisdom.net, Mayor of Houston Proclaims October 12, 1996 "Houston Master Li Hongzhi Day" [Internet]. Available from: http://photo.minghui.org/selected-En/u_master/223707220591.htm [Accessed: 11-January-2017]

[58] Clearwisdom.net, Brief Introduction to Falun Dafa; Photo exhibition dataset of Falun Dafa [Internet]. Available from: http://en.falundafa.org/introduction.html; http://photo.minghui.org/photo/images/exhibition/newest_7.htm [Accessed: 11-January-2017]

[59] Xie FT, Zhu T. Ancient wisdom for modern predicaments: The truth, deceit, and issues surrounding Falun Gong. Cultic Studies Review. 2004;**3**(1):[Internet]. Available from: http://www.icsahome.com/articles/ancient-wisdom-for-modern-predicaments-the-truth–deceit–and-issues-surrounding-falun-gong [Accessed: 11-January-2017]

[60] Epochtimes. Congratulations letter published on (in Chinese, 台灣副總統呂秀蓮致賀法輪功大會(全文)) [Internet]. Available from: http://www.epochtimes.com/b5/2/12/31/n261383.htm [Accessed: 11-January-2017]

[61] Minghui. A report on the effect of Falun Gong in curing diseases and keeping fit based on a survey of 355 cultivators of Falun Gong at certain sites in Beijing, China. 2000-08-03 [Internet]. Available from: http://www.falundafa-pa.net/survey/survey96_e.pdf [Accessed: 11-January-2017]

[62] Minghui. A Summary of Five Independent Studies of the Health Benefits of Falun Gong; Falun Dafa Greatly Improves Health: Survey of over 6,000 Cultivators in Dalian, Liaoning Province [Internet]. Available from: http://en.minghui.org/html/articles/2012/5/6/133105p.html;http://en.minghui.org/html/articles/2001/6/17/11524.html [Accessed: 11-January-2017]

[63] Milner C. Western Counseling Gets a Boost from Eastern Wisdom, written by, at Epoch Times. 2016-07-19 [Internet]. Available from: http://www.theepochtimes.com/n3/2097261-western-counseling-gets-a-boost-from-eastern-wisdom/ [Accessed: 11-January-2017]

[64] McCoy WF, Zhang LJ. Falun Gong Stories: A Journey to Ultimate Health [Internet]. Available from: http://en.minghui.org/emh/download/publications/health_index.html [Accessed: 28-December-2016]

[65] Minghui. Present the truth for history (in Chinese, "向历史展示真實") [Internet]. Available from: http://media.minghui.org/mh/articles/2000/8/3/3315.html [Accessed: 11-January-2017]

[66] Falun Dafa website. Books & Recent Writings of Mr. Li Hongzhi [Internet]. Available from: http://en.falundafa.org/falun-dafa-books.html [Accessed: 01-December-2016]

[67] Bond U, Schlesinger MJ. Ubiquitin is a heat shock protein in chicken embryo fibroblasts. Molecular and Cellular Biology. 1985;**5**:949-956

[68] Bond U, Schlesinger MJ. The chicken ubiquitin gene contains a heat shock promoter and expresses an unstable mRNA in heatshocked cells. Molecular and Cellular Biology. 1986;**6**:4602-4610

[69] Summers C, Rankin SM, Condliffe AM, Singh N, Peters AM, Chilvers ER. Neutrophil kinetics in health and disease. Trends in Immunology. 2010;**31**:318-324

[70] Minghui. Falun Gong practicing can adjust the immune system in both directions (in Chinese, 研究結果：法輪功修煉可雙向調節免疫系統) [Internet]. Available from: http://www.minghui.org/mh/articles/2005/3/18/97594p.html [Accessed: 01-December-2016]

[71] Minghui. San Diego: Scientific Analysis of the Effects of Falun Dafa Presented at International Council of Psychologists [Internet]. Available from: http://www.clearwisdom.net/emh/articles/2007/8/24/88865.html [Accessed: 01-December-2016]

[72] Liu J. Mind Body Medicine & Healthology, Body-mind-spirit Science & Practice. 2014. [Internet]. Available from: https://www.scribd.com/document/300299416/Mind-Body-Medicine-Healthology-Body-mind-spirit-Science-Practice [Accessed: 21-December-2016]

[73] Minghui. A case of 67 years old female suffering vehicle accident (in Chinese, 粉碎性骨折不治而癒) [Internet]. Available from: http://en.minghui.org/html/articles/2016/12/22/160420.html [Accessed: 01-December-2016]

[74] Epochtimes. Practitioners from Tsinghua had re-union in New York after 17 years persecution of Falun Dafa in China (in Chinese, 風雨十七年 十多名清華人紐約聚首). 2016. [Internet]. Available from: http://www.epochtimes.com/b5/16/5/16/n7901085.htm [Accessed: 29-December-2016]

[75] Dong YH, Huang CF, Liao J, Chen CY, Liu J, Hsu KH. An observational cohort study on terminal cancer survivors practicing falun gong (FLG) in China. In: 2016 ASCO Annual Meeting; June 2016; Chicago, USA. (Available from: Journal of Clinical Oncology 34, 2016 (suppl; abstr e21568))

[76] Li H, editor. Falun Gong. Gloucester, MA: Fair Winds Press; 2001. Chapter II.

[77] Minghui. Falun Dafa in Columbia University in November 2016 (in Chinese, 哥倫比亞大學圖片展傳真相) [Internet]. Available from: http://www.minghui.org/mh/articles/2016/11/1/哥伦比亚大学图片展传真相（图）-337103.html [Accessed: 29-December-2016]

[78] Minghui. Over 100 People Join Falun Gong Club at Fresher Fair at the University of the West of England [Internet]. Available from: http://en.minghui.org/html/articles/2016/9/18/159203.html [Accessed: 29-December-2016]

[79] Li H. On Dafa (*Lunyu*) [Internet]. Available from: http://en.minghui.org/html/articles/2012/1/22/130880p.html [Accessed: 29-December-2016]

[80] Minghui. India: Two-Day Program to Introduce Falun Dafa to Primary School Teachers. [Internet]. Available from: http://en.minghui.org/html/articles/2012/1/22/130880p.html [Accessed: 29-December-2016]

[81] Li H. What is Expected of Falun Dafa Assistance Centers, in The Great Way of Spiritual Perfection, Appendix I (version Sep 25, 2014) [Internet]. Available from: http://en.falundafa.org/eng/html/dymf_2014/dymf_2014_4.htm [Accessed: 29-December-2016]

[82] Li H, editor. Falun Gong. Gloucester, MA: Fair Winds Press; 2001. Chapter III.

[83] The Falun Dafa Information Center. Overview of Persecution [Internet]. Available from: http://www.faluninfo.net/category/1/[Accessed: 29-December-2016]

[84] NIH. Mind-Body Medicine Practices in Complementary and Alternative Medicine [Internet]. Available from: https://report.nih.gov/NIHfactsheets/ViewFactSheet.aspx?csid = 102 [Accessed: 29-December-2016]

[85] Emanuel EJ, Daniels ER, Fairclough DL, Clarridge BR. The practice of euthanasia and physician-assisted suicide in the United States: Adherence to proposed safeguards and effects on physicians. Journal of American Medical Association. 1998;**280**:507-513

[86] Chandler E. Spirituality. Journal of Hospital Infection. 1999;**14**:63-74

[87] Post SG, Puchalski CM, Larson DB. Physicians and patient spirituality: Professional boundaries, competency, and ethics. American College of Physicians 2000;**132**:578-583

[88] Tracey KJ. The inflammatory reflex. Nature. 2002;**420**:853-859

[89] Zhengjian. Surveys collection for Falun Dafa practitioners' health status (in Chinese, "中國大陸對於法輪功祛病健身效果醫學調查報告總結") [Internet]. Available from: http://www.zhengjian.org/zj/book/html/yxqj/y183.htm [Accessed: 20-January-2017]

Chapter 3

Quality of Life in Venous Diseases of the Lower Limbs

Mahmut Surmeli and Ozlem Cinar Ozdemir

Additional information is available at the end of the chapter

http://dx.doi.org/10.5772/68002

Abstract

Chronic venous diseases of the lower limbs are one of the common and disabling conditions with clinical signs and symptoms ranging from spider veins, to varicose veins or even to venous ulceration, effecting patient's quality of life negatively, particularly in relation to the domains of pain, physical function and mobility. It is not only a serious medical condition but also an undeniable physical and psychosocial problem and has a severe impact on a patient's quality of life. Quality of life is defined as individuals' perception of their position in life and refers to the patient's ability to enjoy normal life activities. It is a common and subjective term that has a number of dimensions described as patient-reported outcomes. There are numerous factors that affect person's quality of life negatively. Some of those are social relationships, economic status, physical health, psychological status, environmental conditions (living place or work), pain and especially chronic diseases. We conducted a review about quality of life in chronic venous diseases of the lower limbs to identify how person's quality of life is affected. Therefore, this chapter will focus on the effect of the chronic venous diseases on the person's quality of life.

Keywords: quality of life, chronic venous disorders, evaluation, body image, varicose veins

1. Introduction

Chronic venous disorders (or in other name chronic venous diseases) (CVDs) are widespread problem and can vary from asymptomatic insufficiency of venous valves to chronic leg ulcers [1]. The importance of venous disorders is determined by the socioeconomic influence of disease according to its severity and the number of affected person [2]. Although it does vary, it is revealed that the venous disorders have high prevalence in the community. Approximately 23% of the adult population have varicose veins and 17% have chronic venous insufficiency (CVI) (which is one of the CVD), in all age groups. In addition, it clinically varies from minimal lower leg edema to severe leg ulcers [3].

CVI is a condition that the blood flow within the veins is insufficient, resulting in pooling of blood in the veins, especially in the lower limbs. It can be resulted from by several venous disorders, particularly in chronic situations. It is characterized by permanent lower limb venous hypertension as a result of venous reflux and/or occlusion and insufficient calf muscle pump function [4]. It represents severe phase of CVD and denotes some symptoms such as edema, skin changes, and venous ulcer [5].

CVI is commonly characterized by a condition including diseases in the venous system of the lower extremities [6]. There are a lot of reports about prevalence in CVI, and its variation is very large. This difference may result from the differences in the definition of venous insufficiency and in the methods used for clinical evaluation of the signs and symptoms [7]. There is a study reported that CVI varies between 1 and 17% in men and 1 and 40% in women [8]. However, it is predicted to be as high as 50% in the general population especially in industrial countries [2, 9].

To understand effects of CVI on the quality of life and health outcomes, it is necessary to know the venous anatomy and pathophysiological mechanism. Therefore, we first decided to begin giving information about the anatomy and pathophysiology of the disease.

2. Anatomy of the venous system

Functionally, the peripheral venous system is assumed as a reservoir to store blood and as a canal to return blood to the heart [10]. Depending on activating peripheral muscle pumps and a series of valved conduits to return blood against gravity, the venous system in the lower limbs is more complex than the arterial system [11]. The deep, superficial and most of perforating veins include bicuspid valves opening just one direction [12]. Veins are consisted of three layers, which are intima, media, and adventitia as the similar arteries, although having much thinner vessel wall than arteries [13]. In addition, they have a weaker muscular layer and less elastic tissue when compared with arteries [14].

To classify the veins in the lower limbs, it is considered the relationship between veins and the muscular fascia and location in either the superficial or deep compartment. In this context, they are separated into three groups. These include the deep veins, the superficial veins and the perforating veins. The deep veins are located between the large muscle groups of the lower limb inside the myofascial compartments beneath the muscular fascia [15, 16]. The superficial veins are located above the deep fascia and drain the microcirculation of cutaneous tissue. The perforating veins are responsible for connection between the superficial and deep veins. There is also another group of veins named as communicating veins connecting veins within the same system (deep to deep, superficial to superficial) [15]. In normal situations, muscles contraction generates pressure during contractions in the fascial compartment and this is directly transferred to the veins. Whereby, venous blood flow is directed to the heart via competent valves in the venous system [16]. When muscle pump relaxes, blood begins refill to the deep venous system. Veins become swollen as the vein is filled by antegrade flow in prolonged standing positions. Therefore, the valves within the veins begin to open and pressure to rise. By dint of muscle pump contraction, the veins begin to discharge and venous pressure decrease [10].

In the lower limbs, as we mentioned above, contraction of the calf muscle provides a significant mechanism to push and direct blood flow within the veins toward the heart. Blood in the deep veins starts to empty throughout the muscle contraction, and blood flows from the superficial to the deep veins via the perforating veins. Therefore, the pressure in the venous system decreases during ambulation [17]. Venous reflux or obstruction at any time in superficial or deep veins is related to venous disease and the clinical manifestations of CVI. An increased pressure in the venous pressure causes a retrograde elevation of pressure into the venules (thinner veins) of the skin. This leads to continuous elevated ambulatory venous pressures (as known commonly venous hypertension) that have been related to development of leg ulcers in advanced stages [18].

3. Pathophysiology of venous disorders

Rising venous pressure and occurring impairment in blood flow by several mechanisms lead to develop venous pathology [19]. It may resulted from valvular incompetence in superficial, deep or perforator veins, venous tributaries, or obstruction in veins, or a combination of these mechanisms [10]. Congenital causes, repetitive infections, trauma or inflammation resulted from deep vein thrombosis (DVT) may damage the valves. Obesity, pregnancy, a pelvic mass and previous history of DVT may cause obstruction in veins [20]. Abnormal formation of veins and arteries as a congenital disease (arteriovenous malformations) [21] or insufficient of the calf muscle pump caused by fatigue, immobility, or decreased ankle mobility resulting from neuromuscular or orthopedic diseases also contribute to this process [22]. Damage the valves results in reverse flow or leakage in the closed valves [23]. In addition, high pressure begin to enter into superficial veins, when the failure of valves in the saphenofemoral and saphenopopliteal junctions (between to the deep and superficial systems) emerges. It can also enter the superficial system due to malfunction of the perforator valves. As the consequence of this situation, veins start to dilate and varicose veins begin to form and Spread to the extremity [10].

In normal conditions, most of perforator veins are inactive. When the venous hypertension develops, they begin to open, and leakage of the blood occurs from deep to superficial veins [24]. Changing in the microcirculation contributes to the macrocirculatory hemodynamic disturbances [25]. If venous hypertension is not treated, it causes changes in the skin with hyperpigmentation, fibrosis in subcutaneous tissue and eventually ulceration [10].

4. Risk factors

There are some risk factors are defined leading to development of CVI such as heredity, age, female sex, obesity, pregnancy, prolonged standing conditions (**Figure 1**).

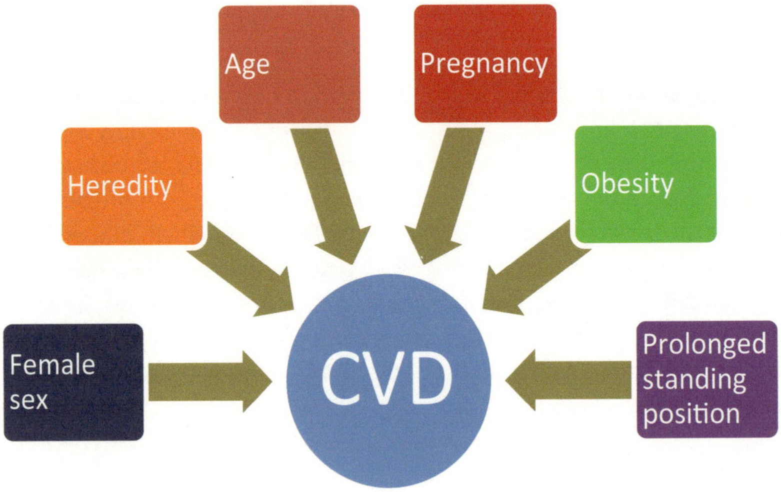

Figure 1. Risk factors of chronic venous disorders.

Several studies have revealed older age as the most important risk factor for CVI. It is more common especially in female adults than male [3, 26]. However, in the Edinburgh Vein research, it has to be reported that varicose veins were more common among males in the general population [27]. Obesity is another risk factor leading to CVI. It is stated that having more than 30 a body mass index (BMI) increases the risk for CVI significantly in both sexes, especially in males [28]. In addition, it is found that a relationship between severe obesity (BMI ≥ 40) and increased limb symptoms without anatomic evidence of CVD. Therefore, it is suggested that the obesity itself is a factor, which contributes to the venous insufficiency [29, 30]. A positive family history, also, is shown to be a predisposing factor for varicose veins or venous disease [28, 31].

5. Symptoms

Symptomatic varicose veins are not assumed as a life-threatening situation, but it is generally progressive, and if progress, it may result in ulceration on the skin [32]. As we emphasized, CVI is a chronic disease seen with the many symptoms ranging from aching of the legs, varicose veins, telangiectasia, muscle cramps, swelling, pruritus, fatigue, throbbing, itching of the skin, to sense of heaviness in lower limbs [33] (**Figures 2** and **3**).

The symptoms of CVI cause considerable morbidity in patients, decreasing on quality of life (QoL) [34]. It is reported that the symptoms are worse especially in women [35]. Some symptoms are exacerbated by the prolonged standing position such as fatigue, heaviness and pain [36]. Therefore, when the patient is being evaluated, these symptoms should questioned with daily activities and work, especially in patients who must be stand for a long periods of time. In general, symptoms are worse at the end of the day. Exercise for lower limbs, mobilization and leg elevation may help relief in symptomatic signs [37].

http://dx.doi.org/10.5772/68002

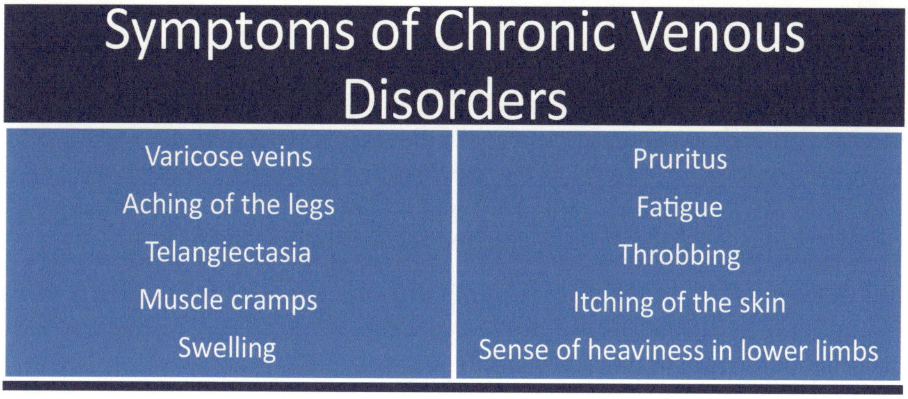

Figure 2. Symptoms of chronic venous disorders.

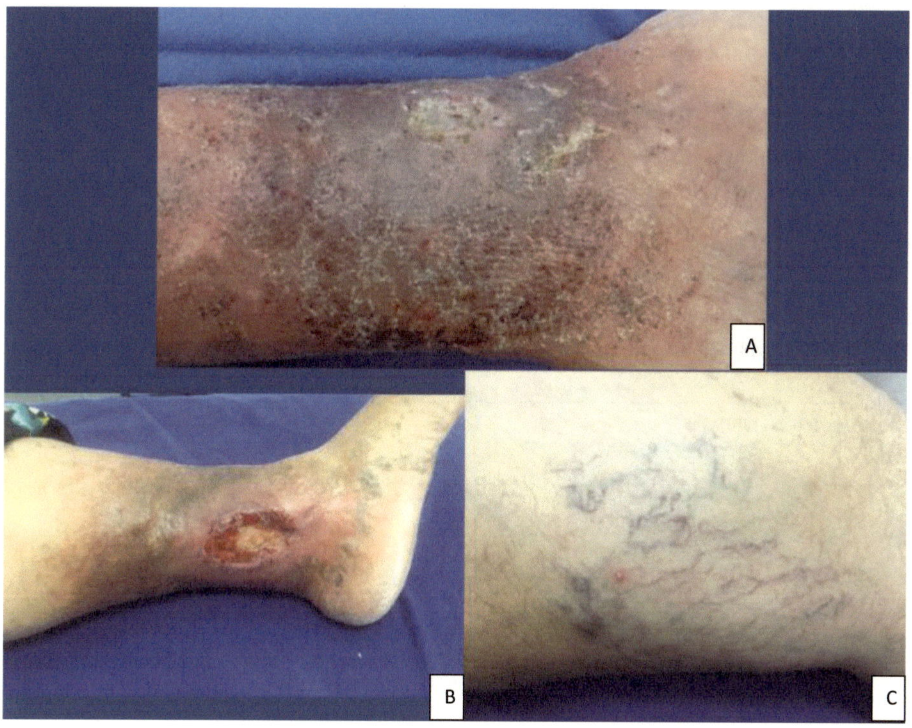

Figure 3. Clinical signs of chronic venous disorders: A: skin changes, B: venous ulcer, C: telangiectasia and reticular veins.

In patients developing chronic outflow obstruction, venous claudication may occur during activities such as walking or climbing stairs. It is also stated that there is an association between QoL scores to severity of disease. Patients indicating the more severe signs and symptoms reported the worse QoL scores [37].

6. Classification

CVI, due to the nature of the disease, presents various signs and symptoms, and this creates a significant challenge to determine severity of disease, to interpret and compare published reports in the literature, objectively. It may be classified using clinical, anatomical, hemodynamic, or patient reported criteria. The challenges resulted from inconsistent reports emerged an applicable and standardized classification system for venous disease [38].

Although there are some classification methods to determine and classify severity of disease, Clinical Etiological Anatomical Pathophysiological (CEAP) classification and Venous Clinical Severity Score (VCSS) are the most using system among them.

6.1. Clinical etiological anatomical pathophysiological (CEAP) classification

The CEAP classification was developed in 1994 the American Venous Forum to describe the severity and etiology of lower limb venous disease, and it was revised in 2004. Thanks to the adoption of this system, correlation among different studies and clinics, and meaningful communication about CVD could be done easily, in a standard way [39]. It serves as a systematic guide in the routine clinical examination of patients with an accurate diagnosis. In the classification of CEAP, there are four parameters. These are clinical indications (C), etiologic factors (E), anatomic dispersion (A), and underlying pathophysiologic manifestations (P) [40]. Detailed information is given in **Table 1**.

There are two version of CEAP: basic version and extended version. The basic version is developed for practical intent. It does not require training for using it and is commonly used in day-to-day care. In addition, its primary aim is using for clinical application; however, it can be used as well as for the purpose of research. The CEAP classification system is an objective and well-constructed

CEAP classification system			
Clinical	**Etiologic**	**Anatomic**	**Pathophysiologic**
C_0 : There is no any signs of venous disease	**Ec** : Congenital	**As** : Superficial veins	**Pr** : Reflux
C_1 : Telangiectases or reticular veins			
C_2 : Varicose veins	**Ep** : Primary	**Ap** : Perforator veins	**Po** : Obstruction
C_3 : Edema			
C_{4a} : Pigmentation or eczema on the skin	**Es** : Secondary (post-thrombotic, post-traumatic)	**Ad** : Deep veins	**Pr,o** : Reflux and obstruction
C_{4b} : Lipodermatosclerosis or atrophie blanche			
C_5 : Healed venous ulcer C_6 : Active venous ulcer	**En** : There is no cause determined	**An** : There is no venous location identified	**Pn** : There is no venous pathophysiology determined

Table 1. CEAP classification system.

instrument, although it has shortage in detecting key symptoms of venous disease. It is reported that the deviations of CEAP scores between different physicians are not significant [41].

In the study, conducted by Kahn et al. [37], it is found that clinical values of CEAP in patients with venous diseases were significantly associated with generic and disease-specific QoL. Their results emphasized that CVD has an adverse effect on QoL, proportionally severity of disease. In addition, it is showed that the variables of sex, age, country, duration of CVD, BMI, education level, and presence comorbidities were associated with QoL scores.

6.2. Venous severity scoring

Although there are many reports that proving the CEAP classification is a useful instrument in classifying venous diseases, there is need more detailed information concerning severity and longitudinal changes of disease in patient during treatment [42]. In the basis of this thought, the American Venous Forum developed Venous Severity Scoring (VSS) from ingredients of the CEAP for measurement of disease severity, in 2000 [43]. The VSS includes three components about scoring of disease. These are the Venous Disability Score (VDS), the Venous Segmental Disease Score (VSDS) and the Venous Clinical Severity Score (VCSS).

VDS is a modification of the original CEAP classification, which is providing disability level, and it evaluates the functional impact of CVD in daily activities. VSDS considers both

The venous clinical severity scoring				
Qualification	**Absent: 0**	**Mild: 1**	**Moderate: 2**	**Severe: 3**
Sense of discomfort, Pain, aching, fatigue heaviness	No	Sometimes, it does not reflect in Daily activities	Daily, it does not affect daily activities	Daily, it restricts most of activitiy
Varicose veins	No	Few and scattered	Confined to calf or thigh	Involve calf and thigh
Venous edema	No	Restricted foot and ankle	Spreads above ankle but don not pass the knee	Extends to knee or above the knee
Pigmentation of the skin	No or focal	Restricted in perimalleolar area	Diffuse over lower third of calf	Diffuse more than lower third of calf
Inflammation	No	Restricted in perimalleolar area	Diffuse over lower third of calf	Diffuse more than lower third of calf
Skin induration	No	Restricted in perimalleolar area	Diffuse over lower third of calf	Diffuse more than lower third of calf
Number of active ulcers	0	1	2	>2
Duration of ulcer	Absent	Less than 3 months	Between 3 months to 1 year	Unhealing ulcer more than 1year
Size of ulcer	Absent	Diameter <2 cm	Diameter 2–6 cm	Diameter >6 cm
Compression therapy	Not used	Intermittant	Most days	Every day, mostly

Table 2. Venous clinical severity scoring.

anatomical and pathological aspects of CVD to provide score about obstruction and reflux. VCSS was designed the aim of obtaining the severity of venous disease. It evaluates totally 10 clinical parameters, and each item scores graded from absent (score 0) to severe (score 3) [43], which is shown in **Table 2**. The VCSS provides more reliable information about severity of disease in patient's routine activities [44]. It is mentioned that the VCSS scores are associated with QoL measurements [41].

7. Quality of life

The term 'Quality of life' (QoL) is a broad multidimensional concept that usually includes subjective evaluations of both positive and negative aspects of life and affected by the culture, individuals' goals, expectations, spirituality, standards and concerns [45, 46]. In another way, it is defined as individuals' perception of their position in life and refers to the patient's ability to enjoy normal life activities [46]. It observes life satisfaction, including everything from physical health, family, education, employment, wealth, religious beliefs, finance and the environment [47]. Therefore, the mean of QoL varies among different people.

There is a consensus in the idea that patients' sights are not stable, and it shifts according to their expectations and perceptions. However, these parameters effecting psychological aspects of disease and degree of symptoms are sometimes underestimated in clinical practice. In order to obtain accurate meaning of QoL in clinical medicine and trials, it is necessary to distinguish between QoL in its more general sense [48]. Consequently, the term 'health-related quality of life' (HRQoL) is derived to obtain more reliable clinical outcomes. Thanks to this idea, clinicians can justify the suitability and cost-effectiveness of the treatment they recommend [49].

The quality of life has subjective and objective indicators, which reflect patient's physical and psychological aspects [50] (**Figure 4**). Objective parameters such as income and physical function can be used in assessing quality of life, but they do not give better opinion about perceiving and experiencing individual's lives. They are better for demonstrating of individual's subjectively perceived QoL, which is critical factor in deciding for treatment and health care policy [51–53]. Subjective evaluations are including psychological domain of pleasure, general happiness and well-being mostly useful to define the experience of life [53] and effectiveness of treatment [54].

There are two basic categories of QoL surveys as generic and disease-specific (**Figure 5**). Both of generic-and disease-specific questionnaires should be used in evaluating QoL, as long as surveys that is reliable, valid, and responsive [55]. For evaluating global sense of well-being and obtaining a subjective measure in efficacy of treatment, generic tools are widely used, and they are appropriate to provide information about QoL in a wide spectrum of disease [56]. Furthermore, in the population with varied circumstances, they enable comparison of measures across populations. However, disease-specific measures are more sensitive in defining effect of treatment and changes in the disease over time [55].

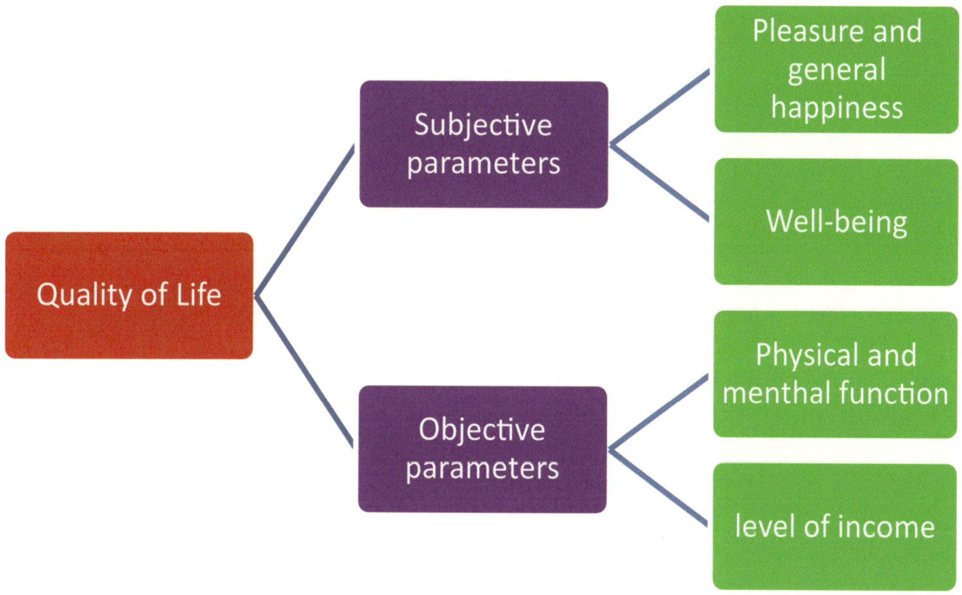

Figure 4. Parameters of quality of life.

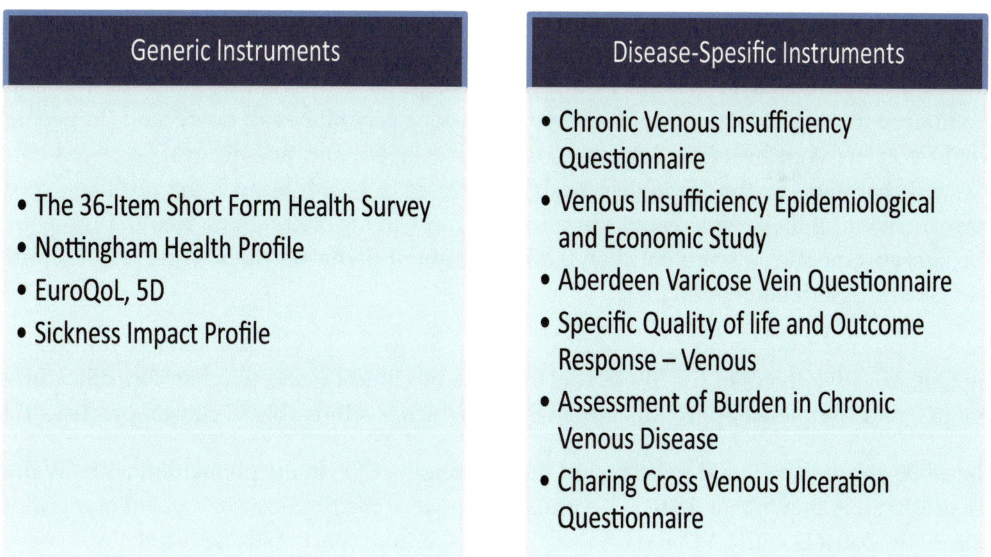

Figure 5. Generic and disease-specific questionnaires.

In particular, in recent years, there is an increasing view that patient-reported QoL is a significant way for evaluating outcomes, especially changes in disease. For chronic conditions such as CVI, evaluation of QoL can provide useful information about burden of disease if it does not captured well by the physician based on other measures [55, 57]. Although mortality rates are low in some diseases such as CVI, patient satisfaction including resolution of symptoms and improvement in HRQoL is recognized as significant treatment outcome measures [58].

As we mentioned above, CVI is a common circulatory disorder that impairs the return of blood to the heart. It mainly affects the legs, causing varicose veins, thrombosis, edema, and ulceration, which affects QoL. There are many generic and disease-specific QoL assessment tools using in CVI. We aimed to explain these instruments, particularly the most widely used in clinical trials.

7.1. Generic instruments

Generic instruments intend for evaluating general QoL, regardless of the diseases or situation of the patient. They usually can be used in healthy population, too [48].

7.1.1. The 36-item short form health survey

A widely used and the most popular well-validated generic quality-of-life instrument is the 36-Item Short Form Health Survey (SF-36) [59]. It is designed to assess generic health status, which is not specific to any age, illness or treatment option [48]. The SF-36 is developed in 1993 using the questions in two categories as physical health and mental health (**Figure 6**). The category of physical health indicates the patient's level of functioning, and mental health indicates of well-being [56]. It contains eight items, which is multidimensional, measuring overall health, functional status and well-being [60]. Besides these items, there are two transition question related to general health and perception of individual's health. These are 'Compared to one year ago, how would you rate your general health now?' and 'In general, would you say your health is: (excellent, very good, good, fair, poor)?', respectively [48]. It is stated that items in the SF-36 detect negative states of health besides the positive aspect; however, some of these dimensions are similar to those in the Nottingham Health Profile [61]. The survey generates a score between 0 and 100, and 0 point indicates worst, 100 indicates best general health perception [60].

Being applicable for a broad range of diseases and thoroughly valid and reliable for measuring QoL are advantages of the SF-36. It is found to be clinically and psychometrically consistent [62] and completed approximately in 5 min, which is admissible in clinical practice [41].

The SF-36 has demonstrated to be a good fit for assessing QoL in the population with CVD. In the studies, it is shown that there is a correlation between the physical component of questionnaire in the patients with CVD and severity of illness. In addition, the strong correlation between all subdomains of the physical components and disease severity measured by the CEAP classification is demonstrated in the literature. However, for the mental component, it is not presented. Correlations between vitality [63] and mental health [64] are poor and inconsistent [38].

Kaplan et al. [63], in their study, used SF-36 to evaluate QoL in patients with CVD, and they found that there was significant correlation between physical component of QoL and severity of venous disease. They also reported that CVD in the lower extremities has a significant effect on physical health.

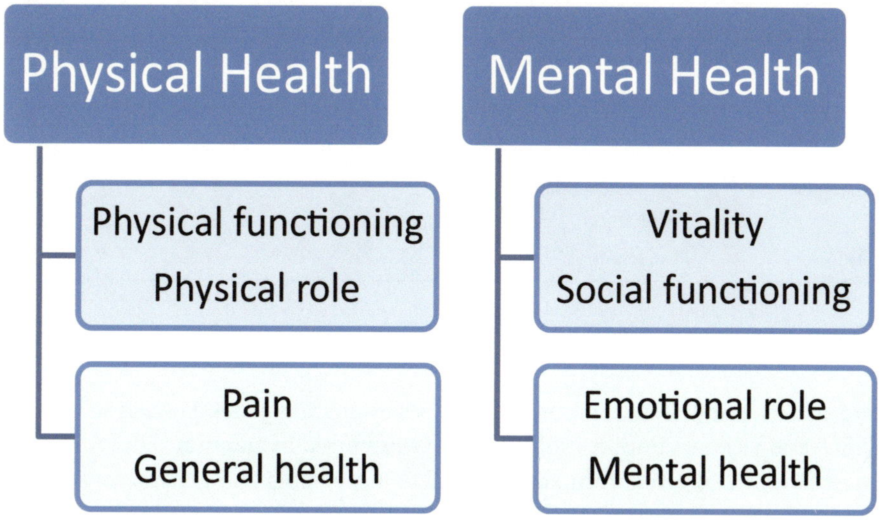

Figure 6. Scales of SF-36.

Darvall et al. [34] conducted a study including 284 patients to determine whether there is any relationship between symptoms in lower limb and generic HRQoL in patients with untreated varicose veins. To obtain data they used SF-12 (shortened version of SF-36) and CEAP. In conclusion, they presented that both physical and mental HRQoL are meaningfully worse in patients with lower limb symptoms regardless of clinical grade of disease. Furthermore, they emphasized that generic HRQoL instruments can be used for comparing effect of treatments.

7.1.2. Nottingham Health Profile

Apart from SF-36, the Nottingham Health Profile (NHP) is another generic instrument designed with a view to measure QoL. Emotional, social and physical health parameters, which are affected by the disease, can assessed with the NHP for a wide spectrum of illness and situation. It measures subjective health status in sleep, emotional reactions, physical mobility, social isolation, pain, and fell of energy. The questions in each area are answered by 'yes' or 'no' [65]. An example for each section is given in **Table 3**.

In a study, the SF-36 is compared with the NHP in patients with of chronic limb ischemia in different degrees. As a result of this study, it is demonstrated that the SF-36 has more internal consistency among patients with mild ischemic symptoms, but the NHP has greater sensitivity to change in patients with more severe ischemia [66].

The NHP is a short questionnaire, and easy to complete. In the NHP, expressions and sentences are not complicated and confounding. It is used frequently both in medical (although it was not designed for that purpose) and non-medical situations to evaluate general health status of individual. However, it is less sensitive in detecting the minor changes in health status and identifying particular problems [48].

Parameters	Examples of questions	Yes	No
Pain	I have pain at night		
Energy	I am tired all the time		
Sleep	I take pills to help me sleep		
Physical mobility	I can walk about only indoors		
Social isolation	I feel lonely		

Table 3. Examples of questions from the NHP.

There are also two other less common instruments evaluating general health status: EuroQoL, 5D (EQ-5D) and Sickness Impact Profile (SIP). We preferred to mention about the most widely used questionnaires to give information related to ingredients and general aspects of them.

7.2. Disease-specific instruments

Disease-specific surveys evaluate factors associated with specific diseases and effect of treatment. In this way, the sensitivity in treatment options and outcomes can increased in comparison with generic instruments, for evaluating QoL in venous diseases [56]. There are four instruments specific to venous disease, which have been used and validated. For evaluating venous diseases with full scope, the Chronic Venous Insufficiency Questionnaire (CIVIQ) and the Venous Insufficiency Epidemiological and Economic Study (VEINES) should be preferred, while the Aberdeen Varicose Vein Questionnaire (AVVQ) and the Charing Cross Venous Ulceration Questionnaire (CXVUQ) are suitable to address particular aspects of venous disease [67].

7.2.1. Chronic Venous Insufficiency Questionnaire

Chronic Venous Insufficiency quality of life Questionnaire (CIVIQ) was developed in France to create a disease-specific QoL instrument should be preferred in place of the generic QoL instruments [68]. Owing to psychometric evaluations and linguistic validity in many countries, it has been the most widely used QoL instrument in venous diseases [66]. According to CEAP classification, the CIVIQ is more sensitive to evaluate QoL for patients in between C0 and C4 clinical grade, because the items in this questionnaire are developed to obtain outcomes about effect of particularly varicose veins and edema on QoL, not venous ulcers [41].

Since CIVIQ was developed, it has undergone some changes. In the first version of the CIVIQ, four areas may evaluate including physical, psychological, social and pain, which is accepted as having effect on QoL. There were different numbers of questions in each category, and this made it difficult to calculate a compound score [68]. Therefore, the idea that there is a need to correction in the CIVIQ is aroused. A new revised version of the questionnaire is named as the CIVIQ-20, which allows weighting the categories equally to obtain a global score.

http://dx.doi.org/10.5772/68002

Nevertheless, it is proved that both versions of the CIVIQ are valid, reliable and consistent for measuring QoL [68, 69]. The items are scored according to the five-point Likert scale [70]. The items of CIVIQ are given in **Table 4**.

There is also the third version of the CIVIQ as CIVIQ-14. The CIVIQ-14 was designed to facilitate its use in clinical areas with the factorial stability of its previous version [71]. It is built by removing 6 items and one dimension causing to the factorial instability in the CIVIQ-20 questionnaire. The 14 remaining items covered three dimensions as physical, pain and psychological [72].

The data obtained with CIVIQ are entirely trustworthy in terms of reflecting the effects of the disease according to self-reporting by the patient with CVD. The CIVIQ was shown to be reliable owing to having good internal consistency, reproducibility and responsiveness [73]. Unlike the AVVQ which are going to be mentioned later, it is more sensitive to making allowance for the more psychological effects on the QoL resulting from CVI [41].

Items of CIVIQ-20	1	2	3	4	5
1. Intensity of pain in legs					
2. Interferes with work or daily activities					
3. Sleeping poorly					
4. To stand for a long time					
5. Climbing stairs					
6. In crunching/kneeling					
7. Walking briskly					
8. Travelling by car/plane/bus					
9. Doing housework					
10. Going to parties					
11. Performing sport activities					
12. Feel on edge					
13. Getting tired easily					
14. Feel like a burden people					
15. Feeling weaker and stiffer					
16. Embarresment to show legs					
17. Easily irritable					
18. Impression of being disabled					
19. Difficulty getting up in the morning					
20. Do not feel like going out					

Table 4. The items of CIVIQ.

7.2.2. Venous Insufficiency Epidemiological and Economic Study

The Venous Insufficiency Epidemiological and Economic study (VEINES Qol/Sym) evaluates the epidemiology and outcomes of CVD [74]. Unlike the generic QoL questionnaires lacking determining detailed information and lower sensitivity for CVD, the purpose of the VEINES QoL/Sym is to present a whole picture of the effect of CVD in the lower limbs [41].

Through the agency of the VEINES, it is evaluated that QoL and symptoms in many conditions in CVD ranging from telangiectasia, varicose veins, skin changes, edema to leg ulcers [38]. Psychometric analyzes revealed that the questionnaire was valid and reliable, as well as manifests correlation with SF-36 and clinical grading according to CEAP. Furthermore, it is stated that the questionnaire is reliable and valid for evaluating of QoL and symptoms in patients with acute deep vein thrombosis (DVT) [75].

The questionnaire includes 35 items in two categories as the VEINES quality-of-life questionnaire (VEINES-QoL) is containing 25 items for evaluating QoL, and the VEINES symptom questionnaire (VEINES-Sym) containing 10 items measuring symptoms of disease. With the VEINES QoL/Sym, physical symptoms can measured more effectively than psychological and social aspects especially in patients with lower limb venous disease ranging from uncomplicated varicose veins to DVT, and the higher scores are associated with the better QoL outcomes [75].

7.2.3. Aberdeen Varicose Vein Questionnaire

Aberdeen Varicose Vein Questionnaire (AVVQ) developed by the Garratt et al. [76] in 1993 as a patient-based questionnaire for measuring especially varicose vein outcomes. It is more specific for disease, particularly focuses on symptoms and questions related to varicose veins [41].

There are some issues that can examined by the AVVQ such as social situation and physical symptoms including pain, edema in lower limbs, venous ulcers, using compression therapy and the impact of varicose veins on routine activities. The questionnaire contains 13 questions and is scored from 0 to 100. The higher scores indicate more severe effect of disease (0 point—no effect, 100 point—severe effect) [76]. It is demonstrated in a study matched the AVVQ scores with SF-36 in the literature that there is significant correlation between the AVVQ and SF-36 outcomes [77]. Additionally, there is evidence reporting the AVVQ is reliable, is associated with patient symptoms significantly, and is considered to be greater responsive and sensitive than generic QoL questionnaires [49].

Being short and easy to complete are some of advantages the AVVQ. It is particularly suitable for assessment of uncomplicated varicose vein and outcomes of treatment. In addition, it allows measurement and evaluation of treatment effects on daily life, if it is used in combination with generic questionnaires, such as the SF-36 or NHP [41].

7.2.4. Specific Quality of life and Outcome Response—Venous

Specific Quality of life and Outcome Response Venous (SQOR-V) is another disease-specific questionnaire separated to five domains about emotional problems, appearance, and limitation in movements, physical discomfort and risk to health. It contains 46 questions that are based on patient reported symptoms not clinical signs [78].

The SQOR-V gives more sensitive assessment in functional impact of CVD in patients with the class C1–C3 according to CEAP classification system and was designed especially for this purpose. There is evidence that the SQOR-V has a significant correlation with the AVVQ [78], but it should be pointed out that the AVVQ more sensitive than SQOR-V in demonstrating ulceration, varicose veins and other skin changes resulted from CVD [79].

7.2.5. *Charing Cross Venous Ulceration Questionnaire*

The Charing Cross Venous Ulceration Questionnaire (CXVUQ) was developed to provide a reliable measure for QoL in patients with venous ulcers. Prior to development of the CXVUQ, there was not available any reliable questionnaire evaluating effect of venous ulcer on QoL and treatment outcomes [67].

7.2.6. *Assessment of Burden in Chronic Venous Disease*

Assessment of Burden in Chronic Venous Disease (ABC-V) was developed by Guex to evaluate directly burden of disease caused by varicose veins. It is a specific tool for evaluating of disease severity and treatments outcomes [80].

8. Discussion

Patients suffering from CVD, causing such as leg ulcers, edema, venous stasis, venous hypertension, etc., are demonstrate worse QoL. This is resulted from sustained discomfort in effected area especially in lower limbs, pain, limited mobility and prolonged healing time. Furthermore, particular treatment options and their financial cost have a significant impact on QoL [81]. It has been preferred to determine the effects of disease and treatment methods on disease from the point of view of the patient, especially in recent years [67]. There are two ways to follow outcomes of effects of venous diseases. One of these is using patient-perceived and reported QoL measurements, another one evaluating clinical signs and symptoms by the physician reports. It is reported that there are many definitions and measuring methods of QoL. Even so, it is defined that there are some key concepts of QoL, which include reliable assessment of the disease's functional effect on daily life activity, focusing on the patient's point of view and feelings, and a taking into account of physical, social and psychological issues [56].

There are many study investigated QoL in patients with CVI. In the study conducted by the dos Santos Crisóstomo et al. [82], using CIVIQ to evaluate effect of manual lymphatic drainage (MLD) on HRQoL and symptoms in patients with CVI, revealed that a four-week period of MLD treatment, which comprised approximately 40 min duration sessions, provide improvement in CVI, especially in terms of clinical severity of disease mostly associated with edema, symptoms and pain HRQoL. Furthermore, thanks to this research, it was demonstrated that the MLD had good effect on CVI and pain HRQoL, and it could still be observed after only 4 weeks of follow-up. On the other hand, it was found that MLD has not any changing effect on the physical, social, or psychological components of HRQOL. Another study carried out by Molski et al. [83] reported that MLD decreased the clinical severity of disease and lower limb volume in patients with CVD, and improved QoL. They also reported that if MLD was started in the preoperative period, there would have been better improvement in quality of life.

Patients with venous ulcers are known as having poor HRQoL [84, 85], and there is a global concern about the link between HRQoL and venous ulcers. For defining health policies, obtaining information about clinical outcomes and determining effect of venous ulceration emerged that the importance of studying by HRQoL in patients with venous ulceration [85]. In a study investigating the relationship between venous ulcer and HRQoL, González de la Torre et al. [86] reported that venous ulcers had negative effect on patients' HRQoL, especially in their emotional status. Additionally, the association with the degree of the wound and decreased HRQoL was found in the study. In another study carried out by Birks et al. [87], it is reported that generic questionnaires were reasonably well to measure HRQoL in patients with venous leg ulcers.

As mentioned before in this chapter, there are some disease specific questionnaires to measure QoL, and each has some advantages and limitations. Kuet et al. [79] conducted a study to evaluate the relationship between the AVVQ and the CIVIQ-14. Their purpose was to compare disease-specific QoL tools with generic QoL and clinician-driven tools. They used the AVVQ, CIVIQ-14 and EQ-5D to measure QoL, and CEAP classification and the VCSS to obtain information about severity of disease. In conclusion, it is reported that there was a strong correlation between the two disease-specific QoL (the AVVQ and CIVIQ-14). There was a significant correlation also found between these disease-specific QoL questionnaires and the generic QoL (EQ-5D). Furthermore, there was existed a strong relationship between the clinical scoring system and disease-specific QoL questionnaires.

There is many factors effect QoL, and evidences demonstrate increasingly significant correlation between questionnaires and clinical severity of disease. Shepherd et al. [78] carried out a study in patients undergoing treatment for symptomatic veins to investigate the relationships between clinical severity of disease and QoL. Assessments were applied by using CEAP and VCSS to evaluate severity of disease, and SF-12 (generic), the AVVQ (disease-specific) and the Specific Quality of life and Outcome Response-Venous (SQOR-V) questionnaires to measure QoL. As a result, strong positive correlation was found between the AVVQ and the SQOR-V. In addition, the correlation between the AVVQ to SF-12 and the VCSS also was found statistically significant.

As mentioned, venous disorders demonstrate a lot of clinical condition, which are effect individual in many ways. Post-thrombotic syndrome (PTS) is one of a widespread clinical condition seen in patients especially after deep venous thrombosis (DVT) and is characterized with symptoms such as pain, edema and skin changes [88]. Effect of PTS on the QoL has been investigated for years and decreased QoL in patients developed PTS and is demonstrated in many studies [89, 90]. Furthermore, in patients having prior DVT history, QoL and clinical severity of disease are worse in comparison with patients having other forms of CVD [91]. Broholm et al. [90] found that PTS was associated with worse both in disease-specific (measured with VEINES-QOL/Sym) and generic (measured with SF-36) QoL. In addition, they stated that there was a significant negative correlation between PTS and the VEINES-QOL/Sym scales.

The fact that increased clinical degree of venous disease was commonly associated with deteriorating health outcomes and QoL has been proven by many studies. Carradice et al. [92] in their study which is conducted with the aim of to explore the impact of venous disease, and assess any gradual direct effect of clinically advanced disease on HRQoL, evaluated 561 patients with the having different clinical degree of venous disease according to CEAP clinical grade. From those, 456 patients had C2–6 and 105 patients had C0–1 clinical degree, which is assumed as a control group. They also used VCSS system to determine severity of disease. To evaluate QoL, they use SF-36 and EQ-5D as generic instruments, and the AVVQ as a disease-specific instrument. As a result, they reported that there was a significant correlation between severity of disease and deterioration in both disease-specific and generic QoL for all clinical grades, in particular in patients with more severe grade had worse QoL scores.

In conclusion, chronic venous disorders are globally common problem, and with the clinical properties, it has great impact on the QoL. To know how disease affects patient's quality of life, and whether there will be any change in the condition of disease in the course of the treatment, it is essential to use valid and reliable disease-specific assessment tools when patients are evaluated by the clinician. Therefore, it should not be considered only physical findings, when evaluating changes in the state of the disease, but also patient's satisfaction and sense of well-being should be assessed. Through the evaluation of all physical and mental aspects of the disease, the effectiveness of treatment outcomes and changes in burden of disease can be assessed more objectively.

To determine level of QoL, there are some general and disease-specific questionnaires, which are developed particularly for patients with venous disorders. The SF-36 and the NHP are most widely used to determine QoL both in patients and healthy population. However, it is proved that using disease-specific questionnaires is more sensitive and reliable for measuring QoL in patients with CVD. The negative effect of CVD on the QoL has been proved in many studies by using different disease-specific instruments, but there is not any consensus which questionnaire is the best among them. The CIVIQ, VCSS, AVVQ, SQOR-V and CXVUQ are mostly used for evaluation QoL in patients having venous disease, and all are sensitive in detecting QoL levels specifically for venous diseases.

Author details

Mahmut Surmeli and Ozlem Cinar Ozdemir*

*Address all correspondence to: ozlemcinar314@hotmail.com

School of Physical Therapy and Rehabilitation, Abant Izzet Baysal University, Bolu, Turkey

References

[1] Fowkes FGR. Epidemiology of Venous Disorders. In: Labropoulos N, Stansby G, editors. Venous and Lymphatic Diseases: New York: Taylor & Francis US; 2006:37.

[2] Beebe-Dimmer JL, Pfeifer JR, Engle JS, Schottenfeld D. The epidemiology of chronic venous insufficiency and varicose veins. Annals of Epidemiology. 2005;15(3):175-84.

[3] Rabe E, Pannier-Fischer F, Bromen K, Schuldt K, Stang A, Poncar C, et al. Bonner Venenstudie der Deutschen Gesellschaft für Phlebologie. Phlebologie. 2003;32(1):1-14.

[4] Labropoulos N, Giannoukas AD, Nicolaides AN, Veller M, Leon M, Volteas N. The role of venous reflux and calf muscle pump function in nonthrombotic chronic venous insufficiency: correlation with severity of signs and symptoms. Archives of Surgery. 1996;131(4):403-6.

[5] Gloviczki P, Comerota AJ, Dalsing MC, Eklof BG, Gillespie DL, Gloviczki ML, et al. The care of patients with varicose veins and associated chronic venous diseases: clinical practice guidelines of the Society for Vascular Surgery and the American Venous Forum. Journal of Vascular Surgery. 2011;53(5):2S–48S.

[6] Raffetto J, Eberhardt R, Cronenwett J, Johnston K. Chronic venous disorders: general considerations. Rutherford's Vascular Surgery. 7th ed. Philadelphia, PA: Saunders Elsevier. 2010:831-40.

[7] Porter JM, Moneta GL, on Chronic AICC, Disease V. Reporting standards in venous disease: an update. Journal of Vascular Surgery. 1995;21(4):635-45.

[8] Robertson L, Evans C, Fowkes F. Epidemiology of chronic venous disease. Phlebology. 2008;23(3):103-11.

[9] Pannier-Fischer F, Rabe E. [Epidemiology of chronic venous diseases]. Der Hautarzt; Zeitschrift fur Dermatologie, Venerologie, und verwandte Gebiete. 2003;54(11):1037-44.

[10] Eberhardt RT, Raffetto JD. Chronic venous insufficiency. Circulation. 2014;130(4):333-46.

[11] Hargens AR, Millard RW, Pettersson K, Johansen K. Gravitational haemodynamics and oedema prevention in the giraffe. Nature.1987;329: 59-60.

[12] van Bemmelen PS, Bedford G, Beach K, Strandness D. Quantitative segmental evaluation of venous valvular reflux with duplex ultrasound scanning. Journal of Vascular Surgery. 1989;10(4):425-31.

[13] Carter C, Anderson F, Wheeler H. Epidemiology and pathophysiology of venous thromboembolism. Venous thromboembolism: an evidence-based atlas. Armonk: Futura. 1996:3-20.

[14] Browse NL, Burnand KG, Thomas ML. Diseases of the veins: pathology, diagnosis and treatment: London: Edward Arnold; 1988.

[15] Caggiati A, Bergan JJ, Gloviczki P, Jantet G, Wendell-Smith CP, Partsch H, et al. Nomenclature of the veins of the lower limbs: an international interdisciplinary consensus statement. Journal of Vascular Surgery. 2002;36(2):416-22.

[16] Bergan J, Pascarella L. Venous anatomy, physiology and pathophysiology. In: The vein book. Elsevier Academic Press, San Diego, CA, 2007:42.

[17] Marston W, Fish D, Unger J, Keagy B. Incidence of and risk factors for iliocaval venous obstruction in patients with active or healed venous leg ulcers. Journal of Vascular Surgery. 2011;53(5):1303-8.

[18] Klode J, Stoffels I, Körber A, Weindorf M, Dissemond J. Relationship between the seasonal onset of chronic venous leg ulcers and climatic factors. Journal of the European Academy of Dermatology and Venereology. 2011;25(12):1415-9.

[19] Burnard KG, Wadoodi A. The physiology and hemodynamics of chronic venous insufficiency of the lower limb. In: Handbook of Venous Disorders: Guidelines of the American Venous Forum Third Edition ed. Gloviczki P. CRC Press, London. 2008:47-48.

[20] Gudjonsson JE, Elder JT. Psoriasis: epidemiology. Clinics in Dermatology. 2007;25(6): 535-46.

[21] Tobón J, Whitney JD, Jarrett M. Nutritional status and wound severity of overweight and obese patients with venous leg ulcers: a pilot study. Journal of Vascular Nursing. 2008;26(2):43-52.

[22] Woo K, Lo C, Alavi A, Queen D, Rothman A, Woodbury G, et al. An audit of leg and foot ulcer care in an Ontario Community Care Access Centre. Wound Care Canada. 2007;5(1):S17-27.

[23] Margolis DJ, Berlin JA, Strom BL. Risk factors associated with the failure of a venous leg ulcer to heal. Archives of Dermatology. 1999;135(8):920-6.

[24] Menon RR. Venous Physiology and Hemodynamics of Lower Limbs. In: Chronic Venous Disorders of the Lower Limbs: A Surgical Approach. Springer India, 2015:18-19

[25] Gschwandtner ME, Ehringer H. Microcirculation in chronic venous insufficiency. Vascular Medicine. 2001;6(3):169-79.

[26] Criqui MH, Jamosmos M, Fronek A, Denenberg JO, Langer RD, Bergan J, et al. Chronic venous disease in an ethnically diverse population the san diego population study. American Journal of Epidemiology. 2003;158(5):448-56.

[27] Evans C, Fowkes F, Ruckley C, Lee A. Prevalence of varicose veins and chronic venous insufficiency in men and women in the general population: Edinburgh Vein Study. Journal of Epidemiology and Community Health. 1999;53(3):149-53.

[28] Rabe E, Pannier F. What have we learned from the Bonn Vein Study? Phlebolymphology. 2006;13(4):188.

[29] Padberg F, Cerveira JJ, Lal BK, Pappas PJ, Varma S, Hobson RW. Does severe venous insufficiency have a different etiology in the morbidly obese? Is it venous? Journal of Vascular Surgery. 2003;37(1):79-85.

[30] Iannuzzi A, Panico S, Ciardullo AV, Bellati C, Cioffi V, Iannuzzo G, et al. Varicose veins of the lower limbs and venous capacitance in postmenopausal women: relationship with obesity. Journal of Vascular Surgery. 2002;36(5):965-8.

[31] Zöller B, Ji J, Sundquist J, Sundquist K. Family history and risk of hospital treatment for varicose veins in Sweden. British Journal of Surgery. 2012;99(7):948-53.

[32] Magnusson M, Nelzen O, Risberg B, Sivertsson R. A colour Doppler ultrasound study of venous reflux in patients with chronic leg ulcers. European Journal of Vascular and Endovascular Surgery. 2001;21(4):353-60.

[33] Staffa R. Chronic venous insufficiency epidemiology. Prevalence. 2002;60(70):55-7.

[34] Darvall K, Bate G, Adam D, Bradbury A. Generic health-related quality of life is significantly worse in varicose vein patients with lower limb symptoms independent of CEAP clinical grade. European Journal of Vascular and Endovascular Surgery. 2012;44(3):341-4.

[35] Bradbury A, Evans C, Allan P, Lee A, Ruckley CV, Fowkes F. What are the symptoms of varicose veins? Edinburgh vein study cross sectional population survey. BMJ. 1999;318(7180):353-6.

[36] Bergan J. Conrad Jobst and the development of pressure gradient therapy for venous disease. Surgery of the Veins. 1985:529-40.

[37] Kahn SR, M'lan CE, Lamping DL, Kurz X, Bérard A, Abenhaim LA, et al. Relationship between clinical classification of chronic venous disease and patient-reported quality of life: results from an international cohort study. Journal of Vascular Surgery. 2004;39(4):823-8.

[38] Writing C, Wittens C, Davies A, Bækgaard N, Broholm R, Cavezzi A. Management of Chronic Venous Disease: Clinical Practice Guidelines of the European Society for Vascular Surgery (ESVS). European Journal of Vascular and Endovascular Surgery. 2015;49(6):678-737.

[39] Onida S, Lane TR, Davies AH. Varicose veins and their management. Surgery (Oxford). 2013;31(5):211-7.

[40] Eklöf B, Rutherford RB, Bergan JJ, Carpentier PH, Gloviczki P, Kistner RL, et al. Revision of the CEAP classification for chronic venous disorders: consensus statement. Journal of Vascular Surgery. 2004;40(6):1248-52.

[41] Catarinella FS, Nieman FH, Wittens CH. An overview of the most commonly used venous quality of life and clinical outcome measurements. Journal of Vascular Surgery: Venous and Lymphatic Disorders. 2015;3(3):333-40.

[42] Meissner MH, Moneta G, Burnand K, Gloviczki P, Lohr JM, Lurie F, et al. The hemodynamics and diagnosis of venous disease. Journal of Vascular Surgery. 2007;46(6):S4–S24.

[43] Rutherford RB, Padberg FT, Comerota AJ, Kistner RL, Meissner MH, Moneta GL. Venous severity scoring: an adjunct to venous outcome assessment. Journal of Vascular Surgery. 2000;31(6):1307-12.

[44] Vasquez M, Munschauer C. Revised venous clinical severity score: a facile measurement of outcomes in venous disease. Phlebology. 2012;27(suppl 1):119-29.

[45] Group TW. The World Health Organization quality of life assessment (WHOQOL): development and general psychometric properties. Social Science & Medicine. 1998;46(12): 1569-85.

[46] Emerson EB. Evaluating the impact of deinstitutionalization on the lives of mentally retarded people. American Journal of Mental Deficiency. 1985;90:277-288.

[47] Barcaccia B. Quality Of Life: Everyone Wants It, But What Is It? 2013 [updated 31.12.2016; cited 2016 26.12.2016]. Available from: http://www.forbes.com/sites/iese/2013/09/04/quality-of-life-everyone-wants-it-but-what-is-it/#3136a6c2c7d9.

[48] Fayers PM, Machin D. Scores and measurements: validity, reliability, sensitivity. Quality of Life: The Assessment, Analysis and Interpretation of Patient-Reported Outcomes, Second edition. Chichester, England. 2007:77-108.

[49] Smith J, Garratt A, Guest M, Greenhalgh R, Davies A. Evaluating and improving health-related quality of life in patients with varicose veins. Journal of Vascular Surgery. 1999;30(4):710-9.

[50] Ferrans CE, Powers MJ. Quality of life index: development and psychometric properties. Advances in Nursing Science. 1985;8(1):15-24.

[51] Briscoe M. Perceived quality of life: problems in the assessment of health and social functioning. Midwife, Health Visitor & Community Nurse. 1985;21(3):98.

[52] Campbell A. The sense of well-being in America: Recent patterns and trends: McGraw-Hill Companies; 1981.

[53] Oleson M. Subjectively perceived quality of life. Image: the Journal of Nursing Scholarship. 1990;22(3):187-90.

[54] Mogos MF, August EM, Salinas-Miranda AA, Sultan DH, Salihu HM. A systematic review of quality of life measures in pregnant and postpartum mothers. Applied Research in Quality of Life. 2013;8(2):219-50.

[55] Patrick DL, Deyo RA. Generic and disease-specific measures in assessing health status and quality of life. Medical Care. 1989;27(3):217-32.

[56] Davies AH, Rudarakanchana N. Qualify of life and outcome assessment in patients with varicose veins. Venous disease simplified: TFM Publishing Ltd., Shropshire, UK; 2006.

[57] Kahn SR, Solymoss S, Lamping DL, Abenhaim L. Long-term outcomes after deep vein thrombosis: postphlebitic syndrome and quality of life. Journal of General Internal Medicine. 2000;15(6):425-9.

[58] McDaniel MD, Nehler MR, Santilli SM, Hiatt WR, Regensteiner JG, Goldstone J, et al. Extended outcome assessment in the care of vascular diseases: revising the paradigm for the 21st century. Journal of Vascular Surgery. 2000;32(6):1239-50.

[59] Jenkinson C, Layte R, Coulter A, Wright L. Evidence for the sensitivity of the SF-36 health status measure to inequalities in health: results from the Oxford healthy lifestyles survey. Journal of Epidemiology and Community Health. 1996;50(3):377-80.

[60] Brazier JE, Harper R, Jones N, O'cathain A, Thomas K, Usherwood T, et al. Validating the SF-36 health survey questionnaire: new outcome measure for primary care. BMJ. 1992;305(6846):160-4.

[61] Ware Jr JE, Sherbourne CD. The MOS 36-item short-form health survey (SF-36): I. Conceptual framework and item selection. Medical Care. 1992;30(6):473-83.

[62] McHorney CA, Ware Jr JE, Raczek AE. The MOS 36-Item Short-Form Health Survey (SF-36): II. Psychometric and clinical tests of validity in measuring physical and mental health constructs. Medical Care. 1993;31(3)247-63.

[63] Kaplan RM, Criqui MH, Denenberg JO, Bergan J, Fronek A. Quality of life in patients with chronic venous disease: San Diego population study. Journal of Vascular Surgery. 2003;37(5):1047-53.

[64] Moura RM, Gonçalves GS, Navarro TP, Britto RR, Dias RC. Relationship between quality of life and the CEAP clinical classification in chronic venous disease. Brazilian Journal of Physical Therapy. 2010;14(2):99-105.

[65] Hunt SM, McKenna S, McEwen J, Williams J, Papp E. The Nottingham Health Profile: subjective health status and medical consultations. Social Science & Medicine Part A: Medical Psychology & Medical Sociology. 1981;15(3):221-9.

[66] Wann-Hansson C, Hallberg IR, Risberg B, Klevsgård R. A comparison of the Nottingham Health Profile and Short Form 36 Health Survey in patients with chronic lower limb ischaemia in a longitudinal perspective. Health and Quality of Life Outcomes. 2004;2(1):1.

[67] Vasquez M, Munschauer C. Venous Clinical Severity Score and quality-of-life assessment tools: application to vein practice. Phlebology. 2008;23(6):259-75.

[68] Launois R, Reboul-Marty J, Henry B. Construction and validation of a quality of life questionnaire in chronic lower limb venous insufficiency (CIVIQ). Quality of Life Research. 1996;5(6):539-54.

[69] Neglén P, Hollis KC, Olivier J, Raju S. Stenting of the venous outflow in chronic venous disease: long-term stent-related outcome, clinical, and hemodynamic result. Journal of Vascular Surgery. 2007;46(5):979-90. e1.

[70] Launois R, Mansilha A, Jantet G. International psychometric validation of the Chronic Venous Disease quality of life Questionnaire (CIVIQ-20). European Journal of Vascular and Endovascular Surgery. 2010;40(6):783-9.

[71] Launois R, Le Moine J, Lozano F, Mansilha A. Construction and international validation of CIVIQ-14 (a short form of CIVIQ-20), a new questionnaire with a stable factorial structure. Quality of Life Research. 2012;21(6):1051-8.

[72] Le Moine J-G, Fiestas-Navarrete L, Katumba K, Launois R. Psychometric Validation of the 14 items ChronIc Venous Insufficiency Quality of Life Questionnaire (CIVIQ-14): Confirmatory Factor Analysis. European Journal of Vascular and Endovascular Surgery. 2016;51(2):268-74.

[73] Biemans A, Van Der Velden S, Bruijninckx C, Buth J, Nijsten T. Validation of the chronic venous insufficiency quality of life questionnaire in Dutch patients treated for varicose veins. European Journal of Vascular and Endovascular Surgery. 2011;42(2):246-53.

[74] Abenhaim L, Kurz X. The VEINES study (VEnous Insufficiency Epidemiologic and Economic Study): an international cohort study on chronic venous disorders of the leg. Angiology. 1997;48(1):59-66.

[75] Kahn SR, Lamping DL, Ducruet T, Arsenault L, Miron MJ, Roussin A, et al. VEINES-QOL/Sym questionnaire was a reliable and valid disease-specific quality of life measure for deep venous thrombosis. Journal of Clinical Epidemiology. 2006;59(10):1056. e1–e4.

[76] Garratt AM, Macdonald LM, Ruta DA, Russell IT, Buckingham JK, Krukowski ZH. Towards measurement of outcome for patients with varicose veins. Quality in Health Care. 1993;2(1):5-10.

[77] Klem T, Sybrandy J, Wittens C. Measurement of health-related quality of life with the Dutch translated Aberdeen Varicose Vein Questionnaire before and after treatment. European Journal of Vascular and Endovascular Surgery. 2009;37(4):470-6.

[78] Shepherd AC, Gohel MS, Lim CS, Davies AH. A study to compare disease-specific quality of life with clinical anatomical and hemodynamic assessments in patients with varicose veins. Journal of Vascular Surgery. 2011;53(2):374-82.

[79] Kuet ML et al. Comparison of disease-specific quality of life tools in patients with chronic venous disease. Phlebology. 2014,29(10):648-653.

[80] Guex J, Rahhali N, Taïeb C. The patient's burden of chronic venous disorders: construction of a questionnaire. Phlebology. 2010;25(6):280-5.

[81] Partsch H. Compression therapy for deep vein thrombosis. Vasa. 2014;43(5):305-7.

[82] dos Santos Crisóstomo RS, Costa DSA, Martins CdLB, Fernandes TIR, Armada-da-Silva PA. Influence of manual lymphatic drainage on health-related quality of life and symptoms of chronic venous insufficiency: a randomized controlled trial. Archives of Physical Medicine and Rehabilitation. 2015;96(2):283-91.

[83] Molski P, Kruczyński J, Molski A, Molski S. Manual lymphatic drainage improves the quality of life in patients with chronic venous disease: a randomized controlled trial. Archives of Medical Science. 2013;9(3):452-8.

[84] International consensus. Optimising wellbeing in people living with a wound. An expert working group review. London: Wounds International, 2012. Available from: http:// www. woundsinternational.com

[85] González-Consuegra RV, Verdú J. Quality of life in people with venous leg ulcers: an integrative review. Journal of Advanced Nursing. 2011;67(5):926-44.

[86] González de la Torre H, Quintana-Lorenzo ML, Perdomo-Pérez E, Verdú J. Correlation between health-related quality of life and venous leg ulcer's severity and characteristics: a cross-sectional study. International Wound Journal. 2016. doi:10.1111/iwj.12610

[87] Birks Y, Nelson EA, Scanlon E, Cullum NA. Quality of life of people with venous leg ulcers: a comparison of the discriminative and responsive characteristics of two generic and a disease specific instruments. Quality of Life Research. 2005;14(7):1705-18.

[88] Prandoni P, Lensing AW, Cogo A, Cuppini S, Villalta S, Carta M, et al. The long-term clinical course of acute deep venous thrombosis. Annals of Internal Medicine. 1996;125(1):1-7.

[89] Kahn SR, Ducruet T, Lamping DL, Arsenault L, Miron MJ, Roussin A, et al. Prospective evaluation of health-related quality of life in patients with deep venous thrombosis. Archives of Internal Medicine. 2005;165(10):1173-8.

[90] Broholm R, Sillesen H, Damsgaard MT, Jørgensen M, Just S, Jensen LP, et al. Postthrombotic syndrome and quality of life in patients with iliofemoral venous thrombosis treated with catheter-directed thrombolysis. Journal of Vascular Surgery. 2011;54(6):18S–25S.

[91] Kahn S, M'lan C, Lamping D, Kurz X, Berard A, Group AftVS. The influence of venous thromboembolism on quality of life and severity of chronic venous disease. Journal of Thrombosis and Haemostasis. 2004;2(12):2146-51.

[92] Carradice D, Mazari F, Samuel N, Allgar V, Hatfield J, Chetter I. Modelling the effect of venous disease on quality of life. British Journal of Surgery. 2011;98(8):1089-98.

Chapter 4

Psychosomatic Interrelations in Cardiovascular Diseases and Their Consequences on Patient's Quality of Life

Maria Suciu and Carmen Cristescu

Additional information is available at the end of the chapter

http://dx.doi.org/10.5772/intechopen.69699

Abstract

At present, the cardiovascular diseases (CVD) are among the major concerns of World Health Organization (WHO), assurance health systems and researchers in the specific field, because they are the number one cause of deaths globally, and according to several perspective studies, they will become the pathology that generates the greatest economic burden worldwide, through morbidity, disability, poor quality of life, and cause of death. Nowadays, the psychic field and the vulnerability of individual mental level to stress are an important link in the development of mental illness, CVD (included in the group of psychosomatic disorders), and also of interrelationship between them. The combined effects of these factors are reflected in the behavioral and cardiovascular (CV) system's pathophysiological changes, which determine the impairment of health-related quality of life for the CV patients, on both short-term and long-term. This chapter aims to address both the interrelationship of psychosocial factors with CVD in terms of its multifactorial etiology and the mark of this bidirectional link on the quality of life of the patients. In several sections, the following issues will be described: general aspects regarding the relationship between homeostasis-stress-pathology; the role of stress as a psychosocial factor in the multifactorial etiology of CVD; the implications of mental disorders in the pathogenesis of CVD; behavioral aspects of CV patient during his illness, and strategies for improving therapeutic adherence and the quality of life of these patients.

Keywords: stress, anxiety, depression, cardiovascular diseases, quality of life

1. Introduction

In 1978, World Health Organization (WHO) defines health not only as a mere absence of disease or infirmity but as a state of complete well-being, physically, mentally, and socially. Also, in 2013, in its report regarding the global health, the same organization specifies that the cardiovascular diseases (CVD) are the leading cause of premature adult death in the world, and one of the major causes of disability worldwide. In Europe, statistics of the European Commission (EUROSTAT) on causes of death identified CVD as occupying the first position in the 2013 ranking [1]. All these determine a cost increase in the health insurance system and a decrease in the individuals' quality of life (QOL). According to the 2016 recommendations of the European Society of Cardiology [2], prevention of CVD is a major public health objective. Due to an increased prevalence of CVD in the general population and a negative effect on both social and economic status today, they are considered "civilization diseases" [3]. Numerous clinical studies have revealed a very close correlation between the occurrence of cardiovascular (CV) pathology, genetic, and life style factors and modifiable physiological and biochemical factors during the individuals' lifetime. Also, it was demonstrated that a better management of the traditional modifiable CV risk factors (through measures of therapeutic education of the patient) may help reduce mortality and morbidity from CVD in the general population, especially in patients who have an increased risk of developing complications [4]. In recent decades, researchers' attention was directed toward other CV risk factors, including nontraditional psychosocial variables. These risk factors, insufficiently approached before, but potentially modifiable, have become object of the INTERHEART study, including one called "psychosocial stress" (and reunited under the name of psychosocial factors) [5]. An association between emotions and CVD was observed since ancient times. Hence, clinical observations of Hippocrates and Galien led to the establishment of the four fundamental human temperaments: sanguine, choleric, melancholic, and phlegmatic. The effect of negative emotions (e.g., anger, anxiety, and depression) on the health of the individual is already a well known, fact sustained by several clinical trials. Nowadays, the anxiety and mood disorders are widely recognized as two psychosocial factors, the presence of which has been observed in a large number of patients with CVD.

2. General aspects regarding the relationship between homeostasis—stress—pathology

2.1. The concepts of homeostasis and stress

The "homeostasis" term was first introduced in medicine by Hippocrates and its dysregulation consequences were called pathologies. "Any dyscrasia or rupture of the normal equilibrium is the cause of diseases" [6]. Walter Bradford Cannon, a Harvard physiologist, connects the concept of homeostasis and the concept of stress in 1915, and 20 years later, in 1935 he recognizes that the connectedness is a secretion of adrenaline produced by the adrenal medulla, consequence of an adaptive hyperfunction of the sympathetic nervous system (SNS), required to maintain normal limits of homeostasis [7]. The structures by which the body controls and regulates its own homeostasis are the interventions of the central nervous system (CNS), autonomic, endocrinal, and immune systems. The main control mechanism at the base homeostasis is a negative "feedback." The concept of "mental stress" was first introduced in 1956 and perfected over many decades of research by Hans Selye, pioneering endocrinologist, recognized as the "father of stress," who defined it as "the nonspecific response of the body to a series of external stressor agents of physical, chemical, biological, and psychological origin" and which can trigger future pleasant or unpleasant events. He depicted the existence of two stress-response syndromes, one local and one general [8]. Local adaptation syndrome is a localized response, limited to the aggression area of the stress agent, immediate and short-termed, and manifested by an acute inflammatory response, while the general adaptation syndrome is defined by a general reaction of the body, it is a long-term reaction and involves the activation of the CNS by neuroendocrine axis, hypothalamic-pituitary-adrenal axis (HPA axis), and the autonomic nervous system (ANS) through the SNS [8]. Nevertheless, this biological theory of Hans Selye is just a description of adaptive physiological reactions to the presence of stress factors and discusses only the behavior of different organ systems and omits one's subjective perception of stressful situation emerging in one's life, the relational, cognitive, behavioral, and affective aspects that can develop after such a situation being ignored. In the context of the new approach, in 1984, R. Lazarus defines stress in a broader sense and comes up together with Folkman with the model "Theory of Cognitive Appraisal," which explained the mental process that influences the stressors [9].

2.2. Biological stress effects on the body

The systemic response to stress is a general nonspecific reaction and has at its base the intervention of numerous biological systems, two of which are major components: local and CNS through neuroendocrine axis and immune axis, but particular roles in modulating it play the basal nuclei, hippocampus, and amygdala [10].

2.2.1. The role of the amygdala and hippocampus in the stress response

The subjective perception of the stimuli is possible due to the intervention of two limbic structures: the amygdala and the hippocampus. The amygdala is a meeting point for neuronal sensory afferents for most perceptions that receive information from the entire cortex it generates efferent to the hypothalamus and thus modulates the ANS and the hypothalamo-hypophyseal axis [8]. The amygdala affects the emotional valence of sensory stimuli playing a very important role in the behavioral response and facial expressions associated with feelings of fear and anxiety. Both of them being involved, the thalamus and the amygdala activate an almost immediate response for fear [11, 12]. The hippocampus is, on the other hand, a structure designed to generate an appropriate behavioral response to a stimulus. Through functional integration of the two structures, it is possible to adapt the individual's vigilance and attention to stress [11, 12].

2.2.2. Neuroendocrine axis response to stress

This type of response is triggered and modulated by the hypothalamus that integrates received afferents and is highly interconnected with other parts of the CNS, limbic system, and ascending reticular activating system (ascending RAS) and is also connected with areas of the ANS through the locus coeruleus-norepinephrine system which is the principal site for brain synthesis of noradrenaline, and the HPA axis by the release of the corticotropin-releasing hormone (CRH) [13]. The role of the cerebral cortex is to increase alertness and attention through incoming afferents by thalamus and ascending RAS, under stress conditions; the role of the limbic system is to imprint the emotional component to the stress response (manifested by states of fear, anger, joy, or elation), while the ascending RAS is the one that induces the cortical alertness state (manifested by vigilance and excitement), increasing the activity of SNS and reactive of the muscle tone [13]. In the periphery, the neuroendocrine axis response to stress is mediated through the effects of increased plasma levels of catecholamines (adrenaline and noradrenaline) and cortisol, hormonal messengers responsible for CV stimulatory effects in the context of organism's adaptive responses to stress [13]. These effects are treated as a whole for preparing the body for *fight and flight* [14]. Besides these two major systems, in the neuroendocrine response of organism to stress, a number of other hormones are involved such as: thyroid's hormones, antidiuretic hormone (ADH), aldosterone hormone, etc.

2.2.3. *The immune response to stress*

There is a bidirectional relationship, of mutual modulation, between the neuroendocrine and immune response to stress. It has been found that stress, both the cognitive (e.g., an emotional shock), recognized by the CNS, and the noncognitive (e.g., secondary to an infection, inflammation) have the capacity to trigger a cascade of biological reactions that will directly or indirectly involve the immune system. Thus, activation of the immune system is considered to be a noncognitive stressor using neural and neuroendocrine circuits identical to those that occur in cognitive stress. The immune response, both local and systemic, to stress may be one of the inhibitions (with anti-inflammatory action) in the context of eustress and one of the stimulations (with pro-inflammatory action) in the context of distress. Two mechanisms underlie the immune response to stress: hormonal and nervous. In the hormonal mechanism, lymphocytes secrete pro-inflammatory cytokines type IL-1β, IL-6, TNF-α that activate both the HPA axis with consecutive hypersecretion of cortisol [15] and contribute to expression of receptors for cortisol and adrenaline with direct consequences in mediating innate and, respectively, acquired immune response; increased circulating levels of above mentioned cytokines IL-1β, IL-6, and IFNγ (interferon gamma) have been emphasized in depression [16]. Cytokine function in modulating the process of organisms adapting to stress and their direct relation with a series of neurotransmitters puts them nowadays in the spotlight, playing a key role in the body's immune response to stress [17]. In the context of the nervous mechanism, the main role is played by the sympatho-adrenergic mechanism and stress neuropeptides with implications both in the periphery on lymph nodes, of the spleen, and thymus, but also centrally through CRH intervention that activates ANS, a situation that leads to summing the action of adrenaline with cortisol and have as final effects the functional inhibition of the immune system.

3. The role of stress as a psychosocial factor in the plurifactorial etiology of CVD

Numerous studies have proven the role of psychosocial factors in increasing incidence of CVD and short- and long-term prognosis of patients with manifest CVD [18]. The impact of psychosocial factors for both QOL and prognosis is becoming increasingly well defined in patients with heart failure (HF) [19], postcoronary bypass surgery [18], and implantation of a cardioverter defibrillator [20]. A distinction between psychosocial stress and psychosocial factors must be made. Defining the scientific concept of psychosocial stress implies the creation of a model for CVD (**Figure 1**) and its integration at various levels of all potential psychosocial factors involved in the development of these pathologies [21].

According to this model, it can be recognized by all experts that not only negative life events (financial loss, loss of partner, workplace, etc.), but also the negative emotions (depression, anxiety, fear, etc.) can become CV risk factors [18]. There is no official definition of psychosocial stress. Thus, we can consider that the psychosocial stress model on the CV system is the released expression by the CV system itself to the action of psychosocial factors on it.

3.1. Mechanisms involving psychosocial stress in the pathogenesis of CVD and psychosomatic disorders induced by it on the heart level

Stress is an important part of our lives, essential in the short term, but it can become dangerous as a long-term presence. Mechanisms through which psychosocial factors can increase the risk of developing or worsening CVD are various and complex.

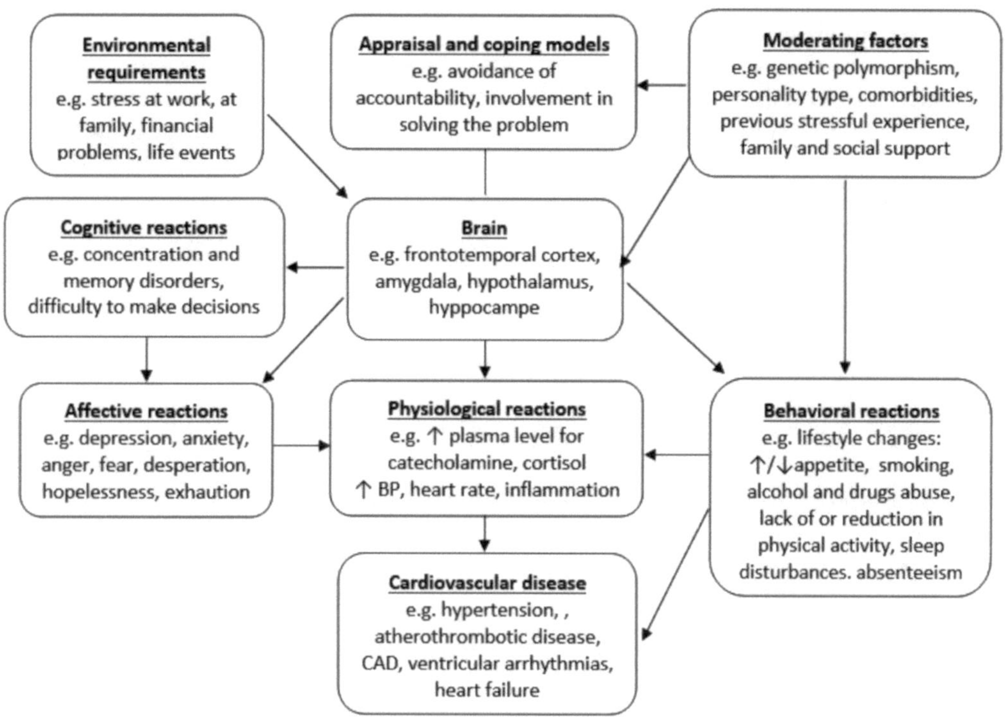

Figure 1. A stress model integrating psychosocial risk factors of CVD (after Ref. [21]).

3.1.1. *An unhealthy lifestyle*

One first impact of long term of psychosocial stress is the individual's recourse to an unhealthy lifestyle. Therefore, these patients smoke and drink more, often adopt an unhealthy diet, reduce their physical activity, etc., in comparison to the patients with no psychosocial factors [18]. Financial barriers of a high-quality healthcare are also contributing to an unfavorable prognosis, as after an acute myocardial infarction (MI), for example.

3.1.2. *Pathophysiological changes*

In CVD, participation of the psychosocial factor in both constitution and modulation of their evolution, together with other etiopathogenic factors is uneven, but it is proven by a series of experimental and epidemiological studies [22]. The leading role is the one of SNS hyperactivities (by disrupting the physiological balance between sympathetic and parasympathetic), being the body's first defense mechanism that not only occurs both in acute conditions (requiring its rapid adaptation), but also in conditions of chronic persistence of stressor factors [23]. CV changes induced by the SNS hyperactivity during acute stress are represented by: short-term increase of blood pressure (BP) values, mediated through the activation of alpha-adrenergic receptors and heart rate (HR) by stimulating beta-adrenergic receptors, transient endothelial dysfunction, lowering the threshold for triggering the cardiac arrhythmias and increasing the risk of MI, left ventricular dysfunction, and even the sudden death [18, 24]. Under normal conditions, the simultaneous growth of both the HR and the BP involves modulation of baroreflex function. Particularly, under stress conditions, the sensitivity of baroreflex mechanisms is reduced, with loss of the major cardioprotective mechanism autonomic reflex [25], and finally, the persistence of this disorder favors the appearance of high BP. The risk of MI induced by psychological stress appears due to the simultaneous increase of both the systemic vascular resistance and the myocardial oxygen requirements, effects of the SNS stimulation, and concurrent discharge of catecholamines. Adaptive response of the CV system under stress conditions occurs as a physiological response, but the generation alterations of autonomic impulses can trigger ventricular arrhythmias especially in the presence of a pre-existing CVD [21, 25].

A series of recent studies indicate that the asymmetrical brain activity plays an important role, too, in the heart vulnerability to ventricular arrhythmias, an aspect proven by the fact that the lateralization of brain activity during emotional stress may stimulate asymmetric heart, and by inhomogeneous repolarization which creates conditions of electrical instability that underlie the generation of ventricular arrhythmias [26]. It has been found that patients with CAD have this exaggerated asymmetric brain activity in several cortical areas during periods of psychosocial stress, such as the left parietal cortex, the left anterior cingulate, the right visual association cortex, the left fusiform gyrus, and the cerebellum [27].

The presence of the long-term stress creates favorable conditions for the persistence of arterial hypertension (HT), initiation, and progression of the atherosclerotic vascular process [17] through both persistence at high levels of sympathetic tonus and as a result of its association with inflammatory process [17, 28] initiated by the elevated plasma levels of cortisol following the entry into action of the HPA axis [14, 28] and the development of resistance of the glucocorticoid receptor [29], also due to activation of the sympathetic renin-angiotensin-aldosterone system (RAA system). A number of behavioral disorders (increased appetite, sedentariness, and excess consumption of energy drinks) can be added, that the individual can acquire even some pathologies developed over a long period of stress, such as metabolic syndrome. The persistence of elevated BP beyond the moment of acute stress and its transformation into chronic stress conditions in a chronic pathology—arterial HT—are both due to the role the CNS plays in part through autonomic regulation of BP, hydro electrolytic balance, and ADH release [22], and entry into action of RAA system and the occurrence of inflammatory process at a vascular level, particularly the adaptive immune response that plays an important role in the pathogenesis of this disease [30]. It was initially revealed that experimental damage to certain regions of the brain like the anteroventral third ventricle, the subfornical organ, prevents many forms of experimental HT [31], but recent studies have shown that these forebrain region plays an important role to peripheral activation of T cells and vascular inflammation during angiotensin II-dependent HT [32]. One of the principal pathways activating the RAA system is the SNS. Both, acute and chronic stress can increase renal renin levels and plasma levels of angiotensin II (ANG II) that acts to maintain the constant extracellular volume and BP by different mechanisms. High levels of renin and ACT II are attenuated by β-adrenergic blockade [33], thereby demonstrating the relationship between the two systems. The local renin-angiotensin systems (RASs) exist at the level of solid tissues (e.g., heart, kidney), but it is also present in the circulating RAS in a variety of circulating cells [34], and the ANG II type 1 receptors (AGTR1) are the most numerous expressed, and they mediate most of the known functions of ANG II [35]. Stimulation of these receptors, by increased plasma levels of AGT II, is responsible for the persistence of HT, pro-inflammatory, and pro-oxidative actions [35]. AGTR1 expression in all immune cells that were observed has contributed to involvement of the RAS in vascular inflammation and the atherosclerotic process [35, 36]. According to the latest discoveries, both CNS and increased levels of ANG II and salt from the body are responsible for T-lymphocytes activation, which once activated, reached the peripheral blood vessels, and kidneys start the release of pro-inflammatory cytokines responsible for vasoconstriction and retention of sodium thereby contributing to the long-term maintenance of high BP values, respectively, arterial HT [31, 32, 36]. Reduction of the inflammatory process at the vascular level both through direct blocking of the production of ANG II by means of pharmacological agents like angiotensin converting enzyme inhibitors (ACE inhibitors) or indirectly by other pharmacological agents such us statins, is associated with the decrease of plasma levels of various pro-inflammatory and pro-coagulant markers from the old, high sensitive C-reactive protein, soluble cellular adhesion molecules of immunoglobulin family, the selectins group, up to long pentraxin 3, a new multimeric inflammatory biomarker, from the family of pentraxins that is nowadays one of the mechanisms of intervention in reducing BP values in hypertensive patients and reduction of the progression of the atherosclerotic process, issue proven by numerous clinical studies [37–39].

3.2. The psychosocial consequences of cardiovascular diseases

Psychosomatic disorders in CV patients are manifested through the appearance of psychical disorders echoing the suffering of the somatic, at the CV system level. Development of atheroma plaque both peripherally and at the cerebral level and also the decrease of cardiac output that has repercussions over the cerebral flow in patients with cardiac insufficiency can induce the process of chronic cerebral hypoxia and is the cause of developing symptoms such as asthenia, fatigue, headache, and sleep disorders, in which over time can lead to diminished cognitive performance (attention, memory, and learning) and even ideational flow. A series of studies have pointed out that the cerebral vascular disease is associated with cognitive function and shows a key role in the decline of memory [40]. It has been found that 25–75% of the stroke patients can develop vascular cognitive impairment [40]. The development of carotid artery stenosis was associated with the decrease in neuropsychological performance due to the chronic drop of cerebral blood flow [40] and by affecting the neurovascular unit increases the likelihood for dementia [41]. It seems that both symptomatic and asymptomatic carotid stenoses are associated with cognitive impairment [42]. Withal, hypertensive individuals are more susceptible to developing cognitive impairments than those whose BP values are within normal limits [43]. Other cardiac conditions such as atrial fibrillation or HF, both reducing cardiac output and increasing the risk for thromboembolism, have been associated by some, but not all, researches with cognitive decline [44, 45]. Frequently, the people with CAD develop illness-related symptoms in the recovery period, such as breathlessness and chest pain, both of which are associated with post-MI fatigue [46]. Majority of patients with heart disease from a hospital reported a poor quality of sleep and that subjective quality of sleep associated with fatigue post-MI and daytime dysfunction were predictors of depressive symptoms [47]. Among 15–20% of CV patients are depressed and as many of those who have suffered a heart attack later developed a major depressive relapse [48]. Approximately, 25% of patients with HF can develop a major depressive disorder and the share of depression symptoms among them can reach 30–50% [48]. CV patients acquire a special social status, starting from their removal from a number of stressful physical activities (household routine) to reduction or suspension of practicing sports activities, hiking or even limiting sexual activity and ending with granting priority to emergency medical services. All these behavioral changes to which a cardiac patient must suddenly undergo summed up with a number of other lifestyle restrictions concerning quitting smoking, alcohol, coffee intake, or the potential risk of relapse of acute MI are actually new sources of mental stress for him and culminate with the prospect of sudden death or possible long-term disabilities (physical, cognitive, emotional, and social) after stroke. Thus, the CV patient develops a state of fear and restlessness, of constant anxiety, as an expression of the presence of a continuous mental stress in one's life with varying degrees of intensity, playing an important and negative role in the development of one's heart disease itself.

4. The implications of mental disorders in the pathogenesis of CVD

In recent decades, scientists began to increasingly focus their attention toward a group of nontraditional risk factors for CVD, also, including many psychosocial variables in their studies. The most studied issues are represented by the impact that negative emotions (e.g., anger, anxiety, and depression) have on the general health of the individual and further in initiating, precipitating, or worsening of pre-existing pathologies [18]. Although various negative emotions have been studied, the results of several researchers have led to similar conclusions regarding the impact of anxiety disorders and mood disorders which appear as two psychiatric factors with a higher incidence among people with CVD [19, 48, 49].

4.1. Depression and cardiovascular disease

Both experimental and epidemiological studies revealed a bidirectional association between depression and CVD. The changes in homeostatic and neuroendocrine function during the CVD create favorable conditions for the development of mood disorders, such as depression [47, 48], and depression, through behavioral and pathophysiological changes produced by it, is a recognized risk factor for CVD-related morbidity and mortality [48–51]. The major depression is the most common mood disorder frequently encountered among patients with CVD [52] and is the one that portends adverse CV outcomes and increased healthcare costs. The prevalence of depression is at least three times greater in patients with CAD or HF than in the general population, and one in five CV patients is depressed [48, 53]. If a series of depressive symptoms are added to the major depression diagnosis itself (present but not enough to meet the diagnostic criteria), which frequently develop in a patient who recently suffered a heart attack, the share among these patients may increase at 40–65% [48]. The depressive symptoms developed by the CHD and HF patients are susceptible to contribute toward too limiting physical activity and impaired their QOL [53, 54]. Also the occurrence of depressive symptoms on the background of CVD is, as well, involved in both relapsed acute cardiac events (stroke, MI) and increased mortality [50]. Since the CV consequences of depressive symptoms continue to generate events with poor prognosis after acute coronary syndrome, the American Heart Association (AHA) issued a 2014 Scientific Statement recommending that depression be elevated to the status of a risk factor in acute coronary syndrome [55]. As regards to the relationship of depression with other CVD, such as arterial HT study, results are controversial [56].

Despite growing evidence of the close link between depression and CVD, the pathophysiological mechanisms underlying these interconnections are still unclear. A series of the neurobiological processes and mediators that are common to both mood and CVD are considered to underlie this interconnection, consequence of their activation by psychosocial stress. These include the following behavioral and pathophysiological mechanisms: (a) neuroendocrine and neurohumoral changes involving dysfunction of the HPA axis and consecutive activation of the RAA system; (b) immune system alterations including activation of pro-inflammatory cytokines; (c) autonomic and CV dysregulation, which include increasing the sympathetic activity, decreasing the parasympathetic tone, HR and rhythm disturbances, and altered baroreceptor reflex function; (d) central neurotransmitter system dysfunction including dopamine, noradrenaline, and serotonin; and (e) behavioral changes including fatigue and physical inactivity [57].

More and more studies have emphasized the connection between depression and ANS and CNS dysfunction. The response of the HPA axis and consecutive activation of the RAA system are similar to the body's physiological responses to stress factors and has been linked directly or indirectly to CV regulation and endocrine changes associated with depression [57, 58]. Dysfunction of the HPA axis was described for the first time by Carroll [59, 60] and the functional consequences at its different levels translate into alterations in the release of CRF, dysregulated adrenocorticotropic hormone (ACTH) (response to CRF's alterations), and elevated circulating cortisol or cortisone levels [61], which have been observed at the depressive patient in the cerebrospinal fluid, hypothalamus, and locus coeruleus [62]. Damage of the hippocampus caused by stress seems to be the key of the dysregulation of the CNS and is responsible for the depressive symptoms, through its consequences generated on HPA axis' activity [63]. Endocrine and immune systems interact with each other and some of their functional disorders are common to mood disorders and CV disease, playing a role in the combination of the two pathologies. In 1991, Smith was the first to propose the macrophage theory, the pathophysiological mechanism of dysfunction of the immune system associated with depression, as a result of an excessive secretion of monokines such as interleukin (IL)-1, tumor necrosis factor (TNF)-α, and interferon [64, 65]. Similar pro-inflammatory cytokines such as TNF-α, IL-1β, and IL-6 are released into the systemic circulation of post-MI patients, demonstrating an activation of the immune system that is directly linked to specific CVD [16, 65, 66]. Once synthesized, they have negative consequences in both the CV and CNS [66], thus being responsible for occurrences of not only depression symptoms but also neuroendocrine and autonomic systems disorders. Also, peripheral cytokines play an important role in the release and metabolism of several CNS neurotransmitters, including monoamine neurotransmitter (dopamine, noradrenaline, and serotonin) [67], alter the glucocorticoid axis and increasing the plasma levels of cortisol [68], aspects directly related to the onset of clinical depression [69]. A different mechanism underlying the link between depression and CVD is that of the combination of hypercortisolemia and platelet function disorders, including enhanced platelet reactivity and release of different platelet products [61, 70], one can say that it creates the theoretical foundations that explain the pro-atherosclerosis effects of depression. An interesting aspect is represented by the increased levels and reactivity of platelets that has been associated with the presence of depression among both healthy individuals and patients with CVD [70]. Possibly this quantitative and qualitative exaggerated response of the platelets could be responsible for the connection between depression and CVD [71]. An influence of depression in reducing HR variability and baroreflex cardiac control in CAD patients [72], and impaired coronary flow reserve that increases risk of acute coronary syndromes have also been observed. All these mechanisms suggest that patients with depression have an increased risk of developing heart rhythm disorders. Recent studies results have revealed that behind these complex interactions between depression and CVD is a genetic predisposition, further research focused on identifying specific genes that link depression/negative affect and CVD. It was thus determined that the serotonin transporter (SERT) gene polymorphism plays an important role in both pathologies [73]. Under the behavioral aspect, depression is associated with decreased adherence to recommended risk-reducing behaviors for different chronic diseases, including CVD [74], diabetes mellitus, etc. For instance, the development of depressive symptoms in patients with CAD has been associated with decreased adherence to taking CV medications, attending cardiac rehabilitation, quit smoking, physical activity, and modified diet [75]. Many studies suggest that the poor adherence to these recommended behaviors is the link between depression and increased mortality in patients, after MI and poor prognosis in general for the CV patient [74, 76].

4.2. The association between anxiety disorders and cardiovascular diseases

A series of study results suggest that anxiety, too, plays an important role in prognosis of coronary heart disease (CHD) patients independent of depression [18, 19, 49, 52, 77], but there are still many unclear aspects regarding its role as an etiologic factor in the pathogenesis of CVD. The main issue derives from the fact that it is difficult to establish a certain diagnosis of anxiety disease with a CV patient, because many symptoms characteristic of a panic attack (one form of manifestation of anxiety disease) overlap with the clinical symptoms of CHD, and in these circumstances, it becomes difficult to make a differential diagnosis [77]. A meta-analysis was carried out by Roest et al., assessing the association of anxiety disorder with risk of CHD [78]. Their results showed that anxiety seemed to be an independent risk factor for incident CHD and cardiac mortality, independent of sociodemographic variables, biological risk factors, and health behaviors. Anxiety takes different forms in relation to the types of CVD. For example, a patient with atrial fibrillation does not particularly associate symptoms of anxiety, but worry and phobic anxiety have the highest rate of association with CHD [79] and posttraumatic stress disorders are frequently linked with risk of stroke and CHD [80]. Most studies have revealed a close link between anxiety disease and CHD. The mechanisms proposed to underlie this link are represented by the fact that anxiety was associated with a higher risk of developing the ventricular arrhythmias, as a consequence of HR variability after MI [81], and a progression of atherosclerosis [82]. Also, anxiety is associated with an unhealthy lifestyle in patients at risk of CHD [75], such as excess of cigarette smoking or alcohol consumption, lower physical activity, and poor adherence to CV medications [74], which increase the risk of CVD. Pathophysiological mechanisms' model proposed for anxiety's intervention in the pathogenesis of CVD is the one for accumulation of anxiety as a chronic stressor factor along with other negative emotions and its intervention at the level of the two systems: ANS and CNS with the activation of the HPA axis and increased release of plasma catecholamines, and endothelial damage, underlying the initiation and progression of atherosclerosis and development the CAD (also known as CHD), and acute coronary events. If this model is viable, it may be expected that anxiety would be frequently associated with increased BP. However, in medical practice study, results are controversial. There is a link between anxiety and HT, but anxiety as a risk factor for incident HT is unclear [83]. Another model which could explain the conflicting results of various epidemiological studies is represented by the emotional triggering model in which the individual's timing of anxiety measurement is essential [84]. Acute anxiety is often associated with hyperreactivity of the CV system to stress [18]. The close connection between anxiety, CVD, and sudden death, suggests that ventricular arrhythmias may be the principal mechanism for cardiac death. The pathophysiological mechanism that underlies this hypothesis is the reduced HR variability by increased SNS stimulation or impaired vagal control [85]. It was also revealed that anxiety alone is associated with higher platelet levels [71], and co-occurring with depression, when the two comorbidities have the highest platelet levels [86], increases the risk of acute events in CAD patients. The direct effects of the SNS together with the activation of HPA axis increase the risk of developing a CVD and lowering the threshold for myocardial ischemia, arrhythmias, and sudden cardiac death in patients with symptoms associated with anxiety [79, 81].

4.3. The importance of personality type

An important role in increasing the risk of CVD is apparently played by the individual's personality type. Thus, two kinds of personality Type A and Type D have been related to increased risk of CVD. Initial studies have revealed that personality Type A, characterized by hostility, ambition, competitiveness, etc., is more prone to CVD [87]. However, subsequent studies have reported that the association between Type A behavior and the risk of developing the CVD is insignificant or reduced to a limited number of cases [87]. In recent years, attention was also directed on the Type D personality, which is characterized by a combination of negative affectivity and social inhibition. This individual accumulates high levels of chronic distress that are not recognized by him and, therefore, are not to be expressed. The patients with CAD and Type D personality profile have associated a risk almost 2–2.5 times higher for developing adverse cardiac events [88] and in this association, Type D personality appears to be an independent risk factor for CVD.

4.4. Pharmacological implications during depression, anxiety, and CVD treatment

Regarding depression, its prevalence among patients with CVD is high (20–25%), both the immediate and the long-term prognosis of a CV patient worsens 2.2 times if one develops depression, but there is also a controversy about the use of antidepressants, selective serotonin reuptake inhibitors (SSRIs), in particular, whether or not it increases the risk [89]. As follows, antidepressant medication is among the most frequently prescribed medication in the world, and its consumption has skyrocketed in the last decade [90]. However, information concerning antidepressants' relationship with CAD is rather less known. Numerous studies have highlighted an increased risk of CHD during use of antidepressants, especially in tricyclic antidepressants [91]; reports have also emerged in this regard for patients who are users of SSRIs [92]. From a theoretical perspective, antidepressants such as SSRIs (which have potent antiplatelet activities) should have good effects in terms of coagulation. Tri- and tetracyclic antidepressants should not be used for the pharmacotherapy of patients with CVD associated with depression too [91], because by increasing the concentration of monoamines in the synaptic space can have multiple CV side effects by increased catecholamine activity on adrenergic and serotoninergic receptors. Therapy with monoamine oxidase inhibitors (IMAO) is contraindicated in CV patients, as antidepressants are, because they irreversibly or reversibly inactivate the enzyme monoamine oxidase and increase concentrations of the noradrenaline in the synaptic space, leading once more to CV stimulation [91]. The US Food and Drug Administration (FDA) issued in 2011 a drug safety communication regarding citalopram (SSRIs) administration not to exceed the dose of 40 mg/day, based on study findings of QT interval prolongation, with different doses of citalopram [93]. A similar warning was also issued by the European Medicines Agency in 2011. QT interval prolongation is a favoring condition for development of arrhythmias including potentially fatal torsades de pointes [94], issue that has also been recognized with other antidepressants such as escitalopram and amitriptyline [94]. Old age along with other factors such as female gender, family/personal history or concomitant use of drugs with a potential for QT interval prolongation, increase the risk of cardiac toxicity of SSRIs antidepressants particularly in conditions of overcoming the therapeutic plasma concentrations [95]. Sertraline, a SSRIs molecule with a favorable cardiac-safety profile, is still considered a safe and effective treatment for recurrent depression in patients with recent myocardial infarction or unstable angina [96]. Because stress and anxiety disorders are prevalent in patients with CVD, in coronary care units, the use of the benzodiazepines as anxiolytics drugs is reasonable [78]. The benzodiazepines are considered as one of the safest groups of drugs in patients with CVD, because they are free of adverse cardiac effects and can be safely used by them even in the coming period after MI. A number of CV medications can, in their turn, exert neuropsychiatric influences with the consequences on the affect and physical status of an individual. The link between the use of β-adrenergic blocker agents and neuropsychiatric consequences, especially fatigue and depression, has long been described [97]. Many case reports and several small reviews support the link between propranolol and depression and, also, the increase in the number of antidepressant prescriptions in patients under treatment with propranolol supports this point [98]. A recent study has highlighted that all antihypertensive agents in monotherapy used in arterial HT treatment were associated with symptoms of depression, anxiety, or both [99]. Only the intensity of effects on mood is differentiated from one class of antihypertensive agents to another: β-blockers and calcium antagonists may be associated with increased risk, whereas ACE inhibitors and sartans may be associated with a decreased risk of mood disorders [100]. Fatigue alone or in combination with sedation occurs with greater frequency than placebo during therapy with calcium channel blockers, without requiring dose reduction. Diuretics may produce neuropsychiatric symptoms mainly indirectly, through electrolyte disorders or vitamin deficiencies (the loop diuretics-associated with thiamine deficiency), but are not frequently associated with fatigue, sedation, or cognitive impairment [101]. During the treatment with methyldopa, Paykel et al. found that sedation occurs in approximately 30–35% of the patients associated with marked fatigue and other studies reported a cognitive impairment [101], both through direct action of methyldopa or following sedation and translated by impaired concentration and decreased cognitive performance. Digoxin, used in the treatment of congestive HF (secondary after MI or other CVD) and as a rate control drug for atrial fibrillation and atrial flutter, can also have considerable central actions and has been shown to induce a variety of neuropsychiatric side effects, both at therapeutic levels and in toxicity, as: fatigue, depression, "mental disturbance" in 5% of patients taking this tonicardiac drug [102], psychosis, and delirium.

5. Behavioral aspects of the cardiovascular patient during his illness

In case of psychosomatic diseases, category that includes CVD, in which the psychological factors play an important role in their triggering or worsening, numerous studies have highlighted that cognitive representation of illness is a significant predictor of patient's recovery and social reintegration and indirectly through its consequences is reflected on the quality of one's life. Also, studies have shown that individuals can directly influence the results of different interventions on their health, both positive and negative, depending on the mental representation of the illness. The cognitive model of the illness (valid for all psychosomatic illnesses) was created by Leventhal et al. [103] and structures the beliefs of the individual about illness in five dimensions, represented by: identity (name and all signs and symptoms of the disease); cause (perception of possible triggers of the disease); time-line (perception about how long the illness might last and how will it evolve); consequences (perception of effects of the illness in the physical, social, emotional, and economic levels); and curability and controllability (perception of the degree to which the disease is curable and manageable).

This model that best explains the relationship between the illness perceptions and emotional and behavioral responses is called the self-regulatory model, and this process is performed through three stages of interpretation, coping, and appraisal. The patients' cognitive representations and perceptions model of their illness, proposed by Leventhal et al. [103], are known to influence patients' motivation to engage in preventive behavior, or curative healthy behavior with healthy outcomes, all expressed in health-related quality of life (HRQOL). This model explains that based on the individual cognitive and emotional perceptions about the illness, the individual reacts to various internal and external stimuli and generates illness-related cognitive and emotional representations that will be the basis of one's decisions on adaptation strategies in the new context of the disease.

5.1. Patients' quality of life

CVD in general and CAD in particular are nowadays the major cause of premature death in Europe [1] and also an important cause of morbidity, contributing substantially to increasing healthcare costs. It is a known fact that health services around the world are more oriented toward improving the quality of patient care than toward prevention measures, and acute care programs are more common than chronic care programs. For these reasons, WHO in The European Health Report from 2015 sets as its main objective the reduction of premature mortality from CVD, cancer, diabetes mellitus, and chronic respiratory diseases by 1.5% annually, until 2020 [104]. By adopting a Health 2020 Monitoring Framework Program in 2012 as the new European health policy framework, member states mandated the WHO Regional Office for Europe to measure and report on the well-being of the European population as a tool to measure health improvement of the European population. The Murray & Lopez study regarding plausible projections of future mortality and disability between 1990 and 2020 [105] highlighted that the leading causes of disability-adjusted life years predicted by the baseline model were (in descending order): ischemic heart disease, unipolar disorder/major depression, and cerebrovascular disease—three pathologies out of total top four of their results. Both, acute and chronic cardiac illnesses are widely recognized to have a negative impact on patient's QOL. Patrick and Erickson define HRQOL as "the value assigned to duration of life, as the patient's subjective perception about the impact of disease, injury, treatment or policy on their everyday life through the impact on physical, functional, and emotional status" [106].

Essential HT is one of the main risk factors of CVD, and at the same time, one of the main causes of death from noncommunicable diseases, still remaining a public health problem despite early diagnosis and the therapeutic advances in recent years. Undiagnosed and untreated or poorly monitored and treated diseases cause a number of CV complications: from CAD to HF, ischemic stroke, etc., with the impact on family, social, and professional life of patients and implicitly on their HRQOL. Many aspects of the relationship between HT and QOL are still unclear and unsolved. The impact of psychological disorders such as depression and anxiety on CV health is well known and studied by the scientific community, but the relationship of these two comorbidities with arterial HT is still controversial and not well understood [56, 83]. Results of a 2010 study emphasized that both comorbidities significantly associate with essential HT, depression more than anxiety, indeed, and that this association had a significant effect on hypertensive patients' worsened QOL [107]. Awareness of their hypertensive status was highlighted as another factor that can exert influence on the hypertensive patients' QOL. According to the results of some studies, awareness of HT was negatively correlated to HRQOL in hypertensive patients [108] and the methods of modifying the attitudes of the hypertensive patient facing illness (from a passive to an active and positive one) have shown that a positive active attitude is an important element of self-efficacy, which is the most important component of social cognitive theory [109]. Self-efficacy requires motivating the patient to perform their duties and obligations deriving from their new condition, especially the chronically ill, in this case, the CV chronic conditions in order to produce the desired results that will lead to improvements in health-related behaviors and medication compliance. We can say, starting from Leventhal's self-regulatory model, that the change (in a negative or positive sense) of cognitive perception of the disease on the mental level of the patient with chronic cardiac condition will help address the patient's reaction to this disease, respectively, increasing self-efficacy expressed by improving HRQOL and cost reduction in health insurance systems. The poor QOL of treated hypertensive patients seems to be attributed to a more intensive drug treatment for controlling BP and changes in lifestyle, also, issues that may negatively impact on HRQOL [110], but this association was not attributed to adverse effects of BP-lowering drugs or any particular class. Analyzing the impact of association of various comorbidities with the appropriate drug therapy and high BP, it has been observed that the number of comorbidities does not significantly affect the hypertensive patient's QOL; however, the number of drugs exerts significant influences on different aspects in their QOL [111]. This issue could be explained by negative perceptions of patients about a large number of co-prescribed drugs as a result of poly pathologies, perception that entails a low adherence to antihypertensive treatment. All this suggests that awareness of being sick and not the disease itself, the need to respect a medical treatment for life, changes in lifestyle that impose restrictions on smoking, alcohol consumption, physical activity limitation, etc., are reasons for lower QOL of the hypertensive patient.

CAD is one of these chronic CVDs characterized by impaired functional capacity and quality of life. QOL of patients with CAD is influenced by a number of risk factors among which the most important are represented by depression, anxiety, and other associated comorbidities. Study results from the last decade revealed a prevalence of depression among 10–40% of the patients with CAD [50, 52, 53]. Since depression is both a primary and secondary risk factor in the development of morbidity and mortality in patients with CAD, independent of traditional risk factors such as smoking, HT, atherosclerosis, the new guidelines recommend a psychological profile screening of these patients for early detection of the tendency to develop depression [55, 112]. Among these patients, the results of various studies have highlighted that depression is strongly related to the presence of angina [49], and angina severity is directly related to their QOL. An interesting aspect is the fact that self-reporting on the perception of cardiac health status is independently predictive of long-term mortality in those with CAD [78]. Depressed patients with CAD vs. nondepressive reported a wider cognitive perception of the illness' burden [53]. Anxiety is another psychological state whose relationship with CAD was analyzed in a smaller number of studies, and it has been observed that its share in those patients was approximately 36% [78, 79]. The impact of CAD on the HRQOL differs according to gender too; thus in women an unfavorable impact is reported in contrast to men [113]. But other factors may influence the QOL of CAD patients such as age, physical limitation due to angina, and angina frequency has an impact on prognosis as well as on HRQOL. Potential unfavorable prognosis of patients with CAD (at risk of developing MI, unstable angina, cardiac arrhythmias, etc.,) has a strong negative impact on the emotional component of the QOL assessment questionnaire, issue supported by the results of clinical studies highlighting the frequent association of depression and anxiety disorder with CAD. The patient's physical function component is also impaired in patients with CAD, but its intervention is in the background of QOL. The fact is that by summing up the consequences of both components, emotional and physical activity, CAD has a significant negative impact on the HRQOL of these patients. According to new discoveries, the healthcare professionals have an obligation to recognize, identify, and treat these variables that interfere with maximizing the benefits of cardiac surgery and pharmacological interventions aiming to improve a patient's physical QOL. The public health interventions, such as cardiac rehabilitation programs promoting physical activity, changes in diet, and quitting smoking are nowadays the main methods to improve CAD outcomes and their benefits are recognized today. These programs should improve perceived health status in the CAD population, especially in women. It is generally agreed today that benefits and risks of different therapeutic strategies of the underlying disease must be viewed from the individual perspective of the patient, an aspect known as personalized therapy. These principles underlie the concept of the "person-centered care," a concept with dynamic development in the CVD management sector in particular [114].

A large number of studies have tracked the impact of HF on HRQOL, given the fact that it is a disorder with consequences for all assessment dimensions of QOL, and this quality alteration of life reflects back on HF further contributing to increasing its severity. HF is a condition with a large share in the general population, particularly in the aged population, with a prevalence of up to 12%, with high financial costs, high mortality that reaches 50% in the 5 years following diagnosis, and frequent hospital admission [115]. HF has become today a major public health problem worldwide because of its incidence, prevalence, and evolutionary prognosis. The impact of HF on the functional component of QOL appears to be the most important and with the most severe consequences, aspect highlighted by the EPICAL study specifying that a QOL score reduction by 10 points is associated with a 25–35% increase of death rate and hospitalization for HF [116]. Limiting physical activity will hinder the patient's participation in a number of social activities that will restrict its network of social support, the patient becoming increasingly dependent on his family, leading to the development of a sense of burden for them. In this phase, the emotional component is added to deterioration in the QOL; patients face an increasing risk of developing anxiety and/or depression symptoms due to anxiety generated by fear of occurrence of decompensation moments and even sudden death, which further limits the autonomy and their activities [51, 53]. Anxiety symptoms are frequently reported by patients with HF, and these patients have a much higher anxiety level than healthy older adults. It has been found that patients with chronic HF and depressive symptoms (which account for 30–50% among them) have a significantly increased risk of death, repeated hospitalizations, and worsened HRQOL [53, 54]. The association of anxiety and depression symptoms was observed in approximately 30–35% of patients with HF, an association that increases the risk for both reductions in QOL concerning health [117] and shortening of the survival rate. Taking into consideration that patients with HF show a high prevalence of physical symptoms, including dyspnea and fatigue, and more than 30% of these patients also have depressive symptoms [48], both with consequences for the subsequent evolution of the disease and also on the HRQOL of these patients. Heo et al. [118] investigated if a relationship between the two variables exists and found that physical and depressive symptoms have a dose-response relationship with HRQOL. Perception of loss of control exercised by these patients over variables such as physical and depressive symptoms and functional status seem to play an important role, hindering their improvement, an aspect that leads to considerable deterioration of their QOL [119]. Family support offered to patients with chronic illness seems to play an important role, too, in general and in particular in this case in relation to chronic HF patients' QOL. Family support must provide encouragement, empathy, and increase feelings of trust of the patient's own resources to regain their autonomy previous to the disease. Depending on the success of changing these perceptions about their disease, the HF patient may recourse to modification of behavioral self-regulatory model of Leventhal, and so these patients can engage in their care regime, choose whether to adhere or not to their prescribed regimen, thus becoming an active part of patient-centered care programs in order to improve health-related quality of their life, reduce hospitalization rate, and costs in the healthcare system and not least the reduced death rate. All these observations of previous studies support the view that better relationships between patient's perspective on family functioning, greater autonomy support, lower family criticism, and greater knowledge of family members and HF patients are associated with lower levels of depressive symptoms and better HRQOL, aspects proven by results obtained from the analysis of the impact of these factors by Stamp et al. [120]. Complex drug therapy that requires daily administration, also, presents increased risks of adverse effects and interactions that implicitly call for careful biological monitoring, respectively, increased cost, and is an added stress factor for this patient. All this calls for greater economic and social resources from the chronic HF patient, and their absence, especially in the elderly, may be related to HRQOL, rehospitalization, and high mortality. The new behavioral changes (dietary salt restriction, weight control, regular physical activity, self-monitoring of symptoms, etc.,) are part of nonpharmacological treatment of these patients and also have an impact on their QOL and entourage.

Of all the previously presented psychosocial factors, being in a stronger or a weaker relationship associated with deterioration of the QOL of the patient with CVD, depression has received the most attention from researchers in the last decade. The results of these studies have brought a clear evidence of depression association with CVD which is why recent clinical recommendations include mandatory screening for depression and its treatment, as a standard of efficient care of these patients [112].

It is already well known that the type D personality, which is a combination of two characteristics, namely negative affectivity and social inhibition, is associated with an increased incidence and risk of mortality due to CVD [88], and only few studies indicate that type D personality is associated with QOL both in health and in disease [121]. Type D personality is considered an independent predictor of CV morbidity and mortality, the worsening QOL in patients with CAD and HF [122]. Type D personality becomes a pre-morbid condition for affective disorders especially among CV patients and hence the consequences in their HRQOL.

5.2. Therapeutic adherence in cardiovascular disease

Analyzing worldwide data on the global burden of disease, it can be observed that HT ranks first, before smoking and obesity: 26.4% of world adult population had HT in 2000, and its incidence is expected to reach 29.2% in 2025 [123]. Furthermore, today one of the major causes of mortality in the world is the CVD, which is expected to be the world's leading cause of death in 2020, the CHD being the first among them [124]. The high BP level is the most important modifiable risk factor of CVD, but despite the large number of potent antihypertensive drugs they succeed in achieving an optimal control of BP values only in 50–66% of patients [125, 126]. One possible reason for this low rate of BP control is that the antihypertensive medication is not taken as prescribed or not taken at all by the vast majority of patients [127]. In other words, the simple nonadherence to existing CV medication becomes a new risk factor for CVD. In the current context, the main objective of WHO in the 2015 European Health Report is to reduce premature mortality by CVD, cancer, diabetes mellitus, and chronic respiratory diseases by 2020 [104] due to the fact that chronic diseases generate high costs in the health insurance system and affect HRQOL, and they require providing services for their better management. Among the many factors that can increase the effectiveness and efficiency of these services, patients' treatment adherence and compliance to medication play an increasingly important role today, counseling patients is the main objective of the new policy in health for 2025 [128].

5.2.1. Adherence to medication and illness perceptions

Nowadays, in a broader sense, the concept of therapeutic adherence stands for the extent to which a person's behavior regarding the use of prescribed medication, respecting a diet, and/or a lifestyle change (reducing salt intake, stopping smoking, weight loss, etc.,) correspond to the therapist's recommendations [129]. This newly introduced term in medical practice tends to replace the previous term of therapeutic compliance that has its origin in physics, given that the latter transmits an authoritative message from medical and pharmaceutical staff, which implies a passive attitude, of subjecting the patient to therapist decisions and failure of noncompliance can be seen from this perspective being only the fault of the patient. The adhesion term also includes the patient in making decisions about the treatment of his own disease, by establishing a partnership between the patient and the specialist, where the patient is a direct, active, and responsible partner on the results of his own treatment, thus becoming a major player in one's disease management [129]. Compliance can be defined as "taking medications as prescribed" [130] and involves an understanding of the correct use of medication and a positive attitude of the patient facing treatment due to a perceived personal benefit from its application. Therefore, patient compliance to treatment is mirrored in the individual's health and QOL.

Regarding arterial HT, clinical practice reveals that around 16–55% of hypertensive patients give up treatment during the first year of diagnosis and initiation of appropriate therapy of the disease. Because many patients are reluctant to voluntarily provide information on their adherence to antihypertensive treatment, the true rate of nonadherence to medication may be even higher. Nonadherence to antihypertensive therapy is responsible for poor treatment outcomes or treatment failure, finally determining an increase in the rate of hospitalization and thus impairment of their medication-therapy-related QOL [131]. A number of studies have shown that various factors, such as demographic, clinical, treatment related, or behavioral, are responsible for adherence/nonadherence of antihypertensive medication, but in older adults, a major role seems to be played by psychosocial factors [132]. A study conducted in the USA highlighted that including a decline in physical and mental quality of the individual, as components for assessing QOL among hypertensive elderly people, is associated with lack of adherence to medication [133]. The results of numerous studies certify that an increased BP controlled by antihypertensive medication can at the same time improve HRQOL for these patients, reduce the risk of complications, and repeated hospitalizations of the hypertensive patients. A systematic 2016 review analyzing the relationship between QOL and treatment adherence in hypertensive patients certainly emphasizes: "nonpharmacological treatment improves the overall QOL and physical domain of people with arterial HT and adherence to pharmacological treatment has a positive impact on the mental and physical domains of patients, as it did on the overall QOL score" [134]. As predictors of nonadherence to antihypertensive medication, the following factors have been identified: poor knowledge of complications of HT, unavailability of antihypertensive drugs in the healthcare facilities, lack of education of hypertensive patients in the healthcare facilities, prior experience of medication side effects, uncontrolled BP, and taking nonprescribed medications (as self-medication) [135]. Another aspect of adhesion is the persistence of the treatment of patients. In case of HT, it has been observed that tolerance and self-perception that the adverse effects of antihypertensive medications play an important role in the patient's motivation to follow this treatment for the rest of his lives [136]. Thus, hypertensive patient's adherence and persistence to pharmacological and nonpharmacological treatment is a key component of HT management.

Despite advances in the field of CV pharmacology and interventional cardiology that contributed to increasing the rate of survival after heart attack, CAD remains worldwide an important cause of morbidity and mortality [137]. In this context, attention and efforts of specialists, in this pathology, must be directed toward secondary prevention, prevention that involves better management of both lifestyle factors and physiologic parameters, most often with medications. Although the number of prescriptions for CV medications among these patients increased significantly in the last 20 years [138], a large percentage fail to meet therapeutic goals aspect highlighted by the results of the European Action on Secondary and Primary Prevention by Intervention to Reduce Events III (EUROASPIRE III) survey [139]. Therefore, the main issue that was raised in this study was if the patient with CAD is compliant with medical therapy or not [140], an aspect appears to be closely related with a paternal clinician-patient relationship and not of the partnership.

The CV medications, according to clinical evidence, are now responsible for more than 40% reduction in mortality in CAD, and premature discontinuation, willingly by the patient, of some classes as beta blockers, antiplatelet agents, or lipid-lowering drugs [141] have shown unfavorable outcomes on morbidity, mortality, and HRQOL of these patients. A supply of more information specific to pharmacological treatment (e.g., the consequences of nonobservance or sudden withdrawal of treatment) when initiating therapy improving the transition between secondary and primary care and a better explanation of the risk of relapse and the development of other possible complications of the disease, may contribute to attitude and behavior changes toward better patient treatment adherence. Under conditions in which HF is characterized by a high rate of hospitalization and has a negative impact on physical activity and on HRQOL, with high mortality, taking CV medications is essential to control HF symptoms and prevent exacerbations. The low adherence to HF pharmacological and nonpharmacological treatment (physical activity, diet modification, and weight control) increases mortality, morbidity, hospitalization rates, and healthcare costs. The negative affectivity and social inhibition, type D personality characteristic features, which are associated with poor health, a greater number of cardiac symptoms, impaired physical and mental health are known predictor factors of nonadherence to prescribed medication regimens, and if encountered among patients with HF, it translates into a poorer adherence to medication, which may lead to adverse health outcomes implicitly and QOL [142]. It seems that an important role in mediating the relationship between type D personality and adherence to medication is played by medication self-efficacy. Self-efficacy or personal efficacy, which is defined as confidence of the individual in their own ability to complete tasks and achieve objectives [143], seems to be associated with medication adherence, thus being an intermediate measure for health outcomes among HF patients. Low levels of self-efficacy have been associated with poor self-care adherence in patients with HF, and a better medication adherence was observed after improving self-efficacy to medication for type D patients with HF [144]. Therefore, the chronic HF becomes the most frequent diagnosis of hospital discharge, of all chronic conditions, generating a medical and psychological burden that impacts on HRQOL of those patients and economic burden on national healthcare systems. Improvement of their outpatient management by good-quality primary healthcare services becomes a necessity and can only be achieved when the two partners, specialist and patient, establish a partnership and understand that a treatment has more benefits than risks or costs, in other words, a strong degree of therapeutical adherence is essential by developing and applying interventions to enhance medication self-efficacy.

Another general problem of nonadherence is the so-called profligate patients, that is, those patients to whom medication was prescribed but is never followed in the long term. It has been found that more than 20% of prescriptions are never picked up from the pharmacy (given the high cost of medication, patient acceptance of the new status, etc.,), others 20% are honored although used incorrectly (errors in dosage, rhythm by omitting doses, etc.,) or are not consumed in 50% of cases by patient-initiated drug holidays [145]. Nonadherence to treatment appears to be the result of patients' decision, more or less thought-out, and translated by a change in the behavior toward their health, following a partial compliance or an infringement of the healthcare practitioners' guidance on therapeutic recommendations. Nonadherence is a complex phenomenon that is characterized mainly by two concepts tightly connected to each other: adherence (denoting the level of patient taking his medication and respecting the indications of healthcare practitioners regarding self-monitoring treatment and periodical controls) and persistence (denoting the time in complying with medication consumption)

[129, 131]. Medication adherence is thus a crucial self-care behavior for all the patients with chronic disease and especially for the CV patients. For these reasons, all needed efforts are made to develop predictive models for screening to better identify prospective patients at risk of nonadherence to treatment [146].

Leventhal's self-regulatory model is the basic explanatory model of patient adherence to treatment [147]. According to this model, when people try to understand the unfortunate event that suddenly appeared in their lives, sometimes with permanent consequences on personal and family life, they develop their own model at the mental level of its individual perception which brings together the five key components: what is the disease; what are its underlying factors; what are its consequences in everyday life; the time duration of disease and not least if they can cure or control it [147]. The uniqueness of illness representation derives from the fact that it is originated in the person's own intrinsic beliefs, they do not necessarily comply with scientific arguments and is in a permanent dynamics, suffering a series of changes, depending on the age and sex of the patient, the disease and one's experiences throughout it [147]. On the basis of their illness representations, patients subsequently develop their own response and coping strategies in order to adjust or minimize the impact of disease on the health of their own organism. Patients' decision to become adherent to medication is one of the health-related coping strategies. An individual's choice to adhere or not to the treatment regimen depends on one's beliefs about the disease and their perception of the importance of taking medication. Patient behavior for continued treatment, the persistence to treatment, once initiated, is dependent on personal assessments of the obtained results.

5.2.2. *Strategies for improving therapeutic adherence*

Detection of nonadherence share among CV patients and medication noncompliant individual profile is very important but not enough to overcome this problem, if not followed by management plans and existing/emerging technologies. As the phenomenon of nonadherence to therapy has grown in recent years, a number of interventions have been proposed to improve medication compliance of patients, which are centered in three directions: the health system, the therapist, and the patient, the three main players on which the therapeutic adherence behavior of an individual depends. The health system should be so organized as to put more emphasis on the quality of care in both primary and secondary prevention, to take initiatives with regard to organizing health education and therapeutic programs for patients, to use counseling strategies in primary care as the patient's motivational interventions for adoption of healthier CV lifestyles and reduction of their CVD risk, to monitor more closely the non-compliance to treatment, and to find solutions to correct the potential problems in a timely manner; all these actions will lead to improvements in outcomes and cost savings. A number of interventions (improvement of physician-patient communication, counting pills, the use of reminder packaging or electronic event monitoring systems, etc.,) have been approached targeting individuals with medication adherence problems. The benefits were modest as specified by the trial results but had significant effects on medication-taking behavior [148, 149]. The use of packaging interventions (like using pill boxes and blister packs, especially recommended for aging adults with multiple chronic diseases) effectively increase medication adherence, aspect highlighted by a series of studies in this direction [150]. A review of different patient-centered interventions to increase adherence to treatment, which analyzed a total of 141 studies, revealed that a major problem is the lack of patient knowledge about the disease and its proper medication, misunderstanding of their active and direct role responsible for the results of therapy, its relationship to their health and implicitly to HRQOL and their physical activity [151]. Since the HRQOL is a multidimensional evaluative concept that is based on the patients' illness perception, the specialists' attention should be centered less on the intrinsic disease outcomes (e.g., CV morbidity or mortality) and more on the changing perception outcomes in patients with CAD. In order to improve patient-centered care, counseling programs and therapeutic education of patients are, thus, necessary to be introduced. The lack or low levels of knowledge (both patients and their family members) is an important and foundational element regarding their HF care regimen aspect supported by one's education and counseling programs that have improved outcomes such as better adherence to a hygienic-dietary regime, medication adherence, and a reduction in the rate of inpatient. All previous study results show that the presence of different CV risk factors in the general population and the emergence of the CVD, also, are related to a decreased QOL in these patients and support the idea that the events on healthcare education by health-related public programs should promote the importance of preventive measures associated with regular physical activity, with a special focus on women [114].

The conclusion derived from here regards the need to implement interventions on patient-centered care and education. Increased awareness, knowledge, and education of CV patients especially on the benefits/risks of adherence/nonadherence to treatment is an essential component in the increased ability of patients to manage their medication but not sufficient to ensure medication adherence and persistence. Alongside these educational programs, others are required to improve the skills of self-control and auto-management of therapy and disease, aspects derived from the fact that in the long term, these patients need to rely on daily personal effort unattended closely by a therapist. When behavioral strategies are associated with continuous therapeutic education programs, chances for adherence and persistence to medication increase greatly for the CV patient and therapeutic results will not wait to happen. Furthermore, implementing individual educational sessions and not the group one may be better suited to the specific needs of each individual, contributing to a better therapy personalization.

6. Conclusion and suggestions for future developments

As a result of current developments in society, stress is a component increasingly present in everyday life of the individual and even a necessary one to a certain point. Due to the close interrelationship with the CV system (which responds first to its presence), excessive stress generates a range of behavioral disorders (e.g., unhealthy lifestyle) and pathophysiological changes which create conditions favoring development of CVD. Through the high mortality rate, they induce in the general population and from the perspective of evolutionary trend analysis studies, in 2020, CVD will become the pathology that generates the greatest economic burden in the world, as morbidity, disability, reduced QOL, and cause of death.

Once triggered by induced pathophysiology changes, CVD also creates the prerequisites for the development of anxiety and mood disorders, and changes in patient behavior in this phase translate into a high rate of nonadherence to treatment. In these new conditions, it is required to implement ever more acutely in primary care practice of the CV patient strategies for the prevention of CVD and increasing/improving therapeutic adherence. An earlier identification of CV patient's psychosocial profile in view of new discoveries and a patient-centered health education can help in reducing CV mortality and morbidity and thus improving their QOL.

Author details

Maria Suciu* and Carmen Cristescu

*Address all correspondence to: suciu01maria@yahoo.com

Faculty of Pharmacy, "Victor Babeş" University of Medicine and Pharmacy, Timişoara, Romania

References

[1] Eurostat Causes of Death Statistics [Internet]. Available from: http://epp.eurostat.ec.europa.eu/statistics_explained/index.php/Causes_of_death_statistics

[2] Piepoli MF, Hoes AW, et al. 2016 European Guidelines on cardiovascular disease prevention in clinical practice the Sixth Joint Task Force of the European Society of Cardiology and Other Societies on Cardiovascular Disease Prevention in Clinical Practice (constituted by representatives of 10 societies and by invited experts). European Heart Journal. 2016;37:2315-2381

[3] Kreatsoulas C, Anand SS. The impact of social determinants on cardiovascular disease. Canadian Journal of Cardiology. 2010;26:8C-13C

[4] Bruckert E. Problématique de la prévention des maladies cardiovasculaires. In: Foucaud J, Bury JA, Balcou-Debussche M, Eymard C, dir. Education thérapeutique du patient. Modèles, pratiques et évaluation. Saint-Denis: Inpes, coll. Santé en action; 2010. pp. 150-152

[5] Yusuf S, Hawken S, et al, INTERHEART Study Investigators. Effect of potentially modifiable risk factors associated with myocardial infarction in 52 countries (the INTERHEART study): Case-control study. Lancet. 2004;364(9438):937-952

[6] Hippocrate, -460, -377. Corpus Hippocratum

[7] Cannon WB. The Wisdom of the Body. New York: W.W. Norton; 1939

[8] Selye H. The Stress of Life. New York: Mc Graw-Hill; 1956

[9] Lazarus R. Stress, Appraisal, and Coping. New York; 1984

[10] Galinowski A, Lôo H. Biologie du stress. Annales Medico-Psychologiques. 2003;161 (10):797-803

[11] Ledoux J. Émotion, mémoire et cerveau. Pour La Science. 1994;202:50-57

[12] Morris JS, Friston KJ, et al. A neuromodulatory role for the human amygdala in processing emotional facial expressions. Brain. 1998;121:47-57

[13] Hammond C, Tritsch D. Neurobiologie cellulaire: les neurotransmetteurs. Paris: Doin Editeurs; 1990

[14] Hery F, Fraincois-Bellan A-M, Grino M, editors. Stress and Neuroendocrine Systems. Paris: John Libbey Eurotext; 1993. pp. 83-95

[15] Connor TJ, Leonard BE. Depression, stress and immunological activation: The role of cytokines in depressive disorders. Life Sciences. 1998;62(7):583-606

[16] Maes M, Bosmans E, et al. Interleukin-1 beta: A putative mediator of HPA axis hyperactivity in major depression?. American Journal of Psychiatry. 1993;150(8):1189-1193

[17] Thurin JM, Baumann N. Stress, pathologies et immunité. Paris: Lavoisier MSP; 2010

[18] Rozanski A, Blumenthal JA, et al. The epidemiology, pathophysiology, and management of psychosocial risk factors in cardiac practice: The emerging field of behavioral cardiology. Journal of the American College of Cardiology. 2005;45(5):637-651

[19] Friedmann E, Thomas SA, et al, Sudden Cardiac Death in Heart Failure Trial Investigators. Relationship of depression, anxiety, and social isolation to chronic heart failure outpatient mortality. American Heart Journal. 2006;152(5):940.e1-940.e8

[20] Matchett M, Sears SF, et al. The implantable cardioverter defibrillator: Its history, current psychological impact and future. Expert Review of Medical Devices. 2009;6(1):43-50

[21] Von Känel R. Psychosocial stress and cardiovascular risk : Current opinion. Swiss Medical Weekly. 2012;142:w13502

[22] Gasperin D, Netuveli G, et al. Effect of psychological stress on blood pressure increase: A meta-analysis of cohort studies. Cadernos de Saude Publica. 2009;25:715-726

[23] Julius S. Sympathetic hyperactivity and coronary risk in hypertension. Hypertension. 1993;21:886-893

[24] Schwartz AR, Gerin W, et al. Toward a causal model of cardiovascular responses to stress and the development of cardiovascular disease. Psychosomatic Medicine. 2003;65:22-35

[25] La Rovere MT, Specchia G, et al. Baroreflex sensitivity, clinical correlates, and cardiovascular mortality among patients with a first myocardial infarction. A prospective study. Circulation. 1988;78:816-824

[26] Kuo CS, Munakata K, et al. Characteristics and possible mechanism of ventricular arrhythmia dependent on the dispersion of action potential durations. Circulation. 1983;67(6):1356-1367

[27] Soufer R, Bremner JD, et al. Cerebral cortical hyperactivation in response to mental stress in patients with coronary artery disease. Proceedings of National Academy of Sciences U S A. 1998;95(11):6454-6459

[28] Chrousos GP. The hypothalamic-pituitary-adrenal axis and immune mediated inflammation. New England Journal of Medicine. 1995;332:1351-1362

[29] Cohen S, Janicki-Deverts D, et al. Chronic stress, glucocorticoid receptor resistance, inflammation, and disease risk. Proceedings of National Academy of Sciences USA. 2012;109(16):5995-5999

[30] Harrison DG, Guzik TJ, et al. Inflammation, immunity, and hypertension. Hypertension. 2011;57:132-140

[31] Brody MJ, Johnson AK. Role of the Anteroventral Third Ventricle Region in Fluid and Electrolyte Balance, Arterial Pressure Regulation, and Hypertension. New York: Frontiers in Neuroendocrinology Raven Press; 1980

[32] Marvar PJ, Thabet SR, et al. Central and peripheral mechanisms of t-lymphocyte activation and vascular inflammation produced by angiotensin II-induced hypertension. Circulation Research. 2010;107:263-270

[33] Aguilera G, Kiss A, et al. The renin angiotensin system and the stress response. Annals of NewYork Academy of Sciences. 1995;771:173-186

[34] Paul M, Poyan Mehr A, Kreutz R. Physiology of local renin-angiotensin systems. Physiological Reviews. 2006;86:747-803

[35] Wright JW, Yamamoto BJ, Harding JW. Angiotensin receptor subtype mediated physiologies and behaviors: New discoveries and clinical targets. Progress in Neurobiology. 2008;84:157-181

[36] Madhur MS, Lob HE, et al. Interleukin 17 promotes angiotensin II-induced hypertension and vascular dysfunction. Hypertension. 2010;55:500-507

[37] Andor M, Suciu M, et al. Correlation analysis between biochemical, functional and structural markers of endothelium damage in patients with essential hypertension. Medicine in Evolution. 2013;XIX(1):188-197

[38] Andor M, Suciu M, et al. Influence of rosuvastatin therapy on biochemical markers, functional and structural parameters of endothelial dysfunction in patients with essential hypertension. Healthmed. 2014;8(6):710-719

[39] Buda V, Andor M, et al. The influence of perindopril on PTX3 plasma levels in hypertensive patients with endothelial dysfunction. Farmacia. 2016;64(3):382-389

[40] Zhong W, Cruickshanks KJ, et al. Carotid atherosclerosis and 10-year changes in cognitive function. Atherosclerosis. 2012;224(2):506-510

[41] Gorelick PB, Scuteri A, et al. Vascular contributions to cognitive impairment and dementia: A statement for healthcare professionals from the American Heart Association/American Stroke Association. Stroke. 2011;42(9):2672-2713

[42] Wang T, Mei B, Zhang J. Atherosclerotic carotid stenosis and cognitive function. Clinical Neurology and Neurosurgery. 2016;146:64-70

[43] Mancia G, Fagard R, et al. 2013 Practice guidelines for the management of arterial hypertension of the European Society of Hypertension (ESH) and the European Society of Cardiology (ESC): ESH/ESC Task Force for the Management of Arterial Hypertension. Journal of Hypertension. 2013;31(10):1925-1938

[44] Almeida OP, Garrido GJ, et al. Cognitive and brain changes associated with ischaemic heart disease and heart failure. European Heart Journal. 2012;33(14):1769-1776

[45] Marzona I, O'Donnell M, et al. Increased risk of cognitive and functional decline in patients with atrial fibrillation: Results of the ONTARGET and TRANSCEND studies. CMAJ. 2012;184(6):E329-E336

[46] Alsén P, Brink E, et al. Fatigue after myocardial infarction: Relationships with indices of emotional distress, and sociodemographic and clinical variables. International Journal of Nursing Practices. 2010;16(4):326-334

[47] Norra C, Kummer J, et al. Poor sleep quality is associated with depressive symptoms in patients with heart disease. International Journal of Behavioral Medicine. 2012;19(4):526-534

[48] Gnanasekaran G. Epidemiology of depression in heart failure. Heart Failure Clinics. 2011;7(1):1-10

[49] Fleet R, Lavoie K, Beitman BD. Is panic disorder associated with coronary artery disease? A critical review of the literature. Journal of Psychosomatic Research. 2000;48(4-5):347-356

[50] Carney RM, Freedland KE. Depression, mortality, and medical morbidity in patients with coronary heart disease. Biological Psychiatry. 2003;54:241-247

[51] Van der Kooy K, van Hout H, et al. Depression and the risk for cardiovascular diseases: Systematic review and meta analysis. International Journal of Geriatric Psychiatry. 2007;22:613-626

[52] Frasure-Smith N, Lesperance F. Depression and anxiety as predictors of 2 year cardiac events in patients with stable coronary artery disease. Archives of General Psychiatry. 2008;65(1):62-71

[53] Ruo B, Rumsfeld JS, et al. Depressive symptoms and health-related quality of life: The Heart and Soul Study. JAMA. 2003;290(2):215-221

[54] Gottlieb SS, Kop WJ, et al, HF-ACTION Investigators. Relation of depression to severity of illness in heart failure (from Heart Failure and a Controlled Trial Investigating Outcomes of Exercise Training [HF-ACTION]). American Journal of Cardiology. 2009;103(9):1285-1289

[55] Lichtman JH, Froelicher ES, et al, AHA Statistics Committee of the Council on Epidemiology and Prevention and the Council on Cardiovascular and Stroke Nursing.

Depression as a risk factor for poor prognosis among patients with acute coronary syndrome: Systematic review and recommendations: A scientific statement from the AHA. Circulation. 2014;129(12):1350-1369

[56] Meng L, Chen D, et al. Depression increases the risk of hypertension incidence: A meta-analysis of prospective cohort studies. Journal of Hypertension. 2012;30(5):842-851

[57] Grippo AJ, Johnson AK. Stress, depression, and cardiovascular dysregulation: A review of neurobiological mechanisms and the integration of research from preclinical disease model. Stress. 2009;12(1):1-21

[58] Sanchez-Gonzalez MA, May RW, et al. Sympathetic vasomotor tone is associated with depressive symptoms in young females: A potential link between depression and cardiovascular disease. American Journal of Hypertension. 2013;26(12):1389-1397

[59] Carroll BJ, Curtis GC, Mendels J. Neuroendocrine regulation in depression: I. limbic system-adrenocortical dysfunction. Archives of General Psychiatry. 1976a;33(9):1039-1044

[60] Carroll BJ, Curtis GC, Mendels J. Neuroendocrine regulation in depression. II. discrimination of depressed from nondepressed patients. Archives of General Psychiatry. 1976b;33(9):1051-1058

[61] Weber B, Lewicka S, et al. Increased diurnal plasma concentrations of cortisone in depressed patients. Journal of Clinical Endocrinology and Metabolism. 2000;85 (3):1133-1136

[62] Bissette G, Klimek V, et al. Elevated concentrations of CRF in the locus coeruleus of depressed subjects. Neuropsychopharmacology. 2003;28(7):1328-1335

[63] Vaidya VA. Stress, depression, and hippocampal damage. Journal of Biosciences. 2000; 25(2):123-124

[64] Smith RS. The macrophage theory of depression. Medical Hypotheses. 1991;35(4):298-306

[65] Dunn AJ, Swiergiel AH, de Beaurepaire R. Cytokines as mediators of depression: What can we learn from animal studies? Neuroscience and Biobehavioral Reviews. 2005;29 (4-5):891-909

[66] Francis J, Zhang ZH, et al. Neural regulation of the proinflammatory cytokine response to acute myocardial infarction. American Journal of Physiology-Heart and Circulatory Physiology. 2004;287(2):H791-H797

[67] Raison CL, Borisov AS, et al. Activation of central nervous system inflammatory pathways by interferon-alpha: Relationship to monoamines and depression. Biological Psychiatry. 2009;65(4):296-303

[68] Raison CL, Borisov AS, et al. Interferon-alpha effects on diurnal hypothalamic-pituitary-adrenal axis activity: Relationship with proinflammatory cytokines and behavior. Molecular Psychiatry. 2010;15(5):535-547

[69] Anisman H. Cascading effects of stressors and inflammatory immune system activation: Implications for major depressive disorder. Journal of Psychiatry Neuroscience. 2009;34(1):4-20

[70] Laghrissi-Thode F, Wagner W, et al. Elevated platelet factor 4 and b-thromboglobulin plasma levels in depressed patients with ischemic heart disease. Biological Psychiatry. 1997;42(4):290-295

[71] Zafar MU, Paz-Yepes M, et al. Anxiety is a better predictor of platelet reactivity in coronary artery disease patients than depression. European Heart Journal. 2010;31(13):1573-1582

[72] Watkins LL, Grossman P. Association of depressive symptoms with reduced baroreflex cardiac control in coronary artery disease. American Heart Journal. 1999;137(3):453-457

[73] McCaffery JM, Duan QL, et al. Genetic predictors of depressive symptoms in cardiac patients. American Journal of Medical Genetics B. Neuropsychiatric Genetics. 2009;150B(3):381-388

[74] Bautista LE, Vera-Cala LM, et al. Symptoms of depression and anxiety and adherence to antihypertensive medication. American Journal of Hypertension. 2012;25(4):505-511

[75] Bonnet F, Irving K, et al. Anxiety and depression are associated with unhealthy lifestyle in patients at risk of cardiovascular disease. Atherosclerosis. 2005;178(2):339-344

[76] Whooley MA, de Jonge P, et al. Depressive symptoms, health behaviors, and risk of cardiovascular events in patients with coronary heart disease. JAMA. 2008;300(20):2379-2388

[77] Tully PJ, Cosh SM, Baumeister H. The anxious heart in whose mind? A systematic review and meta-regression of factors associated with anxiety disorder diagnosis, treatment and morbidity risk in coronary heart disease. Journal of Psychosomatic Research. 2014;77(6):439-448

[78] Roest AM, Martens EJ, et al. Anxiety and risk of incident coronary heart disease: A meta-analysis. Journal of American College of Cardiology. 2010;56(1):38-46

[79] Albert CM, Chae CU, et al. Phobic anxiety and risk of coronary heart disease and sudden cardiac death among women. Circulation. 2005;111(4):480-487

[80] Vaccarino V, Goldberg J, et al. Posttraumatic stress disorder and incidence of coronary heart disease: A twin study. Journal of American College of Cardiology. 2013;62(11): 970-978

[81] Martens EJ, Nyklícek I, et al. Depression and anxiety as predictors of heart rate variability after myocardial infarction. Psychological Medicine. 2008;38(3):375-383

[82] Paterniti S, Zureik M, et al. Sustained anxiety and 4-year progression of carotid atherosclerosis. Arteriosclerosis, Thrombosis and Vascular Biology. 2001;21(1):136-141

[83] Hildrum B, Mykletun A, et al. Effect of anxiety and depression on blood pressure: 11-year longitudinal population study. British Journal of Psychiatry. 2008;193(2):108-113

[84] Edmondson D, Newman JD, et al. Emotional triggers in myocardial infarction: Do they matter? European Heart Journal. 2013;34(4):300-306

[85] Kawachi I, Sparrow D, et al. Decreased heart rate variability in men with phobic anxiety (data from the Normative Aging Study). American Journal of Cardiology. 1995;75(14):882-885

[86] Brydon L, Magid K, Steptoe A. Platelets, coronary heart disease, and stress. Brain, Behavior and Immunity. 2006;20(2):113-119

[87] Sirri L, Fava GA, et al. Type A behaviour: A reappraisal of its characteristics in cardiovascular disease. International Journal of Clinical Practices. 2012;66(9):854-861

[88] Denollet J, Schiffer AA, Spek V. A general propensity to psychological distress affects cardiovascular outcomes: Evidence from research on the type D (distressed) personality profile. Circulation Cardiovascular Quality and Outcomes. 2010;3(5):546-557

[89] Hatcher S, Arroll B. Newer antidepressants for the treatment of depression in adults. BMJ. 2012;344:d8300

[90] Stephenson CP, Karanges E, McGregor IS. Trends in the utilisation of psychotropic medications in Australia from 2000 to 2011. Australian and New Zealand Journal of Psychiatry. 2013;47(1):74-87

[91] Rosenberg LB, Whang W, et al. Exposure to tricyclic antidepressants is associated with an increased risk of incident CHD events in a population-based study. International Journal of Cardiology. 2010;145(1):124-125

[92] Blanchette CM, Simoni-Wastila L, et al. A secondary analysis of a duration response association between selective serotonin reuptake inhibitor use and the risk of acute myocardial infarction in the aging population. Annals of Epidemiology. 2008;18(4):316-321

[93] FDA.gov [homepage on the Internet]. FDA Drug Safety Communication: Abnormal heart rhythms associated with high doses of Celexa (citalopram hydrobromide) [updated March 3, 2012]. Available from: http://www.fda.gov/Drugs/DrugSafety/ucm269086.htm. Accessed January 20, 2017

[94] Girardin FR, Gex-Fabry M, et al. Drug-induced long QT in adult psychiatric inpatients: The 5-year cross-sectional ECG Screening Outcome in Psychiatry study. American Journal of Psychiatry. 2013;170(12):1468-1476

[95] Alvarez PA, Pahissa J. QT alterations in psychopharmacology: Proven candidates and suspects. Current Drug Safety. 2010;5(1):97-104

[96] Glassman AH, O'Connor CM, et al, Sertraline Antidepressant Heart Attack Randomized Trial (SADHEART) Group. Sertraline treatment of major depression in patients with acute MI or unstable angina. JAMA. 2002;288(6):701-709

[97] Ko DT, Hebert PR, et al. Beta-blocker therapy and symptoms of depression, fatigue, and sexual dysfunction. JAMA. 2002;288(3):351-357

[98] Hallas J. Evidence of depression provoked by cardiovascular medication: A prescription sequence symmetry analysis. Epidemiology. 1996;7(5):478-484

[99] Johansen A, Holmen J, et al. Anxiety and depression symptoms in arterial hypertension: The influence of antihypertensive treatment. The HUNT study, Norway. European Journal of Epidemiology. 2012;27(1):63-72

[100] Boal AH, Smith DJ, et al. Monotherapy with major antihypertensive drug classes and risk of hospital admissions for mood disorders. Hypertension. 2016;68(5):1132-1138

[101] Wurzelmann J, Frishman WH, et al. Neuropsychological effects of antihypertensive drugs. Cardiology Clinics. 1987;5(4):689-701

[102] Patten SB, Love EJ. Neuropsychiatric adverse drug reactions: Passive reports to Health and Welfare Canada's adverse drug reaction database (1965-present). International Journal of Psychiatry Medicine. 1994;24(1):45-62

[103] Leventhal H, Meyer D, Nerenz D. The common sense model of illness danger. In: Rachman S, editor. Contributions to Medical Psychology. New York: Pergamon; 1980. pp. 7-30

[104] World Health Organization. The European health report 2015, Targets and beyond – Reaching new frontiers in evidence. 1-150; Available at http://www.euro.who.int/__data/assets/pdf_file/0006/288645/European-health-report-2015-full-book-en.pdf

[105] Murray CJL, Lopez AD. Alternative projections of mortality and disability by cause 1990-2020: Global Burden of Disease Study. Lancet. 1997;349(9064):1498-1504

[106] Patrick DL, Erickson P. Health Status and Health Policy: Quality of Life in Health Care Evaluation and Resource Allocation. New York: Oxford University Press; 1993

[107] Saboya PM, Zimmermann PR, Bodanese LC. Association between anxiety or depressive symptoms and arterial hypertension, and their impact on the quality of life. International Journal of Psychiatry Medicine. 2010;40(3):307-320

[108] Mi B, Dang S, et al. Association between awareness of hypertension and health-related quality of life in a cross-sectional population-based study in rural area of Northwest China. Medicine (Baltimore). 2015;94(29):e1206

[109] Tang S, Bishwajit G, et al. Improving the blood pressure control with the proActive attitude of hypertensive patients seeking follow-up services: Evidence from China. Medicine (Baltimore). 2016;95(14):e3233

[110] Trevisol DJ, Moreira LB, et al. Health-related quality of life is worse in individuals with hypertension under drug treatment: Results of population-based study. Journal of Human Hypertension. 2012;26(6):374-380

[111] Suciu L, Cristescu C, et al. The impact of hypertension and associated comorbidities on quality of life assessment questionnaire SF-36 V2. Farmacia. 2013;61(3):503-517

[112] Lichtman JH, Bigger JT Jr, et al. Depression and coronary heart disease: Recommendations for screening, referral, and treatment: A science advisory from the American Heart

Association Prevention Committee of the Council on Cardiovascular Nursing, Council on Clinical Cardiology, Council on Epidemiology and Prevention, and Interdisciplinary Council on Quality of Care and Outcomes Research: Endorsed by the American Psychiatric Association. Circulation. 2008;118(17):1768-1775

[113] Norris CM, Murray JW, et al. Gender roles in persistent sex differences in health-related quality of life outcomes of patients with coronary artery disease. Gender Medicine. 2010;7(4):330-339

[114] Martinelli LM, Mizutani BM, et al. Quality of life and its association with cardiovascular risk factors in a community health care program population. Clinics (Sao Paolo). 2008;63(6):783-788

[115] Tsuchihashi-Makaya M, Hamaguchi S, et al, JCARE-CARD Investigators. Characteristics and outcomes of hospitalized patients with heart failure and reduced vs preserved ejection fraction: Report from the Japanese Cardiac Registry of Heart Failure in Cardiology (JCARE-CARD). Circulation Journal. 2009;73(10):1893-1900

[116] Alla F, Briançon S, et al, EPICAL Investigators. Self-rating of quality of life provides additional prognostic information in heart failure. Insights into the EPICAL study. European Journal of Heart Failures. 2002;4(3):337-343

[117] Moser DK, Heo S, et al. "It could be worse... Lot's worse!" Why health-related quality of life is better in older compared with younger individuals with heart failure. Age Ageing. 2013;42(5):626-632

[118] Heo S, Moser DK, et al. Dose-dependent relationship of physical and depressive symptoms with health-related quality of life in patients with heart failure. European Journal of Cardiovascular Nursing. 2013;12(5):454-460

[119] Banerjee T, Lee KS, et al. Limited association between perceived control and health related quality of life in patients with heart failure. Journal of Cardiovascular Nursing. 2014;29(3):227-231

[120] Stamp KD, Dunbar SB, et al. Family context influences psychological outcomes of depressive symptoms and emotional quality of life in patients with heart failure. Journal of Cardiovascular Nursing. 2014;29(6):517-527

[121] Pedersen SS, Denollet J. Type D personality, cardiac events, and impaired quality of life: A review. European Journal of Cardiovascular Prevention Rehabiliation. 2003;10(4):241-248

[122] Saeed T, Niazi GS, Almas S. Type D personality: A predictor of quality of life and coronary heart disease. Eastern Mediterranean Health. 2011;17(1):46-50

[123] Kearney PM, Whelton M, et al. Global burden of hypertension: Analysis of worldwide data. Lancet. 2005;365(9455):217-223

[124] Gaziano TA, Opie LH, Weinstein MC. Cardiovascular disease prevention with a multidrug regimen in the developing world: A cost-effectiveness analysis. Lancet. 2006;368(9536):679-686

[125] Kotseva K, Wood D, et al. EUROASPIRE IV: A European Society of Cardiology survey on the lifestyle, risk factor and therapeutic management of coronary patients from 24 European countries. European Journal of Preventive Cardiology. 2016;23(6):636-648

[126] Yoon S, Fryar C, Carroll M. Hypertension prevalence and control among adults: United States, 2011-2014. NCHS data brief, no 220. Hyatts ville, MD: National Center for Health Statistics. 2015. www.cdc.gov/nchs/data/databriefs/db220.htm. [Accessed 18-10-2016]

[127] Bosworth HB, Granger BB, et al. Medication adherence: A call for action. American Heart Journal. 2011;162(3):412-424

[128] World Health Organization. Noncommunicable Diseases and Mental Health, Global Action Plan for the Prevention and Control of Noncommunicable Diseases 2013-2020. Geneva, Switzerland: WHO; 2013

[129] Cristescu C, Voicu M, et al. Pharmacy clinique. Timişoara: Eurobit; 2015. pp. 85-123

[130] Morisky DE. Adherence or compliance behavior, encyclopedia of public health [Online]. Available from: URL: http://www.encyclopedia.com/doc/1G2- 3404000026.html

[131] Kolandaivelu K, Leiden BB, et al. Non-adherence to cardiovascular medications. European Heart Journal. 2014;35(46):3267-3276

[132] Wang PS, Bohn RL, et al. Noncompliance with antihypertensive medications: The impact of depressive symptoms and psychosocial factors. Journal of General Internal Medicine. 2002;17(7):504-511

[133] Holt EW, Muntner P, et al. Health-related quality of life and antihypertensive medication adherence among older adults. Age Ageing. 2010;39(4):481-487

[134] Souza AC, Borges JW, Moreira TM. Quality of life and treatment adherence in hypertensive patients: Systematic review with meta-analysis. Revista de Saude Publica. 2016;50:71

[135] Lulebo AM, Mutombo PB, et al. Predictors of non-adherence to antihypertensive medication in Kinshasa, Democratic Republic of Congo: A cross-sectional study. BMC Research Notes. 2015;8:526

[136] Gregoire JP, Moisan J, et al. Determinants of discontinuation of new courses of antihypertensive medications. Journal of Clinical Epidemiology. 2002;55(7):727-734

[137] Scarborough P, Bhatnagar P, et al. Coronary Heart Disease Statistics. London, England: British Heart Foundation; 2010

[138] DeWilde S, Carey IM, et al. Trends in secondary prevention of ischaemic heart disease in the UK 1994-2005: Use of individual and combination treatment. Heart. 2008;94(1):83-88

[139] Kotseva K, Wood D, et al, EUROASPIRE Study Group. EUROASPIRE III: A survey on the lifestyle, risk factors and use of cardioprotective drug therapies in coronary patients from 22 European countries. European Journal of Cardiovascular Prevention Rehabiliation. 2009;16(2):121-137

[140] Vrijens B, De Geest S, et al, ABC Project Team. A new taxonomy for describing and defining adherence to medications. British Journal of Clinical Pharmacology. 2012;73(5):691-705

[141] Ho PM, Spertus JA, et al. Impact of medication therapy discontinuation on mortality after myocardial infarction. Archives of Internal Medicine. 2006;166(17):1842-1847

[142] Wu JR, Moser DK. Type D personality predicts poor medication adherence in patients with heart failure in the USA. International Journal of Behavioral Medicine. 2014;21(5):833-842

[143] Bandura A. Self-efficacy: Toward a unifying theory of behavioral change. Psychological Review. 1977;84(2):191-215

[144] Jia-Rong W, Song EK, et al. Type D personality, self-efficacy, and medication adherence in patients with heart failure—A mediation analysis. Heart and Lung. 2015;44(4):276-281

[145] Christensen AJ. Patient Adherence to Medical Treatment Regimens. New Haven: Yale University Press; 2004

[146] Nguyen TM, La Caze A, Cottrell N. What are validated self-report adherence scales really measuring? A systematic review. British Journal of Clinical Pharmacology. 2014;77(3):427-445

[147] Leventhal M, Cameron L. Behavioural theories and the problem of compliance. Patient Education and Counseling. 1987;10:117-138

[148] Zolnierek KB, Dimatteo MR. Physician communication and patient adherence to treatment: A meta-analysis. Medical Care. 2009;47(8):826-834

[149] Cutrona SL, Choudhry NK, et al. Physician effectiveness in interventions to improve cardiovascular medication adherence: A systematic review. Journal of General Internal Medicine. 2010;25(10):1090-1096

[150] Conn VS, Ruppar TM, et al. Packaging interventions to increase medication adherence: Systematic review and meta-analysis. Current Medical Research Opinion. 2015;31(1):145-160

[151] Conn VS, Ruppar TM, et al. Patient-centered outcomes of medication adherence interventions: Systematic review and meta-analysis. Value Health. 2016;19(2):277-285

Chapter 5

Quality of Life in Children with Cerebral Palsy

Ozlem Cinar Ozdemir and Sezen Tezcan

Additional information is available at the end of the chapter

http://dx.doi.org/10.5772/67996

Abstract

Cerebral palsy is a complex and serious disease that can affect all age groups around the world. There is no prevalence differentiation between developed and developing countries. Preeclampsia, maternal trauma, low birth weight, gestational age, birth asphyxia, intrauterine infection, etc. can also be included in the list of risk factors. Functional limitations, self-care difficulties, behavioral problems, seizures, cognitive, sensory, social and emotional impairments and difficulties of daily life activities can affect health-related quality of life in children and adolescents with cerebral palsy. So, it is important to evaluate health-related quality of life in cerebral palsy. It is stated that there are many generic instruments to evaluate health-related quality of life. However, using specific instruments for cerebral palsy is more sensitive than generic instruments to evaluate effect of cerebral palsy in children's health-related quality of life. The aim of this chapter is to investigate health-related quality of life instruments in children and adolescents with cerebral palsy.

Keywords: cerebral palsy, health-related quality of life, quality of life, generic measures, condition-specific measures

1. Introduction

Cerebral palsy (CP), in its broadest definition issued by the "Surveillance of Cerebral Palsy in Europe"(SCPE), "is a group of permanent disorders in the development of movement and posture, causing activity limitations that are attributed to non-progressive disturbances that occurred in the developing fetal or infant brain." CP incidence is within the range of 1.5–2.5 per 1000 live births. In premature births and extreme low birth weights, this range elevates to 40–100 per 1000 live births [1].

Between the years 1980 and 2000, CP prevalence was within the range of 1.3–4.4 per 1000 live births, and there is no prevalence differentiation between developed and developing countries. Europe, North America, Australia, Hong Kong and Japan-based studies demonstrated that prevalence ratios were within the range of 1.3–3.6 per 1000 live births [2–5]. In China [6], Turkey [7] and India [8], on the other hand, these ratios varied between the ranges of 1.3 and 4.4 per 1000 live births [9].

Studies have ensured the following statistics, commonly reported in the United States:

- About 764,000 children and adults currently have CP.
- About 500,000 children under the age of 18 currently have CP.
- About two to three children out of every 1000 have cerebral palsy (the United States studies have yielded rates as low as 2.3 per 1000 children to as high as 3.6 per 1000 children).
- About 10,000 babies born each year will develop CP.
- Around 8000–10,000 babies and infants are diagnosed per year with CP.
- Around 1200–1500 preschool-aged children are diagnosed per year with CP [10].

2. Risk factors and known causes of CP

Risk factors in CP have been classified as prenatal, perinatal and postnatal phases (**Table 1**). Prenatal maternal risk factors are associated with delayed onset of menstruation, irregular menstruation cycles, an extended menstrual cycle, and maternal drug use, which have all been associated with an increased ratio of CP risk. Among perinatal causes are preeclampsia, maternal

Prenatal	Perinatal	Postnatal
Maternal disease in pregnancy	Length of labor	Neonatal seizures
Oligohydramnios	Membrane rupture	Respiratory distress syndrome
Polyhydramnios	Induction of labor	Hypoglycaemia
Perinatal infection	Augmentation	Jaundice
Pre-eclampsia	Meconium	İnfections (meningitis, sepsis,malaria)
Placental abnormalities	Abnormal fetal presentation	
Small for gestational age	Mode of delivery	
Large for gestational age	Birth asphxia	
Placenta praevia	Sentinel events(cord prolapse, cord around neck, specifically tight cord, uterine rupture)	
Hemorrage		

Table 1. Risk factors of CP.

http://dx.doi.org/10.5772/67996

trauma, antepartum hemorrhage, Factor V Leiden mutation, gene for prothrombin, placental thrombose, neonatal stroke, autoimmune and coagulation disorders, and multiple pregnancies. Cephalopelvic disproportion, largeness or abnormal positioning of the fetus that are linked with the cord prolapse induced perinatal asphyxia; severe intrapartum hemorrhage; extended or traumatic labor experience, also extension of the second phase of labor; emergency caesarean section; early separation of placenta; abnormal fetal position; chorioamnionitis; meconium presence; tight nuchal cord are among the CP risk factors while delivering the baby [11]. Preterm birth, emboli and thrombose, intrauterine infection, genetic disorders, neonatal seizures, neonatal sepsis, and respiratory diseases are the other critical risk factors of developing CP [12, 13]. CP prevalence is strongly linked with gestational age and birth weight. Literature studies manifested that babies with an extra-low birth weight (below 1500 g) are 20–80 times likely to develop CP when compared to babies with 2500 and higher grams of birth weight [14, 15].

Depending on the time brain damage occurred, it would be illuminating to classify causes for CP as prenatal, perinatal, postnatal (**Table 2**). Prenatal causes are congenital malformations, vascular incidences, and maternal infections detected during the first and second trimesters (rubella, cytomegalovirus, toxoplasma). Rarely experienced causes during prenatal period are metabolic diseases, maternal toxin intake and infrequent genetic syndromes. Perinatal causes are obstructed labor, and antepartum hemorrhage, neonatal ensefalopati that might be certain risk factors for triggering hypoxia of the fetus. Infection and injuries, cerebrovascular events, meningitis, septicemia and malaria can also be included in the list of postnatal causes [11].

Prenatal	Perinatal	Postnatal
Congenital brain malformations	Obstructed labor	Infection
Vascular events	Antepartum hemorrhage	İnjuries
Maternal infections during the first and second trimesters of pregnancy(rubella, cytomegalovirus,	Cord prolapse	Apparent life-threatening events
Metabolic disorders	Neonatal encephalopathy	Cerebrovascular accidents
Maternal ingestion of toxins		Following surgery for congenital malformations
Rare genetic syndromes		Meningitis
		Septicaemia
		Malaria

Table 2. Known causes of CP.

3. Classification of CP

3.1. Classification on the basis of neurological findings

1956-dated classification system issued by the American Cerebral Palsy Academy is still commonly utilized system of today: four motor types have been classified as spastic, dyskinetic, ataxic, and hypotonic (**Figure 1**) [16, 17]. Spastic type has been subcategorized as: spastic hemiplegia, spastic diplegia, spastic quadriplegia, and dyskinetic type has been subcategorized as: chorea, athetosis, ballismus, tremor, rigidity, dystonia [18]. Among children with CP, the most prevalent type is spastic (58% bilateral and 30% unilateral), and the second most prevalent type is dyskinetic type (7%). Ataxic type is common in a ratio of 4%, and there are certain unclassifiable forms with a ratio of 1% [1, 19].

Spastic CP: This is the most common form of CP [20]. It has been detected that 35% of spastic CP cases are hemiparetic, 28% are diparetic, and 37% are quadriparetic [16, 20]. In some studies, hemiparetic, diparetic, quadriparetic terminologies are replaced with the terms unilateral and bilateral [21, 22].

Among children with spastic quadriplegia that integrates four extremities, the severity of influence is extremely high (**Figure 2**). Retardation in physical development, severe mental retardation, seeing, hearing, chewing, swallowing and speaking impairments, pseudobulbar palsy and accompanying orofacial impairments, unmanageable epileptic seizures are often frequently experienced among these children [23].

Spastic hemiplegia is a unilateral paresis in which upper extremity is much more impacted than lower extremity (**Figure 3**). Its prevalence in term infants is 56 and 17% in preterm infants [24]. In upper extremity grasp action of the thumb, wrist extension and supination are the most impacted functions, while in lower extremity, dorsiflexion and eversion are among the most impacted functions. Flexor tonus has risen in parallel with hemiparetic posture, elbow and wrist flexion, harvesting position of the knees and feet. Sense anomalies are frequent incidences in the affected extremities. Two-point discrimination and position sensation have also been deteriorated in these cases. Visual field defect, homonym hemianopsia, cranial nerve anomalies, and facial nerve palsy at most are likely to be observed [24].

In spastic diplegia, lower extremities are more severely impacted than upper extremities (**Figure 4**). Among mildly impacted children, in effect of increased tonus of the muscles surrounding the ankle, emerging impaired dorsiflexion may lead to walking on the balls of the

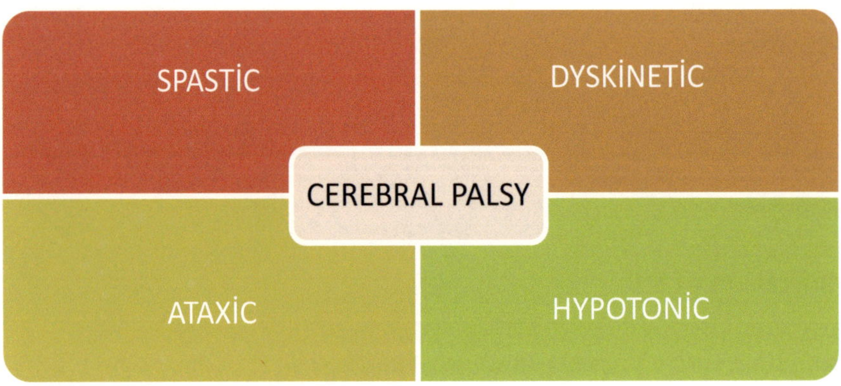

Figure 1. Classification on the basis of neurological findings.

http://dx.doi.org/10.5772/67996

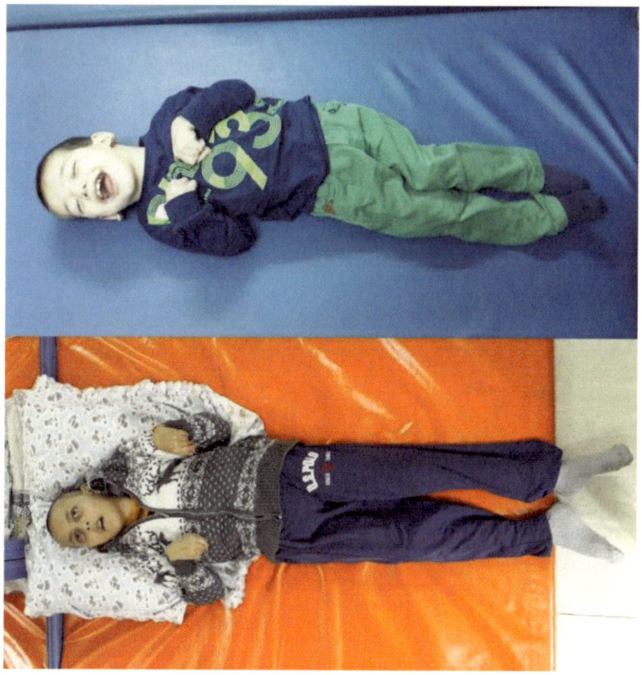

Figure 2. Spastic quadriplegia.

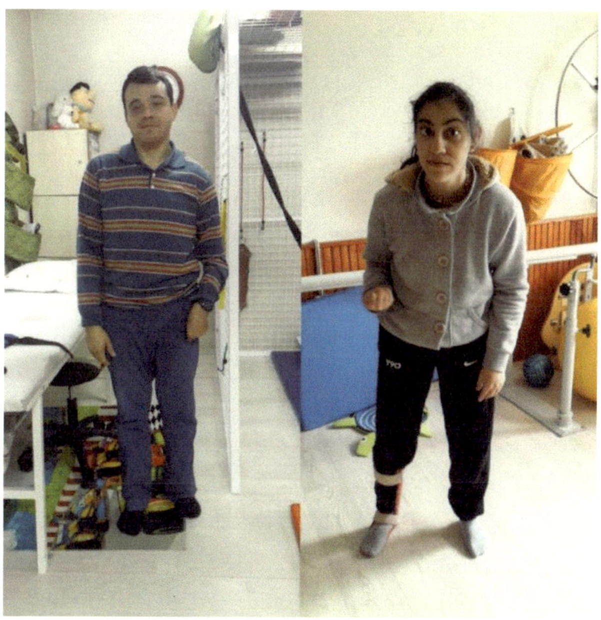

Figure 3. Spastic hemiplegia.

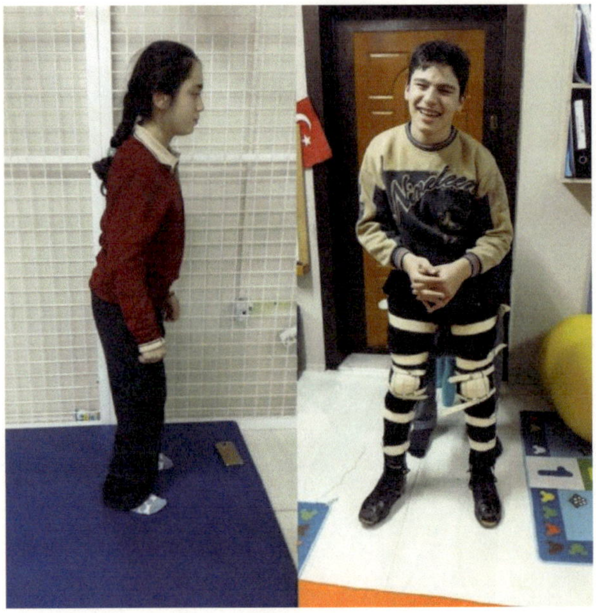

Figure 4. Spastic diplegia.

feet, while in severely affected cases, it is also common to see hip and knee flexion as well [24]. Spastic diplegia is basically triggered by prematurity and low birth weight [25]. When the child stands upright, rigidity in lower extremities is quite visible and due to adductor spasm, there is the scissoring of legs. Sense-perception impairments, epilepsy, mental retardation and strabismus are likely to be detected in such cases [25, 26].

Dyskinetic-type CP: Dyskinetic CP is characterized with unmanageable and unintended bodily movements, and dyskinetic-type CP includes subtypes such as chorea, athetosis, ballismus, tremor, rigidity, and dystonia [27].

Athetosis refers to unintended slow bodily movements such as convulsion. It is mostly common in major joints as major movements and in such movements problems such as timing, distance measuring or movement control are frequently problematic. Rigidity is less prevalent, and there is resistance against passive and active movement. Tremor is rhythmic and small actions mostly prevalent in minor joints. It is usually accompanied with athetosis or ataxia. Dystonia refers to slow, torsional movements that could either impact a single joint or the whole body. Ballismus is the least-prevalent movement disorder. They are random movements in a broad and speedy pattern and commonly observed in a single joint. Chorea includes irregular movements [28].

This type of CP is characterized with birth asphyxia. Severity of dystonic postures is likely to change with respect to body position, emotional state and sleep. On dyskinetic CP, primitive reflexes are more accentuated and sustained for longer periods. These movement patterns diminish in sleep when tonus in impacted extremities is comparatively weaker. It is also common to see posture control and coordination anomalies [24].

http://dx.doi.org/10.5772/67996

Ataxic CP: Children with ataxic CP, balance and coordination system are the most affected one. These children are able to walk only by keeping wide the support surface, and they have severe tremors adversely affecting performing well in daily activities that call for good motor functions [29]. It is common to see cerebellum damage at most [28]. In ataxic CP, the overall condition may be accompanied with spasticity, athetosis, nystagmus, speaking problems, mental retardation, and epilepsy [30].

Hypotonic CP: The basic negation among children with hypotonic-type CP is hypotonia in all muscles [28]. In the advanced levels spasticity, dyskinesia and ataxia in particular may develop in these children. In effect of hypotonia, there is growth retardation, abnormal tendon reflexes and absence of primitive reflexes (**Figure 5**) [23].

3.2. Classification based on motor function

In children with CP, walking ability is strongly affected [31]. Gross Motor Function Classification System (GMFCS) is a classification system developed by Palisano et al. for children with chronic disability and based on the movements initiated by the child him/herself such as sitting, moving and acting. Since children's motor functions are subject to change with the range of age, for each level, functions have been categorized as below age 2; within the age range of 2–4, within the age range of 4–6, and within the age range of 6–12. In the past, the system was only employed for children below age 12, and with its extended version, it can be used for the adolescents of age group 12–18 presently (**Table 3**) [32].

Figure 5. Hypotonic CP.

	GMFCS
Level I	Walks without limitations; limitations in more advanced gross motor skills
Level II	Walks with limitations; limitations walking outdoors and in the community
Level III	Walks with adaptive equipment assistance; limitations walking outdoors and in the community
Level IV	Self-mobility with use of powered mobility assistance; children are transported or use power mobility outdoors and in the community
Level V	Self-mobility is severely limited even with the use of assistive technology

Table 3. Gross Motor Classification System (GMFCS) [30].

Other classification systems employed among children with CP are Manual Ability Classification System (MACS) and Bimanual Fine Motor Function (BFMF) (**Table 4**) [33, 34]. MACS is a classification system that analyzes the way aged 4–18 children with CP use their hands while carrying objects in their daily activities. System does not measure maximum capacity of a child with

	MACS	BFMF
LEVEL I	Handles objects easily and successfully. At most limitations in the ease of performing manual tasks requiring speed and accuracy. However, any limitations in manual abilities do not restrict independence in daily activities.	One hand manipulates without restrictions. The other hand manipulates without restrictions or has Limitations in more advanced fine motor skills.
LEVEL II	Handles most objects, but with somewhat reduced quality or speed of achievement. Certain activities may be avoided or achieved with some difficulty; alternative ways of performing might be used, but manual abilities do not usually restrict independence in daily activities.	(a) One hand manipulates without restrictions. The other hand has only ability to grasp or hold. (b) Both hands have limitations in more advanced fine motor skills.
LEVEL III	Handles objects with difficulty; needs help to prepare and/or modify activities. The performance is slow and achieved with limited success regarding quality and quantity. Activities are performed independently if they have been set up or adapted.	(a) One hand manipulates without restrictions. The other hand has no functional ability. (b) One hand has limitations in more advanced fine motor skills. The other hand has only ability to grasp or worse.
LEVEL IV	Handles a limited selection of easily managed objects in adapted situations. Performs part of activities with effort and limited success. Requires continuous support and assistance and/or adapted equipment for even partial achievement of the activity	(a) Both hands have only ability to grasp. (b) One hand has only ability to grasp. The other hand has only ability to hold or worse. The child needs support and/or adapted equipment.
LEVEL V	Does not handle objects and has severely limited ability to perform even simple actions. Requires total assistance.	Both hands have only ability to hold or worse. The child requires total assistance, even with adaptations.

Table 4. Description of the five level classifications of bimanual fine motor function (BFMF) [34] and manual ability classification system (MACS) [33].

CP, but rather it aims to evaluate individual hand performance of the child carrying objects in his/her daily activities [33]. BFMF was defined by Beckung et al. in 2002. The system classifies grasping and manipulation skill in both hands by using a five-level system [34].

4. Prevalent problems among children with CP

Brain damage among children with CP leads to impaired postural control, retarded balance and movement development and pattern. As a result of the consequential weakness, hypotonicity or hypertonicity, and abnormal inactivation of muscles requiring co-contraction are visible in these cases. Musculoskeletal problems accompany these neuromuscular problems. Seeing, hearing, speaking anomalies, and behavioral problems such as attention deficit and hyperkinesia and epilepsy and communication problems accompany the overall picture [35]. In the reports published by SCPE, it was communicated that there is epilepsy story of 1 out of each 3 child with CP [36].

5. Health-related quality of life (HRQOL) in children with cerebral palsy

*World Health Organization (WHO) provided this definition for quality of life (QOL); "the way an individual perceives his/her own state within the context of one's culture and value system." In the habitat of a person, the concept of QOL that includes one's personal goals, expectations, standards and interests; items such as physical health, mental health, level of independence, social relations, environmental factors and personal beliefs are based on subjectivity reference [37, 38].

WHO cites that QOL does not simply indicate being free of ailments but rather QOL accentuates physical, mental and social well-being of an individual. QOL is evident in four domains: personal inner domain (values, beliefs, aspirations, personal goals, coping with problems etc.), personal social domain (family structure, income level, employment status, social opportunities etc.), external natural environment domain (air, water quality etc.), and external social environmental domain (cultural, social and religious institutes, social opportunities, school, healthcare services, security, transportation, shopping etc.) [39].

QOL and general meaning of QOL require the requirements of clinical medicine and clinical studies to be distinguished, and for this reason, health-related quality of life (HRQOL) concept is mostly used in order to remove the uncertainties [40]. HRQOL refers to a patient's subjective perception on the contentment level of his/her own health status [41]. HRQOL is a study field that that focuses on the awareness of an individual on the fact that a person's physical, psychological and social welfare level and treatment for a disease render certain effects on his/her daily life [42, 43]. QOL and HRQOL are two concepts that are inextricably intertwined.

In relevant literature, the latest studies focused on the assessment of HRQOL in children with CP and directed at increasing the level of HRQOL are significantly critical [44]. Functional deficits that develop among children with CP due to the accompanying physical, cognitive, sensory, emotional, and social impairments block these children's capacity to perform their assigned-social roles, thereby leading to major collapses in the level of HRQOL [44, 45].

Depending on the severity of impact, children with CP are subject to different types of functional limitations and these functional limitations pose adverse effects on their HRQOL level. Further to that, even among children with analogue functional limitation, it is feasible to detect a different level of HRQOL impact [46].

5.1. HRQOL measures for children with cerebral palsy

Usage of both generic and condition-specific QOL scales is accepted as a standard for determining the changes in QOL. Generic scales focus on the wide perspective of QOL and health situation and are used in general population or in a wide scale consisting of various illnesses [40]. General evaluation scales evaluate the general state of well-being and achieve the subjective measurement of treatment results in case of various illnesses. Disease-specific scales evaluate a specific diagnosis group or a patient population. Many researchers think that disease-specific or condition-specific scales are more sensitive compared to general scales [47].

5.1.1. Generic measures

KINDL: KINDL is a generic HRQOL measure for children and adolescents with CP. It was initially developed in German [48]. There are different versions of the scale for different age groups. Among these, Kiddy-KINDL is used for children aged 4–7 and is a version applied through the interviewer (person providing care for child). Other versions are Kid-KINDL used for children aged 8–12 and Kiddo-KINDL used for adolescents aged 13–16. Adult forms of these versions are also available. The scale consists of 24 items and 6 dimensions (physical well-being, spiritual well-being, self-confidence, family, friends, daily activities at school). High scores show that the HRQOL is good [41, 49, 50].

KIDSCREEN: KIDSCREEN is a generic QOL instrument for aged 8–18 children and adolescent with CP. Instrument available in three versions; original long version, 27-item version, and 10 item index version. Original long version consists of 52 items in 10 dimensions; physical well-being, psychological well-being, moods and emotions, self-perception, autonomy, parent relations and home life, social support and peers, school environment, social acceptance (bullying), financial resources. 27-item version evaluates QOL in five dimensions; physical well-being, psychological well-being, parent relations and autonomy, social support, and peers and school. Both instruments self-report and parent-report forms are available [51].

Child Health Questionnaire (CHQ): The Child Health Questionnaire (CHQ) is a generic QOL measure for children for children from 5-to-18 years of age. The scale consists of 14 physical and psychosocial domains: general health perceptions, physical functioning, role/social physical functioning, bodily pain, role/social emotional functioning, role/social behavioral functioning, parent impact-time, parent impact-emotional, self-esteem, mental health, behavior, family activities, family cohesion, and change in health. While CHQCF87 (the child-report questionnaire) consists of 87 items, the long parent-report questionnaire (CHQ-PF50) consists of 50 items, and the short parent-report questionnaire (CHQ-PF28) consists of 28 items [52, 53].

http://dx.doi.org/10.5772/67996

TACQOL: TACQOL is a general scale developed in order to evaluate the QOL of children aged 6–15 with chronic disease, about health. It has two versions named Parent-form (TACQOL-PF) and Child-form (TACQOL-CF). TACQOL-PF is used for children aged 6–15, and TACQOL-CF is used for children aged 8–15. Both scales have seven domains containing 56 items [54, 55].

Pictured Child's Quality of life self Questionnaire (AUQUEI): AUQUEI is a generic measure, evaluates the child's subjective QOL. Two versions are available as ages 3–5 and ages 6–11 and explore following domains; family and social relations, activity (play, schoolarity, leisure), health, functions (sleeping, alimentation), separation [56, 57].

Young Adult Quality of life (YAQOL): YAQOL is an instrument for young adults aged 18–25. The measure consists of five domains; physical health, psychological well-being, social relationship, role function, environmental context and takes approximately 25 min. Answer options include four-point Likert scales (1 = not at all, 2 = somewhat, 3 = usually, 4 = definitely; 1 = false, 2 = mostly false, 3 = mostly true, 4 = true; 1 = never, 2 = seldom, 3 = sometimes, 4 = often) [58, 59].

DİSABKIDS Condition-Generic Module: The scale is a condition-generic module for aged 8–16 years children and adolescents diagnosed with different chronic conditions, and scale has two versions. The long version consists of 37 Likert-scaled items in three dimensions (DCGM-37): mental (independence: 6 items, emotion: 7 items), social (social inclusion: 6 items, social exclusion: 6 items), and physical (limitation: 6 items, treatment: 6 items), and the short version consists of 12 items [60].

Child Health and Illness Profile (CHIP): CHIP is a generic health status and HRQOL measure with 2 forms: CHIP-CE (Child Health and Illness Profile-Child Edition) and CHIP-AE (Child Health and Illness Profile-Adolescent Edition). CHIP-CE is a generic health status and HRQOL instrument for children 6–11 years old or their parents. CHIP-CE consists of four domains: satisfaction, comfort, resilience, and risk avoidance [61]. There is also a parent report form of CHIP-CE. The CHIP-AE is a self-administered (for 11–17 years old) generic health status and HRQOL instrument. CHIP-AE includes six domains (satisfaction, discomfort, resilience, risk avoidance, achievement, disorders), and 20 subdomains consist of 107 items plus an additional 46 (disease or injury specific) items [62, 63].

Exeter Health-Related Quality of Life Measure (EHRQOL): EHRQOL is a generic computer-delivered measure assesses self-reported HRQOL in 6–11 aged children. The scale includes seven domains (activity limitation, physical symptoms, negative feelings, self image, relation with friends, scholar functioning, interaction with family) and consists of 16 pictures, and it takes approximately 20 min [56, 64].

Generic Children's Quality of Life Measure (GCQ): GCQ assesses perceived QOL in children aged between 6 and 14. Scale explores seven domains and consists of 50 questions (25 × 2); physical functioning, physical (sport), positive emotions, self-image, cognitive functioning, relation with friends, interaction with family. In the first section, the child choose the character in the story that he/she feels the most like (Perceived-Self), and in the second section, the child choose the character that he/she would most like to be (Preferred-Self). The measure has girl specific or boy specific version, the only differences being the color and gender of the characters [56, 65].

'Vécu et Santé Perçue des Adolescents' (VSP-A): VSP-A is a generic self-administered measure for healthy and ill adolescents aged 11–17 [66].

Duke Health Profile-Adolescent Version (DHP-A): DHP-A is a 17-item generic self-report measure and consists of four dysfunction dimensions (anxiety, depression, pain, disability), and six health dimensions (physical, mental, social, general health, perceived health, self-esteem). It is used in adolescents ged between 13 to 18 [56, 67].

5.1.2. Condition-specific measures

PedsQOL Cerebral Palsy Module: PedsQOL Cerebral Palsy Module is a condition-specific HRQOL instrument for toddlers, young children, children and teens with CP. The module has parent-report and self-report forms. Parent-report for toddler (ages 2–4) composed of 22 items comprising five dimensions, parent and child report for young children (ages 5–7), children (ages 8–12), teens (ages 13–17) composed of 35 items comprising seven dimensions. The seven dimensions include daily activities (9 items), school activities (4 items), movement and balance (5 items), pain and hurt (5 items), fatigue (5 items), eating activities (5 items), speech and communication (5 items) domains. There are no school activities and speech and communication dimensions in parent-report for toddler form. High scores indicate lower problems [68].

Caregiver Priorities and Child Health Index of Life with Disabilities (CPCHILD): This is a disease-specific instrument for children and adolescents aged 5–18 years with CP. CPCHILD consists of 36 items in six domains. These are personal care (8 items), positioning, transfer, and mobility (8 items), communication and social interaction (7 items), comfort, emotions, and behavior (9 items), health (3 items), and overall quality of life (1 item). The degree of difficulty of accomplishing activity was rated on a seven-point ordinal scale (0: no problem at all, 6: impossible). The level of assistance required was rated on a six-point ordinal scale from 0 ('Independent') to 5 ('Total assistance'). For Comfort, Emotions, and Behavior domain, the frequency of discomfort was rated on a six-point ordinal scale from 0 ('None of the time') to 5 ('All the time'). In the health and overall quality of life domain, items were rated on a six-point ordinal scale. Last, Section 7 (Importance of items) determines the caregivers' rating of the importance of each of the questionnaires' items toward their child's overall quality of life by using a six-point ordinal scale, from 0 ('Least important') to 5 ('Most important'). The questionnaire took approximately 20 min to complete [69].

DISABKIDS CP disease module: DISABKIDS CP disease module is a condition-specific instrument for children and adolescents aged 8–18 years with CP, consists of 14 disease-specific questions and 2 further items on communication about the condition. High scores remark better quality of life [70, 71].

Cerebral Palsy Quality of Life Questionnaire for Children (CPQOL-Child): CPQOL has two forms: primary caregiver-proxy report for children aged 4–12 years and self-report form for children aged 9–12 years. Primary caregiver-proxy report includes seven domains (social well-being and acceptance, functioning, participation and physical health, emotional well-being, access to services, pain and impact of disability, and family health) and consists of 66 items. Self-report form includes 5 domains and 52 items, except for pain and impact of disability, and family health domain [72].

Lifestyle Assessment Questionnaire-Cerebral Palsy: Lifestyle Assessment Questionnaire Cerebral Palsy includes 46 items in six dimensions: physical independence, mobility, clinical burden, schooling, economic burden, and social integration. Total score of scale (Lifestyle Assessment Score-LAS) expressed in percent. The classification of the HRQOL with regard to LAS: good (<30%); mildly affected (30–50%); moderately affected (51–70%); and severely affected (>70%) [73, 74].

Functional limitations, self-care difficulties, behavioral problems, seizures, cognitive, sensory, social and emotional impairments and difficulties of daily life activities can affect health-related quality of life in children and adolescents with cerebral palsy. So, it is important to evaluate health-related quality of life in cerebral palsy. It is stated that there are many generic instruments to evaluate health-related quality of life. However, using specific instruments for cerebral palsy is more sensitive than generic instruments to evaluate effect of cerebral palsy in children's health-related quality of life.

Acknowledgements

We would like to thank our patients for the permission to use their photographs.

Author details

Ozlem Cinar Ozdemir[1]* and Sezen Tezcan[2]

*Address all correspondence to: ozlemcinar314@hotmail.com

1 Abant Izzet Baysal University, School of Physical Therapy and Rehabilitation, Bolu, Turkey

2 Pediatric Rehabilitation Center, Bolu, Turkey

References

[1] Krägeloh-Mann I, Cans C. Cerebral palsy update. Brain and Development. 2009;31(7):537-44. doi:10.1016/j.braindev.2009.03.009

[2] Johnson A. Prevalence and characteristics of children with cerebral palsy in Europe. Developmental Medicine & Child Neurology. 2002;44(09):633-40. doi:10.1017/S0012162201002675

[3] Yeargin-Allsopp M, Braun KVN, Doernberg NS, Benedict RE, Kirby RS, Durkin MS. Prevalence of cerebral palsy in 8-year-old children in three areas of the United States in 2002: a multisite collaboration. Pediatrics. 2008;121(3):547-54. doi:10.1542/peds.2007-1270

[4] Oskoui M, Coutinho F, Dykeman J, Jetté N, Pringsheim T. An update on the prevalence of cerebral palsy: a systematic review and meta-analysis. Developmental Medicine & Child Neurology. 2013;55(6):509-19. doi:10.111/dmcn.12080

[5] Tsui K, Yiu B, Cheng C, Chan C. Prevalence study of cerebral palsy in Hong Kong children. Hong Kong Medical Journal. 2006;12(3):180-4.

[6] Liu J-M, Li S, Lin Q, Li Z. Prevalence of cerebral palsy in China. International Journal of Epidemiology. 1999;28(5):949-54. doi:10.1093/ije/28.5.949

[7] Serdaroglu A, Cansu A, Ozkan S, Tezcan S. Prevalence of cerebral palsy in Turkish children between the ages of 2 and 16 years. Developmental Medicine & Child Neurology. 2006;48(6):413. doi:10.1017/S0012162206000910

[8] Banerjee TK, Hazra A, Biswas A, Ray J, Roy T, Raut DK, et al. Neurological disorders in children and adolescents. The Indian Journal of Pediatrics. 2009;76(2):139-46. doi:10.1007/s12098-008-0226-z

[9] Panteliadis CP. Epidemiology of cerebral palsy. In: Cans C, Sellier E, Mermet M-A, editors. Cerebral Palsy. Germany; 2011. pp. 17-27.

[10] Prevalence of Cerebral Palsy 2016 [cited 03.12.2016]. Available from: http://www.cerebralpalsy.org/about-cerebral-palsy/prevalence-and-incidence.

[11] Reddihough DS, Collins KJ. The epidemiology and causes of cerebral palsy. Australian Journal of Physiotherapy. 2003;49(1):7-12. doi:10.1016/S0004-9514(14)60183-5

[12] MacLennan AH, Thompson SC, Gecz J. Cerebral palsy: causes, pathways, and the role of genetic variants. American Journal of Obstetrics and Gynecology. 2015;213(6):779-88. doi:10.1016/j.ajog.2015.05.034

[13] McIntyre S, Taitz D, Keogh J, Goldsmith S, Badawi N, Blair E. A systematic review of risk factors for cerebral palsy in children born at term in developed countries. Developmental Medicine & Child Neurology. 2013;55(6):499-508. doi:10.1111/dmcn.12017

[14] Platt MJ, Cans C, Johnson A, Surman G, Topp M, Torrioli MG, et al. Trends in cerebral palsy among infants of very low birthweight (<1500 g) or born prematurely (<32 weeks) in 16 European centres: a database study. The Lancet. 2007;369(9555):43-50. doi:10.1016/S0140-6736(07)60030-0

[15] Gladstone M. A review of the incidence and prevalence, types and aetiology of childhood cerebral palsy in resource-poor settings. Annals of Tropical Paediatrics. 2010;30(3):181-96. doi:10.1179/146532810X12786388978481

[16] Dodd KJ, Imms C, Taylor NF. What is cerebral palsy? In: Imms C, Dodd KJ, editors. Physiotherapy and Occupational Therapy for People with Cerebral Palsy. 1st ed. London: British Library Cataloguing; 2010. pp. 7-31.

[17] Rosenbaum P, Paneth N, Leviton A, Goldstein M, Bax M, Damiano D, et al. A report: the definition and classification of cerebral palsy. Developmental Medicine & Child Neurology Supplement 2007;109(suppl 109):8-14. doi:10.1111/j.1469-8749.2007.tb12610.x

[18] Takahashi R, Yamada M, Takahashi T, Ito T, Nakae S, Kobayashi Y, et al. Risk factors for cerebral palsy in preterm infants. Early Human Development. 2005;81(6):545-53. doi:10.1016/j.earlhumdev.2004.11.007

[19] Bax M, Goldstein M, Rosenbaum P, Leviton A, Paneth N, Dan B, et al. Proposed definition and classification of cerebral palsy. Developmental Medicine & Child Neurology. 2005;47(08):571-6. doi:10.1017/S001216220500112X

[20] Baxter P, Morris C, Rosenbaum P, Paneth N, Leviton A, Goldstein M, et al. The definition and classification of cerebral palsy. Developmental Medicine & Child Neurology. 2007;49(s109):1-44. doi:10.1111/j.1469-8749.2007.00001.x

[21] Tarsuslu T, Livanelioğlu A. The effect of pain on activity independence and health-related quality of life in cerebral palsied individuals. Fizyoterapi ve Rehabilitasyon Dergisi. 2008;19:117-22. doi:10.5505/agri.2011.93695

[22] O'Shea M. Cerebral palsy. Seminars in Perinatology. 2008;32:35-41.

[23] Ferrari A, Cioni G. Tetraplegic forms. In: Ferrari A, Lodesani M, Muzzini S, Pascale R, Sassi S, editors. The Spastic Forms of Cerebral Palsy. Italy 2010. pp. 241-73.

[24] Sankar C, Mundkur N. Cerebral palsy-definition, classification, etiology and early diagnosis. The Indian Journal of Pediatrics. 2005;72(10):865-8. doi:10.1007/BF02731117

[25] Mewasingh LD, Demil A, Christiaens FJ, Missa A-M, Cheron G, Dan B. Motor strategies in standing up in leukomalacic spastic diplegia. Brain and Development. 2002;24(5):291-5. doi:http://dx.doi.org/10.1016/S0387-7604(02)00059-1

[26] Panteliadis CP. Clinical characteristics. In: Papavasiliou A, Panteliadis CP, editors. Cerebral Palsy. Germany: Dustri-Verlag; 2011. pp. 89-105.

[27] Hou M, Zhao J-H, Yu R. Recent advances in dyskinetic cerebral palsy. World Journal of Pediatrics. 2006;1:23-8.

[28] Tecklin JS. Pediatric Physical Therapy. 3rd ed. United States of America: Lippincott Williams & Wilkins; 1999.

[29] Krigger KW. Cerebral palsy: an overview. American Family Physician. 2006;73(1):91-100.

[30] Jan JE, Lyons CJ, Heaven RK, Matsuba C. Visual impairment due to a dyskinetic eye movement disorder in children with dyskinetic cerebral palsy. Developmental Medicine & Child Neurology. 2001;43(02):108-12. doi:10.1111/j.1469-8749.2001.tb00725.x

[31] Beckung E, Hagberg G, Uldall P, Cans C. Probability of walking in children with cerebral palsy in Europe. Pediatrics. 2008;121(1):e187–e92. doi:10.1542/peds.2007-0068

[32] Palisano R, Rosenbaum P, Walter S, Russell D, Wood E, Galuppi B. Development and reliability of a system to classify gross motor function in children with cerebral palsy. Developmental Medicine & Child Neurology. 1997;39(4):214-23. doı:10.1111/j.1469-8749.1997.tb07414.x

[33] Eliasson A-C, Krumlinde-Sundholm L, Rösblad B, Beckung E, Arner M, Öhrvall A-M, et al. The Manual Ability Classification System (MACS) for children with cerebral palsy: scale development and evidence of validity and reliability. Developmental Medicine & Child Neurology. 2006;48(07):549-54. doı:10.1017/S0012162206001162

[34] Beckung E, Hagberg G. Neuroimpairments, activity limitations, and participation restrictions in children with cerebral palsy. Developmental Medicine & Child Neurology. 2002;44(5):309-16. doı:10.1111/j.1469-8749.2002.tb00816.x

[35] Levit S. Treatment of Cerebral Palsy and Motor Delay. 5th ed. Chichester: Wiley-Blackwell Publication; 2010.

[36] Sellier E, Uldall P, Calado E, Sigurdardottir S, Torrioli MG, Platt MJ, et al. Epilepsy and cerebral palsy: characteristics and trends in children born in 1976-1998. European Journal of Paediatric Neurology. 2012;16(1):48-55. doı:10.1016/j.ejpn.2011.10.003

[37] Organization WH. WHOQOL: Measuring Quality of Life. Switzerland: World Health Organization; 1997.

[38] Kaya M. Ankara" da 112 Acil Yardım Hizmetlerinde Çalısan Sağlık Personelinin Öznel Yasam Kalitelerinin Değerlendirilmesi [thesis]. Ankara: Ankara Üniversitesi; 2003.

[39] Basaran S, Güzel R, Sarpel T. Qualıty of life and health outcome assessment questionnaıres. Romatizma. 2005;20(1):55-63.

[40] Fayers P, Machin D. Quality of Life. 2nd ed. Chichester: John Wiley & Sons Ltd.; 2007.

[41] Eser E, Yüksel H, Baydur H, Erhart M, Saatli G, Özyurt BC, et al. The Psychometric Properties of the New Turkish Generic Health-Related Quality of Life Questionnaire for Children (Kid-KINDL). Türk Psikiyatri Dergisi. 2008;19(4):409-17.

[42] Erdoğanoğlu Y, Günel MK. Investigation of health-related quality of life of families with have children with cerebral palsy. Toplum Hekimliği Bülteni. 2007;26(2):35-9.

[43] Erdoğanoğlu Y. Serebral Paralizili Çocuklar ve Ailelerinde Sağlıkla İlgili Yasam Kalitesinin Değerlendirilmesi [thesis]. Ankara: Hacettepe Üniversitesi; 2006.

[44] Bjornson K, McLaughlin J. The measurement of health-related quality of life (HRQL) in children with cerebral palsy. European Journal of Neurology. 2001;8(s5):183-93. doı:10.1046/j.1468-1331.2001.00051.x

[45] Schneider JW, Gurucharri LM, Gutierrez AL, Gaebler-Spira DJ. Health-related quality of life and functional outcome measures for children with cerebral palsy. Developmental Medicine & Child Neurology. 2001;43(09):601-8. doı:10.1111/j.1469-8749.2001.tb00242.x

[46] Liptak GS, O'Donnell M, Conaway M, Chumlea WC, Worley G, Henderson RC, et al. Health status of children with moderate to severe cerebral palsy. Developmental Medicine & Child Neurology. 2001;43(6):364-70. doi:10.1111/j.1469-8749.2001.tb00223

[47] Wiebe S, Guyatt G, Weaver B, Matijevic S, Sidwell C. Comparative responsiveness of generic and specific quality-of-life instruments. Journal of Clinical Epidemiology. 2003;56(1):52-60.

[48] Questionnaires 1998. Available from: http://www.kindl.org/english/questionnaires/.

[49] Ravens-Sieberer U, Bullinger M. Assessing health-related quality of life in chronically ill children with the German KINDL: first psychometric and content analytical results. Quality of Life Research. 1998;7(5):399-407. doi:10.1023/A:1008853819715

[50] Ravens-Sieberer U, Bullinger M. News from the KINDL-questionnaire: a new version for adolescents. Quality of Life Research. 1998: 653-653.

[51] Ravens-Sieberer U, Herdman M, Devine J, Otto C, Bullinger M, Rose M, et al. The European KIDSCREEN approach to measure quality of life and well-being in children: development, current application, and future advances. Quality of Life Research. 2014;23(3):791-803. doi:10.1007/s11136-013-0428-3

[52] Hullmann SE, Ryan JL, Ramsey RR, Chaney JM, Mullins LL. Measures of general pediatric quality of life: Child Health Questionnaire (CHQ), DISABKIDS Chronic Generic Measure (DCGM), KINDL-R, Pediatric Quality of Life Inventory (PedsQL) 4.0 Generic Core Scales, and Quality of My Life Questionnaire (QoML). Arthritis Care & Research. 2011;63(S11):S420–S30. doi:10.1002/acr.20637

[53] CHQ: Child Health Questionnaire™ 2016. Available from: https://www.healthactchq.com/survey/chq.

[54] Verrips EG, Vogels TG, Koopman HM, Theunissen NC, Kamphuis RP, Fekkes M, et al. Measuring health-related quality of life in a child population. The European Journal of Public Health. 1999;9(3):188-93.

[55] Vogels T, Verrips G, Koopman H, Theunissen N, Fekkes M, Kamphuis R. TACQOL Manual: Parent Form and Child Form. Leiden: Leiden Center for Child Health and Pediatrics LUMC-TNO; 2000.

[56] Viehweger E, Robitail S, Rohon M-A, Jacquemier M, Jouve J-L, Bollini G, et al., editors. Measuring quality of life in cerebral palsy children. Annales de Réadaptation et de Médecine Physique; 2008: Elsevier. doi:10.1016/j.annrmp.2007.12.007

[57] Pictured Child's Quality of Life Self Questionnaire (AUQUEI) 2016. Available from: https://eprovide.mapi-trust.org/instruments/pictured-child-s-quality-of-life-self-questionnaire #basic_description.

[58] Jiang B, Walstab J, Reid SM, Davis E, Reddihough D. Quality of life in young adults with cerebral palsy. Disability and Health Journal. 2016. doi:10.1016/j.dhjo.2016.04.006

[59] Chen H, Cohen P, Kasen S, Gordan K, Dufur R, Smailes E. Construction and validation of a quality of life instrument for young adults. Quality of Life Research. 2004;13(4):747-59. doi:10.1023/B:QURE.0000021700.42478.ab

[60] Schmidt S, Debensason D, Mühlan H, Petersen C, Power M, Simeoni MC, et al. The DISABKIDS generic quality of life instrument showed cross-cultural validity. Journal of Clinical Epidemiology. 2006;59(6):587-98. doi:10.1016/j.jclinepi.2005.09.012.

[61] Riley AW, Forrest CB, Rebok GW, Starfield B, Green BF, Robertson JA, et al. The child report form of the CHIP–Child Edition: reliability and validity. Medical Care. 2004;42(3):221-31.

[62] Riley, Anne W., et al. "Reliability and validity of the adolescent health profile-types." *Medical care* 1998; 36(8): 1237-1248.

[63] Starfield B, Riley AW, Green BF, Ensminger ME, Ryan SA, Kelleher K, et al. The adolescent child health and illness profile: a population-based measure of health. Medical Care. 1995; 33(5): 553-566.

[64] Eiser C, Cotter I, Oades P, Seamark D, Smith R. Health-related quality-of-life measures for children. International Journal of Cancer. 1999;83(S12):87-90. doi:10.1002/(SICI)1097-0215(1999)83:12+<87::AID-IJC16>3.0.CO;2-O

[65] Collier J, MacKinlay D, Phillips D. Norm values for the Generic Children's Quality of Life Measure (GCQ) from a large school-based sample. Quality of Life Research. 2000;9(6):617-23. doi:10.1023/A:1008915700210

[66] Simeoni M, Auquier P, Antoniotti S, Sapin C, San Marco J. Validation of a French health-related quality of life instrument for adolescents: the VSP-A. Quality of Life Research. 2000;9(4):393-403. doi:10.1023/A:1008957104322

[67] Vo TXH, Guillemin F, Deschamps J-P. Psychometric properties of the DUKE Health Profile-adolescent version (DHP-A): a generic instrument for adolescents. Quality of Life Research. 2005;14(10):2229-34. doi:10.1007/s11136-005-7021-3

[68] Varni JW, Burwinkle TM, Berrin SJ, Sherman SA, BA KA, Malcarne VL, et al. The PedsQL in pediatric cerebral palsy: reliability, validity, and sensitivity of the Generic Core Scales and Cerebral Palsy Module. Developmental Medicine & Child Neurology. 2006;48(6):442-9. doi:10.1111/j.1469-8749.2006.tb01293.x

[69] Narayanan UG, Fehlings D, Weir S, Knights S, Kiran S, Campbell K. Initial development and validation of the Caregiver Priorities and Child Health Index of Life with Disabilities (CPCHILD). Developmental Medicine & Child Neurology. 2006;48(10):804-12. doi:10.1111/j.1469-8749.2006.tb01227.x

[70] Baars RM, Atherton CI, Koopman HM, Bullinger M, Power M. The European DISABKIDS project: development of seven condition-specific modules to measure health related quality of life in children and adolescents. Health and Quality of Life Outcomes. 2005;3(1):1. doi:10.1186/1477-7525-3-70

[71] Content of the DISABKIDS Cerebral Palsy Module. 2012. Available from: http://www.disabkids.org/questionnaire/disabkids-disease-modules/cerebral-palsy-module/.

[72] Waters E, Davis E, Mackinnon A, Boyd R, Graham HK, Kai Lo S, et al. Psychometric properties of the quality of life questionnaire for children with CP. Developmental Medicine & Child Neurology. 2007;49(1):49-55. doi:10.1017/S0012162207000126.x

[73] Dobhal M, Juneja M, Jain R, Sairam S, Thiagarajan D. Health-related quality of life in children with cerebral palsy and their families. Indian Pediatrics. 2014;51(5):385-7. doi:10.1007/s13312-014-0414-z

[74] Mackie P, Jessen E, Jarvis S. The lifestyle assessment questionnaire: an instrument to measure the impact of disability on the lives of children with cerebral palsy and their families. Child: Care, Health and Development. 1998;24(6):473-86. doi:10.1046/j.1365-2214.1998.00083.x

Chapter 6

Purpose in Life for Adolescents with Oncological Diseases

Eva Žiaková, Karolina Barinková and

Katarína Šiňanská

Additional information is available at the end of the chapter

http://dx.doi.org/10.5772/68001

Abstract

Under common circumstances, adolescent age is the period of a young person having to cope with many life problems (resolving issues if personal maturity which increases, respect decreases self-confidence, changing quality of interests, etc.) We can claim that adolescent age itself involves existential character. If this process involves a serious disease, the situation is really demanding. The aim of the chapter is to describe the process of seeking their own identity in the situation of coping with occurrence of an oncologic disease.

Keywords: Adolescents, oncological diseases, purpose of life, life, treatment

1. Introduction

In Slovakia, since 1930 occurrence and mortality rates for cancer are increasing. According to the National oncological register of SR, occurrence has an increasing trend since 1968–1970, when about 12,000 new cases were annually registered in 2003, the number reached 22,000 and according to other data, last year it was 24,000 cases more, one-tenth thereof made up by adolescents [1].

There are specific issues of occurrence, progression and treatment of oncologic disease in adolescents and also issues regarding their re-integration in common adolescent life. Just in this age group, an important role is played by social problems and related issues of social care [2].

Also this period of life is characterized by emotional instability, frequent and striking changes of moods, impulsiveness in acting and instability which make the situation worse. Moreover, it involves increased fatigue, alteration of being apathetic and short phases of increased activity [3]. This study is mainly about the theoretical insight into the problems of oncological diseases in adolescence with some results of research in this field of study.

2. Theoretical background

The issue of the purpose in life is granted more and more importance in the last decades. The cause being that a human finds himself all alone in post-modern times. Not knowing animal instincts that could direct him what he must do and also give up traditions that would determine his obligations [4].

Křivohlavý [5] claims that in creating the purpose in life, it is a process when people evaluate and revaluate the importance of one or several life events. They try to find the sense of it. If people get to stressing situations bringing about great suffering, there is an increased need to find the sense in something that is going on.

Halama [6] states that the purpose in life is not a simple phenomenon. It depends on several aspects, above all ideals and objectives. Those ideals and objectives, experiencing the feeling of having the reason to live, teach mutual interactions. The purpose in life is perceived as a personal system of objectives, ideals and values that enables a person to experience, to realize and to control his life as valuable, purposeful and satisfying.

Seriously ill patients can have some doubts about the purpose in life or they do not lose the hope that such reason exists at all. Just logo therapy provides help in looking for the purpose, it assists the patient to find the reason to live, helps the patient and awakens the desire for satisfaction. It determines that the person is essentially full of will to live.

At present, the issue of such purpose has transferred from searching for "the true values" to the analysis of the role, experience and conditions, under which the purpose in life can develop. Halama [6] presents studies which deal with the relation between the purpose in life and mental wellbeing of a person, the impact of purpose in life to cope with stress and the relation between senselessness and addictions. Debats [5] talks about three theoretical approaches to purposefulness of life. The principal representatives advocating the approach to purposefulness of life are considered Frank, Maslow and Yalom. Everyone of them viewed the issue from a different prospective.

The outlook of the American psychologist Maslow [7] regarding the purpose in life is affected by his definition of superior and inferior needs. He considers the purpose as an inner property which is demonstrated as a power of motivation, when the inferior needs are satisfied. The purpose can be experienced by a self-actualized person who uses creativity to develop his own potential. However, the actual representative of psychotherapy Yalom [8] claims that finally life has no purpose at all. He says that the person must create such purpose and actively be involved in it to cope with such senselessness.

Křivohlavý [5] pursues an opinion that a verbal phrase "purpose in life" belongs to the human subjective—personal and initial experiences. The feeling of a purpose in life depicts a deep motivation, which is significant for any human activity, that is, it is a superior need, which involves the rest of all needs.

Waren [9] adds that to know the purpose of our life is very important because we came in this world with a certain intention. If we have purpose and objective in our life, the life will have a certain direction, because it will help us to concentrate our efforts and energy to important issues. He also says that the purpose will provide motivation for our life, because nothing can give a person more enthusiasm than his own objective. The final declaration is that the purpose in life will prepare us for eternity. The purpose in life of different people is so different, that many specialists tried to classify it and give it some measurable form [2].

A significant person who dealt with the purpose in life was Frankl [7] who states that "To be a human means to be devoted to something, to be concentrated and set for something or someone, by which he asserts himself, develops, is himself and reaches some purpose which should be achieved."

Drapela [10] presents Frank's opinion that searching for purpose in life is the principal responsibility of a human, because any situation in life makes sense, however, it just cannot be introduced in the situation. He also describes that an individual can find such purpose in three ways: (a) By performing an act, where every activity involves a hidden purpose; (b) by experiencing a value, it means experiencing any real human experience that enriches an individual. According to Frank it is love, which is (c) fulfillment of a human purpose by the form of suffering.

Bačová [11] describes, that Frankl considers searching for purpose in life to be the principal motivation and driving force of a human life. One of the alternatives of existential psychology is logo therapy and existential analysis. Frankl survived a concentration camp and his theory is also based on that, which stands on the term Wille zum Sinn, which can be translated as "the will for a purpose". Frankl depicts it as an inborn desire sleeping in a human [7].

Bačová [11] defines such desire as the deepest desire of a human, where it is necessary to respond to day-to-day challenges and questions in life. Frankl [7] adds that if such desire for the purpose is not fulfilled, existential frustration develops that leads to doubts in a human about the purpose of his existence and depression and neurosis of all kinds develop.

Another significant author who faced the issue of the purpose in life and was the founder of individual psychology was Adler. He worked as a general practitioner and later he orientated at psychiatry. An important aspect of his theory of the purpose in life is the origin and the roots of the purpose which he can see in early childhood. A child creates verbal and notional conclusions from the point of his feelings and strong experiences. According to it the child creates general form of his behavior, such as roles in life, purposes in life and general life style, and then the model of an adult personality develops. The human psychic is created by setting a certain goal or an ideal at the age of 15, overcoming the actual state and shortcomings. According to that goal an individual imagines successful future. Without the particular goals the activities of an individual would not find purpose [12, 13].

Analytic psychologist Jung in his concept of the purpose in life includes a personal view of the world. In his opinion a human by being born wakes up in the world, which he does not understand, tries to explain it and discovers it to understand the purpose in life. Explaining the purpose in life is dangerous, because such efforts spring out of the deepest mental misery, as the need to discover the unity of life and purpose. Life itself asks a human the question of his existence which is just being hard to understand. Just that is according to Jung the elementary super-personal life role of every human being [6, 7].

The author who is ranked to the stream of the social psycho-analysis or to the existentialism-orientated authors is Fromm. In relation to the purpose in life those characteristics are important for a human, which separate him from animals, for example, self-confidence, imagination or brain. A human creates an imagination of the world which is internally compact and structured. This imagination has orientation framework and it is ranked among the basic existential needs. An individual sets a goal which he should be set for and go for it. The term "subject of worship" is used by Fromm as a certain elementary value, at which a human can concentrate his vital energy. That value is elementary for other values in life and lifts a human above his existence, doubts and uncertainties and gives purpose to his life [6].

Nákonečný [14] defines the opinion on regarding purpose in life, where growing individualization of a human means growth of his loneliness, and consequently growth of his own purpose in life. Plháková [15] adds that according to Fromm, a human can become an active creator in life, for example, a woman delivers a child and a man production or culture. Humans can also create true love by living for each other. Love is divided as fraternal, maternal and erotic. This is characterized by Fromm as the basis of human existence.

One of the most influential humanist psychologists is Maslow, who is known by his concept of personal growth and hierarchic concept of the needs. Such hierarchy of needs was set according to necessity and order, how human needs demonstrate. Elementary needs are physical needs, after their satisfaction there comes the need of safety, followed by the need to belonging somewhere and to be loved. Another level involves the need for self-esteem, knowledge, followed by the need for beauty, harmony and self-assertion. Superior needs are satisfied after satisfying the inferior need [15, 16]. Maslow [7] considered purpose in life "an internal property of a human which emerges like a vigorous motivational power, when inferior needs are satisfied."

Křivohlavý [5] adds that everyone can choose his purpose in life either according to his own motivation, free will or natural character. However, if it is not fulfilled, a human falls into depression and disease.

According to Yalom [8] life has no final purpose or a goal. He points out four existential dynamics, that is, death, liberty, isolation and senselessness, which he deals with in the issue of purpose in life. Purpose in life is just creative response of a human to the total senselessness of the world. If a human wants to cope with this senselessness, he must with all of his heart dedicate to resolve and experience the purpose in life and actively engage in it. Yalom also pursues two opinions, how a human should find his own purpose in life. The first way is being searching for cosmic purpose, where a human responds to the questions such as "Has

http://dx.doi.org/10.5772/68001

life a purpose in general, if yes, is my life involved in it?" This way of searching for a purpose in life mostly occurs in religious or spiritual ideologies. Another way is searching for a secular purpose, which is more defined by searching for goals and values. We know altruism (life in favor of the others, to be useful for the others), devotion (obligation to the others), creativity (creating something new), hedonism (effort for joy, comfort and happiness) and self-actualization (considering own options). According to Yalom, these values are not about contents, but intensity of engagement, which is a human tool to fulfill his purpose in life.

3. Adolescent age in the context of specifics of an oncological disease

Adolescent age is a temporary phase between childhood and adulthood. It involves one decade of life from 10 to 20 years. This period involves a complete personality transfer in all aspects: somatic, mental and social. Many changes are primarily biologically conditioned, however, always significantly affected by mental and social factors which involve their interaction. Initial sexual maturation involves physical changes (as a result of a complicated process of alternating stimulation and suppression of internal secretion glands, whose hormonal activity starts physical changes and controls them), their being new and intensive impacts mental experience.

Adolescent age is definitely a psycho-social phenomenon. The principal problem of an adolescent is his "social uncertainty". He does not feel as a child, however, yet not as an adult. An adolescent asks himself questions: Who am I? (real self-reflection), What do I want myself to be? (ideal self-reflection) and How other people see me? (assumed reflection by the others). The answers in this age is lability of feelings of self-value, an adolescent oscillates between extremely euphoric expectations and depressive under-estimation. This uncertainty produces conflicts incurring in puberty and adolescent age mostly between children and their parents. Adolescents separate themselves from their parents or other former authorities and get involved in a group of people of the same age, where they are perceived as "adults". A culture of youngsters develops, which is often presented as a culture of rebellion against ruling culture of adults.

However, adolescent age brings also positive development. There is self-reflection (an adolescent is capable to cope with his environment and take critical standpoints to it, he discovers himself as an independently thinking and acting individual, etc.) and to the social separation of adolescents, that is, separation from the family; however, even though an adolescent is emancipated from the family, still some bonds to the family last [17]. Adolescent is also a period when changes in life events occur in a different sight, which has in many aspects existential character, resulting from the character of adolescent problems. Discovering own limits as a normal part in life involves also thinking of death. Research in this area is important, directly related to the issue of oncological disease. Rationality of modern and post-modern human pushes out the ultimate reality of death from mind and it reflects it less. Death is tabooed at present and is extinct as death reflected, and its individual and personal concern is hidden behind anonymous dying in hospitals. Also adolescents realize death and many think about it. Coming across their own mortality is a strong experience. Adolescents thinking about death are more mature than adolescents, for whom this issue was absent or it was supressed [18].

For a child, an oncological disease presents a major change in his previous way of life. After discovering diagnosis he undergoes many examinations, sometimes painful ones, therapy and hospitalization. He must get used to impersonal hospital environment, exposed to unknown situations, adapt to the new regimen, be people and engage in new daily activities, all of that in isolation from the family and friends [19].

It applies also to an adolescent that disease does not affect only individual organs and their functions, but the overall human personality as well. On the top of physical stress, for an adolescent such disease presents a changed life situation which he must cope with. Certainly, an adolescent realizes his disease and his close social environment shares this experience; he reflects to the changed social situation and tries to cope with it. Uncertainty of the parents and other close people, their moods, standpoints, unusual conduct and sudden behaviors changes, all of that can affect the patient severely [20].

Adolescent age is a period of human development, his preparation for adulthood. It depends on the closes surroundings. An adolescent is trying to achieve his own place in the society, that is, the social status and his role in it. However, he needs for that satisfaction of his necessities and requirements in the real extent, time and space and his rights to be respected. From this aspect, we perceive adolescent age as a social phenomenon. And it is true that of all the cases of occurrence of cancer in European population adolescent and infant patient make up only about 1%. It seems that it is unnecessary to deal with this issue if this disease is relatively rare in infants and adolescents. However, cancer is the second most frequents cause of death in infant age. In the period of the last 30 years, we can observe permanent mild increase of its incidence. At present annually 14 of 100,000 children in the age under 15 years fall ill. Annual incidence of cancer in the Slovak republic is 13.0–13.6 per 100, 000 children under 15 years of age. That means that annually in this age group, 170–180 new cases of cancer occur. Other 70–80 involve the group of 15–18 years of age. Under conditions of SR, cancer is on the second place of death causes of children (after accidents) which corresponds to the world statistics. Oncological diseases in children present a serious medical, ethic and social problems [21]. Although the percentage of children and adolescents which get cancer, does not seem so alarming at the first sight in comparison with the percentage of adult population, it is necessary to pay attention to it. Above all because a human society perceives a child as a human youngster which must be long time cared for, it needs a lot of love and understanding and protection and its development must be directed the way to make it able to get involved in the society and which ensures its further growth.

Cancer in children is serious but curable and at present the percentage of curable cases is high. In 1960s, 97% of children with cancer died. Ever since there is growth of cured children rare. At present 75% of infant patients are cured at specialized department [22].

In treatment of cancer patients an unfavorable role has been played for centuries by lasting and still live "myth of cancer". These are false imaginations of disease accompanied by great suffering and inevitable death. Moreover, a part of this negativist attitude is supported by the natural and deeply rooted imagination that children and adolescents should not be fatally endangered; they should not be dying, but living. Such imagination is suddenly ruined, because the disease often comes like a flash of lightning, without previous warning; it affects a kid that had always been healthy before. To "wipe it out" is as much important as to recognize it early and to cure it. Only this way it will be possible to prove objectively and definitely that cancer in children and adolescent is a serious disease, but it makes no difference from other, also serious diseases, which are not surrounded by such a myth. There is a lot of work left still for the specialists in this area [23].

Life quality of children is significantly different from that one of the adults, because disease itself as well as the treatment damages normal psycho-motoric development of the child. An important role is played by his growth and development, immunologic and hormonal immaturity. The disease reduces his life quality by intellectual, emotional and physical impairment. Physical and functional deficiency leads to emotional unbalance of the child, to the feeling of depending on others, to the feelings of inferiority. As he cannot come with the environment, this often involves problems to be a part of desired children's life. All of that can lead to the feeling of fear, loneliness and isolation. Anger with all the surroundings appears with the feeling of crying. Every child must be approached individually [24].

Successful treatment of an adolescent with cancer (with long-term life prospective in future) assumes management of the three areas, which are mutually conditioned (a) biological treatment (a cured child will die as an adult in unlimited time period and for other reason), which is a condition for (b) mental treatment (personal coping with cancer, its treatment and eventual complications and consequences, also unlimited capability to cope with different life situations). This conditions (c) social treatment (equal involvement in society).

Cancer brings an extraordinary interference in the life of a child and an adolescent, their families and all the surroundings. Disease affects not only the patient himself, but all the members of his family. Every family has its own way of common functioning, which is seriously affected by a serious and long-term disease and it is necessary to create a new, temporary model. To keep the family functioning in this period, it is important not only to divide obligations and roles flexibly but to include the common rituals and activities (meals, play, tales before sleep) which should involve the patient as well. The patient must have a space in the family to be involved in the family in spite of his limitations caused by disease. The patient himself should control how much family activities he can stand and when he needs some rest. It is a painful feeling to be a burden.

It is important for his family to understand and know the natural needs of an adolescent considering his age and find a way to satisfy him maximally in spite of the limitations. Communication quality with an adolescent diagnosed with cancer (besides others also the one concerning providing or concealing information regarding his disease) can significantly affect experiencing of such a period by the young person. A young person can feel that something serious is going on with him anyway. He can observe it also from conduct of his environment, atmosphere in the family and non-verbal speech of others. If such conduct is in accordance with the verbal information the patient receives, this will make him feel confused and scared.

Sometimes disease cannot be stopped. If an adolescent's life cannot be saved, we do not decide on death which is coming. However, we can decide on life which is left. In spite of the painful fact that we cannot save the young person, we can do a lot for him to avoid insupportable suffering, to be able to feel joy and life to the fullest. In such case, we concentrate on treatment of the problems caused by the disease with the objective to improve life quality as much as possible. We do not treat the cause of disease, because it is not possible. We treat the problems caused by the disease to avoid his suffering. We cannot change duration of life but its quality. Adolescents, who feel their life is coming to an end, always feel that they walk to the unknown and their way cannot be changed. Uncertainty, fear from departure, loneliness, suffering of the close people cause sadness. They often detach themselves, do not wish to go out and to meet friends and lie or sit alone in the room for hours. Sometimes they reject food and pills. To accept for a human that his life is coming to an end, he must be sometime left alone to cope with the anxiety and internal pain. We should respect his wish, but should stay close to be ready to help when he needed.

To consider it, we must try to sup up the adolescent period in the context of cancer, having regard to the fact that there are significant individual differences between adolescents.

Under standard conditions an adolescent is gradually involved in the society of adults as an equal partner. He has an intensive feeling of adulthood and wants this adulthood to be recognized also by the others. However, adult people often are unable or do not want to respect the efforts of an adolescent to be independent. They often impose their superiority, care and wisdom against him. An adolescent longs for independence and freedom. To become a mature, independent and responsible person, he needs to be free from dependence of the family and parents' authority. Life in the family therefore becomes stressful and uninteresting. He looks for an appropriate position in his life, thinks about future, forms his life goals and thinks about purpose in life. In the moment when a serious oncological disease appears in his life, this process of achieving independence is interrupted. Building his own independence is reverted by the disease to the position of maximal dependence. Attitude to adolescent must be very sensitive, because that period typically involves tendencies to isolation, negativism, feeling inferior and pessimism, supported by severe disease and aggressive treatment, which is long term, cyclic, unpleasant and sometimes even painful. An adolescent becomes dependent of his environment, above all of the parents and relatives. New authorities anger his life in the form of physicians, nurses and specialists. Again he must subordinate to the requirements of adults, severe regiment and discipline.

This period is characterized by emotional liability, frequent mood changes, impulsive action and instability which worsen the situation. Moreover, it involves increased fatigue, alternating and being apathetic and short phases of increased activities. This state can be accompanied by neurovegetative disorders such as sleeping disorders, impaired sleep, appetite disorders, etc. The fact of disease with eventual permanent limitation for future is processed by adolescents with more difficulties than younger children or adults. Endangered life, reality of death is understood as shocking, unjust, destructive and traumatizing fact. They react with anger, confusion, disgust, destructive conduct (externalizing) or escaping from reality and depression (internalization). Sometimes different escapades occur as substance abuse or other auto-destructive conduct as suicidal risks. These are important facts which may affect significantly treatment and rehabilitation process.

Even under normal circumstances the adolescent age is the period when a youngster must cope with a lot of life problems. He should, for example, select and prepare himself for future profession. At the end of this period he may start a career. Disease and following treatment will often cause interruption of studies which may affect the whole process of gradual professionalization of a youngster and totally ruin his imaginations of his own future.

Own body becomes for the adolescent a subject of continual attention and physical maturity increases his self-confidence. An adolescent is often dissatisfied with his body; he would like to look different—which is the source of negative evaluation. An adolescent cares for his appearance, carefully selects clothes, hair style, etc. Self-reflection concerns also interest in his personality. A youngster creates an imagination what he should look like.

He develops self-confidence. Sock adolescents are especially worried what will they look like, if they stay alive and become adult. However, they are afraid how they will be accepted by the people of their age if they are different. Their self-confidence suffers their self. Adolescents create their spiritual orientation; spirituality plays an important role for a human. Such questions are most frequently asked in adolescent period looking for one's individuality and orientation, mainly in the situation when all of a sudden an adolescent comes across a disease facing his own mortality.

Adolescents qualitatively change their interests (regarding intensity, level, depth and diversity). Quantity of interests is reduced, but selected interests are improved and extended and become permanent. According to the researches, most interest adolescents show for sports, trips to the nature, tourism and camping in the nature. Sports are good recreation, ventilation of fight and aggression, occasion to stay in a good team. Adolescents with cancer are, however, endangered by their physical and functional deficiency. Their interest must be adjusted to the new situation. Instead of activities of sport character they may develop their cultural and artistic interests, as music, fine arts, drama, (film and theatre) and literature, although in passive form. A youngster often shows interest in social occasions involving dancing, which provide a good opportunity to meet and know other people of the same age. Disease and impaired immunity system, related different physical changes (as hair loss and emaciation) make it possible to a cancer patient to visit a cinema, theatre and library, to go to a concert or disco.

This period involves a strong interest in discussion, to which an adolescent dedicates enough time. It is necessary as it resolves the issue of purpose in life and helps the adolescents to gain experience through conversation. The experts should use this fact and via conversations let out negative feelings (as fear, anger, anxiety, etc.) of an adolescent/a patient.

They should not hide behind a mask of "a hero", which patients often do. An adolescent is able to oppress and heroically stand the pain, physical or mental. However, this does not mean, that inside he has coped with it. Work of assisting professionals can be made difficult by ambivalent attitude of adolescents to authorities. Besides a strong interest of an adolescent in conversations, one fact makes it difficult that they are demanding partners for communication. They expect partnership attitude, respecting their space, privacy and intimacy; however, on the other hand they need much support and understanding. During treatment they should have a chance to participate actively in it, make decisions and also be responsible for its course.

Social relations issues are also significant—time of first love and of their own discovered sexuality. They break ties of social contacts based on personal ties in the family, but he just cannot remain cut off. He find it with the people of the same age and they spend lots of time together. It is an important source of social experience. He takes over their ways of conduct, appearance. He needs to be accepted and recognized by the people of the same age. In case of rejection by a group of the people of the same age, he would get into a complicated life situation. He cares for good relations with people of the same sex and partners of another sex. Friendship of adolescents is more stable, deeper, more confidential and more sincere. Friendship provides support to a young person in time when he takes his life in his own hands and feels alone and hapless. Pair relations of the people of different sexes are a true need. Awakening sexuality can take over their minds and bodies; sexual issue becomes crucial. Media, conduct of adults and other facts in actual society present sex as an important part of life. Age at the end of adolescent age is suitable (although not optimal) to get married. Theoretically adolescents can take a role of a parent. Emotional unbalance of a patient leads to a feeling of dependence from others, an inferiority feeling. They often cannot cope with their environment and to get involved in desired healthy life of the people of the same age. All of that can lead to a feeling of fear, loneliness and isolation, separation from the surroundings, depressive moods, communication breakdown, eventually aggressive and rejecting conduct.

An important fact which is also reflected in the life of an adolescent with cancer even in case of successful treatment are eventual later effects of anti-cancer treatment. As a result of increasing number of cured children and adolescents these later consequences of anti-cancer treatment come forward to the attention and also related quality of survival. For example, it involves damage of lungs, disorders of growth and development, damaged bones, thyroid, breasts, liver, muscles, senses, immunity system, skin, fertility disorders and also psychosocial problems. At present the members of multidisciplinary team caring for an infant or adolescent patient must concentrate not only on the treatment of the patient but also how to minimize treatment consequences.

Life brings many changes, some are small and expected (e.g. changes of life periods, including adolescent age) and those can take place without even noticing them, not being a burden. However, others are more demanding; we can see them as a burden and look for the ways to cope with them. These ways are called coping strategies. Although adolescent period is considered expecting change; in spite of that it is a sort of burden. As described above, it involves looking for social and personal identity. Disturbing the process of looking for own identity with cancer is very serious. An adolescent getting affected in the phase of looking for an identity, being confronted with a serious life problem, has huge impacts on his whole identity.

4. Results of study

The objective of research performed in January 2014–2015 was to find out which statistically important differences exist in perception of the purpose in life and its components between healthy and sick adolescents studying at grammar schools and apprentice schools according to individual social-demographic aspects. The research was realized in quantitative study by means of a standardized questionnaire PIL [25, 26]. After the evaluation of results by means of *t*-tests for two independent selections in determining the level of significance $p \leq 0.05$ statistically important results were proven, of which we choose the most significant ones. From the aspect of school type in the component of affirmation of living sick students of grammar schools achieved significantly higher score in comparison with healthy students of apprentice schools (015). Similarly in the components of perception of objectives healthy grammar students achieved significantly higher score in comparison to sick grammar school students (012). Regarding gender, in the component self-accepting, healthy adolescents reached significantly higher score in comparison to healthy female adolescents (024). Also in the component of perceiving future, healthy male adolescents achieved significantly higher score than healthy adolescents (039).

In qualitative research, we have found from no standardized interviews that adolescents deal with issues related to the topics of purpose in life even more frequently than healthy ones of the same age. They look for the answers of the kind: "Why do I live?", "What have I achieved?", "What else I am to achieve in my life?", etc. They are trying in maximum extent to fulfill the rest of their life that they are left with. Here we can see the positive aspect of the suffering brought by the disease. In spite of the well-known fact that just adolescent age is the period of the first experiences with addictive substances, the respondents have negative attitude to their use; they do not wish to make easier the solution of difficult situation in life by substance abuse and appreciate the rest of their life they are left with. They do not think about suicide and they appreciate life they are left with. They feel responsible to their families and environment and religious adolescents. Moreover, they do not want to act against laws of God, prefer to pray, ask God for help, redeeming and salvation.

5. Application of results

Assisting workers should provide help in searching for the purpose in life and awaken desire for its fulfillment, use disease for personal development and maturing, to create the correct attitude to the disease and eventually suffering that accompanies the client. Accepting disease is related to experiencing high meaningfulness of life.

In social practice and consulting, it is possible to utilize in favor of sick male and female adolescents above all the results of research targeted above all to reasons to live and experiencing depression related to difficult treatment of oncologic diseases.

- The higher the overall score in perception of their meaningfulness of life is, the lower is the overall score in the scale of experienced depression [27].
- In social work and consulting in favor of adolescents, we recommend to apply purpose-orientated attitude in building resilience according to P.T.P. Wonga [28, 29], above all module PURE and module ABCDE.
- In the interest of resilience of youth, we emphasize above all the need for integrated attitude.

6. Conclusion

The potential contribution of this chapter arises from the findings about the facts of diagnosis of oncological disease in adolescence in the Slovak adolescents. These adolescents try to find some purpose in this difficult life situation.

These findings provide the basis for the possible future studies in this field and for the all people which work with adolescents in the situation of oncological diseases (teachers, medical staff, parent, family members, etc.). The attention is given to help to find some solution in the field of study and in the real life situation.

Author details

Eva Žiaková[1], Karolina Barinková[2]* and Katarína Šiňanská[1]

*Address all correspondence to: karolina.barinkova@gmail.com

1 Social Work Department, Faculty of Arts, P. J. Safarik University, Košice, Slovak Republic

2 Ministry of Interior of Slovak Republic, Košice, Slovak Republic

References

[1] Valovičová, E. 2008. Onkologická výchova pre stredné školy. 2. vyd. Bratislava: Liga proti rakovine. ISBN 978-80-89201-38-9. Podkapitola Výskyt rakoviny v Slovenskej republike, s. 22-23; Kapitola Liečba rakoviny, s. 36-37.

[2] Žiaková, E. 2013. Cesta sociálneho pracovníka k zmyslu jeho práce. In Sociálna práca: cesta k zmyslu života. Košice: Katedra Sociálnej práce Filozofickej fakulty UPJŠ.

[3] Vašutová, M. a kol. 2010. Základy biodromální psychologie. Ostrava: Filozofická fakulta Ostravskej univerzity v Ostravě. ISBN 978-7368-934-6.

[4] Frankl, V. E. 2006. Vule ke smyslu. Vybrané přednášky o logoterapii. Brno: Cesta. ISBN 80-7295-084-3.

[5] Křivohlavý, J. 2006. Psychologie smysluplnosti existence : otázky na vrcholu života. 1. vyd. Praha: Grada. ISBN 80-247-1370-5.

[6] Halama, P. 2007. Zmysel života z pohladu psychológie. Bratislava: SAP-Slovak Academic Press. ISBN 978-80-8095-023-1.

[7] Tavel, P. 2004. Smysl života podla Viktora Emanuela Frankla. Bratislava: Iris.

[8] Yalom, I. D. 1980. Existential psychotherapy. New York: Basic Books.

[9] Waren, R. 2005. Život s jasným cielom. 1. vyd. Bratislava: Porta Libri.

[10] Drapela, V. J. 2008. Přehled teórií osobnosti. 5. vyd. Praha: Portál.

[11] Bačová, M. 2006. Logoterapia a existenciálna analýzy v liečbe závislosti. In Empatia, 2006, roč. 13, č. 2, s. 21-28.

[12] Adler, A. 1999. Porozumění životu: úvod do individuální psychologie. 1. vyd. Praha: Aurora.

[13] Výrost, J., Ruisel, I. 2000. Kapitoly z psychológie osobnosti. Bratislava: Veda.

[14] Nákonečný, M. 2009. Psychologie osobnosti. 2. roz. a preprac. vyd. Praha: Academia.

[15] Plháková, A. 2007. Učebnice obecné psychologie. Praha: Academia.

[16] Říčan, P. 2007. Psychologie osobnosti : obor v pohybu. 5. vyd. Praha: Grada.

[17] Kern, H. a kol. 2000. Přehled psychologie. Praha: Portál. 287 s. ISBN 80-7367-121-2.

[18] Hvozdík, Stanislav. 2001. Psychologické pohlady na edukáciu a postmodernizmus. Prešov: Grafotlač. 343 s. ISBN 80-8068-072-8.

[19] Andrášiová, Mária. 2009. Ked do života vstúpi rakovina. Bratislava: Landy, s.r.o., 2009. 249 s. ISBN 978 80-970127-0-0.

[20] Dunovský, Jiří. 1999. Sociální pediatrie. Praha : Grada, 1999. 279 s. ISBN 80-7169-254-4.

[21] Kubíková, Eliška. 2001. Etické problémy sociálnej starostlivosti u detí s onkologickými ochoreniami. In Medicínska etika a bioetika. ISSN 1335-0560, roč. 8, č. 3-4, s. 9-10.

[22] Koutecký, Josef, Cháňová, Markéta. 2003. Děti s nádorovým oněmocněním I. Rady rodičum. Praha: Triton. 224 s. ISBN 80-7254-332-6.

[23] Koutecký, Josef. 2002. Detská onkológia pre prax. Praha: Triton, 2002. 179 s. ISBN 80-7254-288-5.

[24] Škanderová, Jana. 1996. Starostlivost o dieta s onkologickým ochorením. In Revue profesionálnej sestry. ISSN 1335-1753, 1996, roč. 3, č. 5, s. 6-7.

[25] Crumbaugh, J., L. Maholick. 1964. Purpose in Life Scale: Psychometric Properties for Social Drinkers and Drinkers in Alcohol Treatment.

[26] Crumbaugh, J., L. Maholick. 1964. An experimental study of existentialism: The psychometric approach to Frankl's concept of noogenic neurosis. Journal of Clinical Psychology, 20, 200-207.

[27] Šiňanská, K., E. Žiaková. 2013. Stratégie zvládania a zmysel života onkologicky chorých adolescentov. Možnosti využitia logoterapie vo vztahu pomáhajúci profesionál – pacient. In: Aké princípy vládnu zdravotníctvu? : právne, etické, ekonomické a psychosociálne aspekty zdravotnej starostlivosti : zborník z medzinárodnej vedeckej konferencie: Košice, 11. máj 2013. Košice: Medius. ISBN 9788081550072. s. 303-323.

[28] Šiňanská, K., E. Žiaková. 2014. E-learning as a Significant Part of Education Increasing the Possibilities of Coping with Oncological Disease in Adolescents. In: Procedia - Social and Behavioral Sciences [elektronický zdroj]. ISSN 1877-0428. Vol. 132 (2014), s. 715-721.

[29] Šiňanská, K., E. Žiaková. 2012. Edukácia ako jedna z možností zvládania onkologického ochorenia u adolescentov z aspektu sociálnej práce. In: Aktuálne výskumné otázky pomáhajúcich profesií - nové výzvy pre prax: zborník príspevkov z konferencie mladých vedeckých pracovníkov s medzinárodnou účastou 30. apríla 2010 v Prešove. Prešov: Filozofická fakulta Prešovskej univerzity v Prešove, 2012. ISBN 9788055505800. s. 182-187.

Chapter 7

Well-being in Old Age: A Question of Both Continuity and Change

Ulla Hellström Muhli and Ann-Marie Svensson

Additional information is available at the end of the chapter

http://dx.doi.org/10.5772/intechopen.68680

Abstract

In this chapter, we are concerned with the well-being of people in old age, living at a residential care home (RCH), and how well-being can be supported in gerontological social work and care at the RCH. Based on empirical data consisting of well-being narratives with elderly residents (average age of 91), a dialogical performance analysis was undertaken about their experiences of well-being at the RCH. The findings of importance are reported through three themes: (1) childhood memories as a source of well-being, (2) family and work as a source of well-being, and (3) opportunities for the well-being of the elderly at the RCH. To be an individual with others is a phenomenon of a personal sense of self and a phenomenon of sociality. Well-being is also found in the individual's self-renewal. *Well-being is about a sense of both individual continuity and change.* Well-being is created in social situations with others (including caregivers) in daily interactions and in human contacts at the RCH. This kind of individual self-renewal is about human growth and is a human need regardless of age. Consequently, the human growth in (and despite) old age at RCH should be the main target of gerontological social work and care.

Keywords: well-being, old age, residential care home, sense of self, sociality, human growth

1. Introduction

How do we recognize well-being in old age? How do we know when elderly people feel well-being? How can we support people in old age to experience well-being? Old age here refers to the characteristic profile of the most elderly people (older than 85 years). These kinds of questions are expected to be raised, in order to fulfill the elderly care policy of welfare and to provide support for the well-being of persons in old age. This is especially true for elderly who are dependent on gerontological social work and care [1].

2. Factors affecting the well-being of the elderly residing in the care home

2.1. Well-being related to health

One possible answer to the above questions is that surely there is no problem: well-being is observable, at least if we understand well-being as being an experience of health and of being in good health in old age. This presupposes that we understand the essence of health as well-being, which means to feel good and thereby have a good life. To experience health and well-being requires the ability of "being able to" fulfill small and large life projects [2]. Consequently, the experience of well-being is linked to health and the ability to fulfill one's own life projects [3], regardless of whether they are small or big, regardless of the complexities of life, and regardless of the person's age.

However, such an answer is in a certain way not entirely satisfactory because an understanding of well-being, or to actually experience well-being, is not always a question of having good physical health and being free of disease. A person's health can alternate between the poles of objective health and disease and the subjective experience of feeling bad and feeling good. This means that a person can experience health and well-being despite illness and despite having an objective disease [4]. Well-being is thus not only linked to biological health but also to how people feel in relation to their zest for life, vitality, courage, and experience of meaning and a meaningful life. The meaning of well-being can therefore be related to a person's inner experiences and can be described in terms of a condition in which people experience an ability to "be well" (well-being).

2.2. Well-being related to both health and quality of life (QoL)

Literature on well-being in old age has emphasized well-being as being related to both health and quality of life. With reference to Graham and Shier [5], well-being has to do with how people make sense of their lives. This refers to satisfaction, lack of depression and anxiety, and positive moods and emotions. The meaning of well-being can thus be linked to another concept, namely the concept of quality of life. Just like the concept of health, QoL is an elusive and controversial concept that usually involves a subjective experience and external measurable conditions. Research termed health-related quality of life states that assessments of objective functioning and subjective well-being convey different information and present different problems in relation to validation [6]. The definition of QoL goes back to the World Health Organization´s [7, 8] definition as a "state of complete physical, mental and social well-being, and not merely the absence of disease and infirmity". The definition also includes the individual's level of independence, social relationships, and spiritual issues related to the life context. QoL thus refers to both positive and negative aspects of life, with an association between QoL and well-being. According to Walker [9], QoL is a somewhat amorphous, multi-layered, and complex concept, with a wide range of components—objective, subjective, macro-societal, micro-individual, positive, and negative—which interact with each other (p. 573). Walker also states that QoL is a concept that is difficult to pin down scientifically and that there are competing disciplinary paradigms.

2.3. Well-being related to external and intra-individual conditions

With reference to QoL in the discipline of nursing science, Sarvimäki [10] notes that QoL usually refers to external conditions as well as the areas of intra-individual conditions and experiences (**Figure 1**) and thus also supports the meaning of both internal and external factors in the concept of QoL. It is therefore not sufficient to say that well-being is about having good physical health or being free of disease. Instead, we must also describe how we view the well-being of persons and how best to go about it when we want to describe and support the well-being of persons in old age.

As shown above, the meaning of well-being includes the whole person and is an experience of being in balance, in relation to one´s fellow human beings and to life in general. Well-being is thus not a constant, whose value can be determined and calculated in a context-free formula or table. Instead, well-being is something that can only be experienced and lived out in different ways during the various phases of life and, as such, it can be described to oneself and others. Such an understanding of well-being has in general a humanistic foundation of science and is based on an existentialist philosophy of what matters in human life [11]. This kind of philosophy places particular value on authentic human contact between people and authentic human experiences. This approach indicates that when helping people to feel well-being, humanistic knowledge is important and is a significant factor for effective helping for those who professionally intend to support the well-being of people in old age, and it is an important skill to be able to concretize this knowledge in gerontological social work and care in practice. In this chapter, we are concerned with what actually matters to people in old age who live their lives at a residential care home (RCH) and, specifically, what matters in their lives are connected to well-being. This chapter is an abbreviated and rewritten analysis of a larger project entitled *Elderly people's experience of well-being at nursing homes* [12].

2.4. The purpose of the chapter

The purpose of this chapter is to gain a deeper understanding of experiences concerning well-being as well as to gain a deeper understanding of how to support people's ability to feel

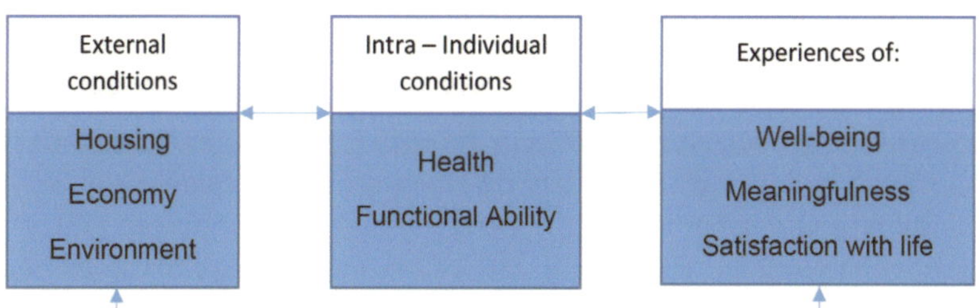

Figure 1. Some key life areas and potential causal relationships between them [9].

and experience well-being in old age. The ability to support well-being is a challenge even for care practitioners and, in this context, in gerontological social work. It demands insight into human life and its processes of change and development. A life incident at one point in time will have a different significance at another. Consequently, the ability of care professionals to support well-being requires that they have an opportunity to understand more about the person in question and his/her life.

However, little research has been performed on the well-being of persons in old age within the discipline of gerontological social work and care for those living at an RCH. Following the distinction between objective well-being and well-being as authentic human experience during the various phases of life, this chapter does not deal with scientific facts of well-being but rather material representation of abstract experiences and concepts of well-being. This kind of humanistic approach can be widely applied in effective support and help by emphasizing on self, on non-directedness, on the inner drive of the individual to find solutions for themselves, on the value of human encounters for well-being, and for its own sake [11]. Knowledge is lacking when it comes to how old persons experience well-being and how care professionals can support well-being in gerontological social work [13, 14]. Referring to Lundin, Berg, and Hellström [15], for example, due to limited autonomy, an elderly person's ability to experience well-being is very dependent on the care professionals. This is true not least when the person in question has moved to an RCH.

3. Caring relationships at the RCH

In gerontological social work and care at an RCH, the caring relationship has a given position. However, one question that needs to be asked is what we do mean by relations and relationships? According to Ohnstad [16], the following definition of relationships is given: "Relationships are the lifeblood and the lifeline of the people. We need to be loved (or if necessary hated) and to feel close (or at worst get to distance)". The caring relationship, however, will enable space to create a good caring communion, which may be a part of the perceived experience of well-being. The care relationship is thus the prerequisite for providing good care, while the caring relationship in itself constitutes care and well-being.

For a caring relationship to be developed, it requires attendance (presence) between the care professional and the care receiver in order to respond to the care needs [17]. As the elderly resident may undergo her/his emotional and reflective work about well-being, the care professional has to undergo her/his emotional labor by listening to the resident's utterances about well-being, for example, in narratives. This is a challenging task for which the care professional has to open herself/himself to be receptive to the elderly resident while being present [18]. Caring is thus a matter of paying attention to the resident—to be there and to listen to, talk to, and understand the other person. Following Brannigan [18], we state that gerontological social work and caring is a state of actions that requires an authentic contact and relationship between the care professionals and the elderly resident. It also requires attendance and not just as a physical presence but rather in terms of a deeper understanding of

http://dx.doi.org/10.5772/intechopen.68680

presence that means to be there (being-there), to be with (being-with), to be for (being-for), to be in relationship (being-in-relation), and to be in the transcendental or in a sensuous and delicate presence (being-in-transcendence) [18].

Receptive attention [17] is thus a key concept when it comes to understanding what authentic contact and relationship is, as well as a basic condition in the care relationship within the gerontological social work and care. As shown, it has to do with an inter-subjective kind of relationship at the micro-level which requires both presence and reciprocity. *Caring* is then about initiating the well-being processes, which can be done by providing room for the elderly resident's inner growth and development [11, 19, 20], despite physical and cognitive limitations. Growth and development can thus be achieved by supporting the elderly resident's self-expression, self-actualization or self-reflection, authenticity, and ability to get in contact with her/his own identity, even given a weak ability for own activity [21]. If the conditions are right, this is what happens; if, however, the conditions are wrong, then the growth will go in other directions [11]. Sudbery [11] also adds that this potential is highly individual but distinctively human.

4. Expressing experiences of well-being through narratives

In order to know something about well-being in the later stages of adult life, we need to ask people who are there and who are actually experiencing it. What does a person's living space look like in old age, within which that person is assumed to be able to experience well-being? The concept of *self* in this context stands for helping the development of self on non-directedness and on the value of human encounter for its own sake, in contrast to didactic methods and targeted objectives [11].

In this chapter, the answers to these questions were obtained in conversations with elderly people living in an RCH. To gain knowledge about what matters to people in old age who are living their lives at an RCH, and about what matters in their lives that is connected to well-being—in this case, what and how well-being is experienced—the language and the narratives have a given place. The narratives, which are later transformed into texts, can be seen as important sources of knowledge. One approach to support this narrative methodology is that people should be understood as narrative human beings rather than mere rational actors [22].

A narrative about well-being can be conveyed to the listener in different ways, and in storytelling, the person can not only say something about their well-being but also can say something about themselves as a person, her/his self-actualization by the way to tell about their well-being [23, 24]. The language therefore plays a key role, not only in the understanding of well-being as communicated but also on how it is understood and created the meaning of human encounters and in conversational interaction with others [25, 26]. Within this approach there is also housed the notion that people structure their experience in different ways to make their world intelligible, primarily through narratives [22]. For this reason, it is possible to study well-being and the social life through and by narratives of persons in old age at an RCH. Similarly, it is possible to study the support for well-being that is provided in the context of care given at the RCH. Narratives are thus a form of knowledge and a form of communication whereby people's actions and words (language use) can be understood in a frame of the narrative. Consequently, the main concern here is to show how a narrative-based approach in the context of the care of persons in old age at an RCH provides a basis for reflecting on well-being. Recorded, transcribed, and translated (from Swedish to English) data has been produced from talk with residents, and an analysis of language use in telling and meaning-making about well-being has been performed.

However, the elderly residents' narratives should be understood in relation to their whole life events. From a gerontological social work and caring perspective, this means that the life course is a dynamic process, which is integrated into various life events. The means for understanding the well-being experiences of an elderly resident's life must be seen in relation to other phases of the life [26, 27]. Life-course perspective assumes that events that occurred earlier in life are understood and considered in another way in old age, through a grid of the lived life. The lifelong ageing is an ongoing process whereby the events from the past, present, and future are intertwined. Important events earlier in life and the person's own ability for self-reflection and managing the situation can affect the experience of well-being. In narrative analysis, there also exists a final phase, namely that of obtaining a new understanding, which means that it is here that the text and the story is completed, not only through the reader (listener) of the new understanding being able to understand the text but also that the text can change the reader, in that the reader alters her/his view of that which has been told [28].

4.1. Participants

This study was conducted at three different RCH facilities within two municipalities in western Sweden. In this chapter, our interest was directed toward five elderly residents' (all female) narratives about how they, through their own lives, experienced well-being at the RCH. The criteria for inclusion in the study included (1) having lived in a RCH for at least 6 months, (2) the ability to speak the Swedish language, and (3) the ability to hold a conversation. Individuals with dementia disabilities and/or impaired decision-making skills were excluded. All interviews took place at the participants' homes at the RCH. All five participants were cognitively well-functioning and could carry on a conversation. The average age of the group was 91, and they had been living at the RCH for a period of 2–7 years. The length of the interviews ranged from 25 to 50 min, depending on each resident's capacity. None of the residents needed to be stopped for any reason, and all of the residents were in good spirits when their interview was finished. The interviews were recorded on an MP3 player and began with open-ended questions such as: (1) Can you tell me about your life—who is Laura? (Please note: Laura is a fictitious name that we are using to represent a summary of the findings for all five participants in the study). The opening question led to the participants beginning to tell us about their lives. In order to capture the well-being aspects in the narratives, supplementary questions were asked such as "How did you do it?" and "What was the well-being for you then?" Another main question asked was "How do you experience the well-being here at the RCH?" All questions asked were aimed at gaining a deeper understanding of the story and, specifically, an understanding of what well-being means to the person of old age

http://dx.doi.org/10.5772/intechopen.68680

who lives permanently at the RCH. The narratives were transcribed verbatim. The purpose of the analysis was merely to capture what was said (content) and to show the dynamic aspects of the narratives in order to be able to study and analyze the narrative function, context, and consequences. Ethical aspects have been discussed and considered in accordance with the Helsinki Declaration of 1975, as revised in 2008 [29], as well as in accordance with the Swedish Government directive [30].

5. Findings

The underlying data collection is comprised of narrative interviews with elderly women living at an RCH. This will be presented in the findings as one case, in terms of a fictitious woman named Laura. The findings of importance that are connected to experiences of well-being are reported here through three themes: (1) childhood memories as a source of well-being, (2) family and work as a source of well-being, and (3) opportunities for well-being of the elderly at the RCH. The analysis shows that well-being seems to be a link between past, present, and future identity. Other important factors for well-being include fellowship and social interaction, as well as being able to enjoy a sense of freedom and independence. The narrations of well-being begin with a story taking place in the person's childhood, continue into adult life, and conclude with a contrast of how life is now at the RCH. Consequently, well-being in the present is reflected through a filter of the life one remembers, and the residents choose and evaluate different events and experiences that have occurred during the course of their lives. This comprises the main structure of the narratives. In the following results, this chronology is shown—1) partly through what the idea of well-being is for the resident and (2) partly through how well-being is described.

5.1. Childhood memories as a source of well-being

When Laura talks about her time as a child, this is done in terms of portrayals of the old days, when children were allowed to be children and there was time to play. Even if there was a shortage of money, as Laura describes it, she had a nice time together with other children and with her family. In the story, Laura is using the temporal dimension of visible time. Laura places the fun experiences of childhood on a time axis, and the meaning becomes clear when comparing how she, as a child, could have fun despite the poor economy: "Of course we didn't have as much money as what the children of today have", she says, and "there wasn't the same extravagance as there is today". This event that Laura describes ends with the narrative points (the plot), that is, that despite this "We had such a nice time all together," and it is clear that the need for money and abundance does not necessarily create well-being but rather it has to do with *being free and without obligations, and being able to have fun in the social interaction with others.*

Her childhood home is portrayed as an ideal place where Laura produced her own subjective experience of well-being by being proud of her father, who was very musical, and she says "oh…father was so musical it didn't matter what instrument you gave him…he could play it.

So, there was always song and music at home/…/I have always felt happy and we had such a nice parental home". This interpretation of the story reinforces the self-identity of Laura and a sense of self-worth by describing not only the father who was musical and in great demand but also herself. Her description of *togetherness* and affiliation with the family represents continuity and wholeness as reflected in the current identity and shown partly by an interest in music that continues to, this day, be and is a source of well-being in situ at the RCH.

Relations, as communions with other people, are connected to experiences of well-being. Relations involve identity-supportive social resources and contribute to the creation of an identity and meaning in the life of resident Laura, not only in childhood but also in the current situation at the RCH. Her story also highlights what Tornstam [26] means by the inner core of social networks, that is, the relationship to friends (and family). In the narrative here, Laura exposes her identity. This is done not only through a story of relations with family members but also through a story and portrayal of a care-free time long ago and through a story of how Laura retains her identity by recalling past events in her life for the listener in the human encounter. What is interesting here is how Laura allows the listener (the interviewer) to understand the event as a dynamic description of well-being by contrasting her childhood moments of joy with the present. To be a kid, to have fun, and as a child to be carefree and without worries depict well-being in contrast to the present where everything is about musts. According to Mead and Morris [31], this model of identity creation occurs during an interaction between how an individual sees himself or herself and how others see him or her.

6. Family and work as a source of well-being

Although the period of childhood seems to be fundamental to the feeling of well-being in the present, Laura describes the period when life was all about building a family and working at a time when her life was filled with joy and happiness. However, this part of her life mostly consisted of hard work, although it was a life period filled with joy as well. She says "it was full up" and "it was a lot of work". Having a family forms a relationship in life and comprises a unit where everybody lends a hand with everyday work and tasks. The family also molded togetherness in life. She concludes by stating that "it went well". In fact, it seems that the everyday work was an important factor that generated happiness and well-being in Laura's life.

When Laura talks about this phase of her life, two analysis levels emerge in her story. (1) She creates two dynamic forces that shape the subjective experience of well-being, that is, the joy of positive development and direction of life, "it went well," while at the same time life, was marked by a laborious existence—"it was full up" and "it was a lot of work." The analysis level (2) deals with the basic core of the story of adulthood, that the family helped out together, and suggests the desire for intimacy and relationships with family members, a sort of fusion of self and others on emotional and working ways. This form of intimacy expresses well-being. In Laura's narration, this analytical model of her identity as commitment to the care of the family is expressed in statements like "it was full to the brim" and "there was work", but at the same time they were happy for the work and it went well. Thus, the commitment during this period in life was to build up resources for future generations and to bring up their children.

7. Well-being at the RCH

In just a few sentences, Laura moves from life events when she was at home with her children to the period of her life when she worked as a switchboard operator and then to the time when she became a pensioner. She concludes her story in the present and the in situ moment at the RCH. With drama and relationships with other people, her account concludes with the statement "so it became my life". With these words, Laura binds together meaningful events in her life and creators of happiness and well-being as well as her own role in these events.

Years are added to years in the story and Laura's conversation returns to the subject of what gives her joy and well-being at the RCH. Laura says that the things that matter for her sense of well-being now are visits by family members and friends. Even if the family members are the closest ones, the contact with friends is also important for her well-being. The relationship with family members, and especially the children, is very important for Laura. She also expresses her longing for her children, despite frequent visits by them to the RCH. This longing gives expression to what Tornstam [26] describes as the interrelations' qualitatively most important quality and that it is the degree of depth or presence that characterizes the relationship.

Although family is the closest thing to her, keeping in touch with friends is also a key factor for Laura's well-being. The visits by friends from the past to the RCH are important for the continuation of her identity, and when she talks about friendship relations, she says "I'm like I am anyway". With this statement, Laura makes a summary of herself, and she is expressing her understanding of the continuity of her own identity, despite ageing and her circumstances and conditions in situ at the RCH. She provides a description of the time or the temporal dimension from the past compared to now and the importance of visits by her friends that "they do not forget me". The talks with friends are described as important because they enable her to talk about current events as well as about the old times. Through friendship relations and conversations with friends, Laura's existence is confirmed by both as the person she had been before and also as she is today. All of this identity construction identifies and confirms Laura's existence and preserves her identity. Apart from the visits by family members and friends, her days are otherwise filled with various activities, not, however, like in the old days, when family and work filled her days. According to Tornstam [26], the consequence of the importance of the friendship relations and social integration relating to well-being is interwoven. All of this identity construction, confirmation of existence, and retention of own identity is noticeable in Laura's narratives.

Even the relationship with the staff at the RCH and the care relationship are described by Laura as a subjective experience of well-being. In this relationship, the staff members provide nursing and care. Laura praises the staff and describes them as kind and helpful: "I have to praise the staff; they are very kind and helpful…they ask how I would like things, what they can do to help me…so they are very accommodating and are kind and competent, providing help and such". Laura says that the conversations between them touch on most things, and it is the act itself which is considered important. The staff members provide help, which Laura confirms by describing how she herself wants to be helped.

In addition to visits by family members and friends, life at the RCH is characterized by trying to fill the days with "meaningful" activities, as opposed to earlier in life when family and work filled Laura's days, and she says: "we have different programmes…today it was going to be…gymnastics". Laura appreciates having something to do and looks forward to such activities. As illustrated, activities can also be a source of well-being. The activities can include music, reading newspapers, bingo, quizzes, or gymnastics. The institutional order and its routines emerge in the description of activities at the RCH in the fact that these activities are organized by the care professionals and not by those who live there. Consequently, the activities are not tailored to the individual's specific needs of well-being.

Based on the findings described, qualitative values of well-being can be summarized as shown in **Table 1**.

As shown, we have presented descriptions of well-being and the *qualitative values of well-being* that are constructed in the elderly residents' narratives. In this way, this chapter contributes to the understanding of what matters for a person's sense of well-being and what enables well-being for the elderly residents at the RCH and in caring relationships. Furthermore, it is also shown how well-being can be supported in the gerontological social work and care at the RCH, primarily through the *caring relationship* and by *being adaptive present*. We can thereby state that gerontological social work and care aimed at supporting well-being is a state of actions that requires an authentic human contact and relationship between the care professionals and the elderly residents. This kind of value on authentic human contact between people and authentic human experience provides knowledge about how it is important and a significant factor for the achievement of effective helping for those who professionally intend to support the well-being of people in old age as well as how it is also an important skill to be able to concretize this knowledge in gerontological social work and care in practice.

Well-being	**The well-being aspects that matter in life at the RCH**	
Qualitative values	Experience of continuity in the identity	Experience of being an individual with others
Content	To talk about memories To be visited by relatives To be visited by friends To converse with the staff To continue to develop their own interests To be considered from the life-course perspective	To retain one's uniqueness as an individual as well as in interactions with others To benefit from affinity with others To be social Getting the chance for self-renewal within RCH interactions and relationships

Table 1. Data organization.

http://dx.doi.org/10.5772/intechopen.68680

8. Discussion

8.1. Substantial considerations: Qualitative values of well-being

In this chapter, we have chosen to use close analysis to study in detail the five participants' narratives as one case (fictitious name Laura). This narrative's methodological approach enables research into the changed meaning in life which influences a disproportionate number of women, for example, the fact that 70% of the most elderly population living at RCHs in Sweden are female [1]. The number of interviews upon which the resulting case is based may be considered to be few. Nevertheless, it is important to point out the difficulty in interviewing suitable informants in a study such as this, as the elderly are often physically and mentally frail. Many women living at the RCH suffer ailments such as dementia and therefore have difficulty in holding conversations. Meinow, Parker, and Thorslund [32] noted in their research that fewer than 5% of those who lived in an RCH managed to conduct an interview and understand the information which also confirms this situation. Furthermore, it should be noted that the women who were interviewed at the RCH were of quite advanced ages, which made interviews last longer than 30 min and hence tiresome for them. Such circumstances were therefore considered and handled with sensitivity and respect by the interviewer.

So what, then, are the qualitative values of well-being? Based on the elderly residents' perceived experience, well-being in this context is described as a qualitative value of (1) a continuity of the own identity and (2) being an individual with others at the RCH. Laura, as a case in point, states that "I'm like I am anyway" and makes a synthesis creation of herself, despite redefinitions of herself throughout the course of her life and due to the ageing process and her circumstances in situ at the RCH. Laura expresses her sense of self and her need for continuity of her own identity which, in turn, is confirmed by her former friends. This is probably also why Laura wants to stay in touch with these friends even though she is now living at the RCH. To be an individual with others is a phenomenon of a personal sense of self and a phenomenon of sociality, which shows the importance of maintaining social networks where relationships with related family members and friends form the innermost core of well-being. However, well-being is also found in the individual's self-renewal, which occurs in interaction with other people and means that well-being is not only a phenomenon in the event that the elderly resident talks about but also may be related to the sociality in the authentic human interaction with other people and in harmony with the environment and society. Well-being is therefore not something that just simply exists but rather something that is also created in social situations with others (including caregivers) in the daily interaction and in human contacts at the RCH. This kind of individual self-renewal has to do with human growth and is a human need regardless of age. Consequently, the human growth in (and despite) old age at the RCH should be the main target of gerontological social work and care.

This chapter contributes knowledge about how self-expression and connection with the elderly resident's identity can be developed within the practical care work at the RCH in order to provide help in increasing the residents' sense of well-being during the last part of their lives at the RCH. This type of specialization in gerontological social work and care is not currently a priority for the benefit of the physical care of elderly residents, but it should be in order to fulfill the elderly care's policy of welfare [1].

9. Theoretical implications

As the residents go through their lives at the RCH, they need to make sense of what matters and, more specifically, what matters in relation to their well-being. This chapter aims to contribute knowledge (albeit based on a fairly small data corpus of empirical data) about this area—a contribution that can be used in further studies and in the practical implications for gerontological social work and care at the RCH. In terms of theoretical implications, the concept of well-being has attracted attention from a variety of academic disciplines, but it has been quite vaguely defined and described. From the results of this chapter, we can learn that this concept is *directly linked to sociality in human interaction* and more precisely, *to social situations and meetings between family members, friends, the care professionals, and the residents.* Well-being is thereby understood in this context as a *dynamic category* and, consequently, we understand well-being in a broader sense than simply experiences of health. However, there is also a need for a continuum or continuous unit of the sense of self throughout the life that is of importance to well-being. In other words, the *well-being category is about both continuity and change*. This is knowledge which highlights the importance of not only supporting physical health care knowledge interventions for people in old age (here at the RCH) but also on the more psychological and social level. The evidence for activity theory [33] in this chapter is that small advantages have been identified for people in old age who are active (or at least interested in activities) and socially engaged if possible at the RCH. However, there is also a small amount of evidence for the continuity theory [34], in the efforts to understand and maintain internal and external structures in life as they progress. Circumstances may change, for example, by moving to the RCH but the residents develop strategies in new circumstances that provide continuity. In this chapter, by sticking to visits by old friends at the RCH as important values for well-being, the residents thereby link the past, as it is perceived, to their current situation *in situ*. The humanistic theories of development [20] of well-being in old age at the RCH set out to emphasize the importance of sense of self and the sociality of human encounters, as well as the struggle to create meaning and the need to renew oneself in order to lead a worthwhile life, all of which are characteristically human. Thus, "being well" refers to experiencing continuity in the self-identity as well as in sociality with others. This kind of emotional experience gives meaning to life, which in turn promotes well-being [14]. This approach arises, in particular, from a life-course perspective as well as from a humanistic model's perspective.

10. Practical implications

In practical terms, and in order to create conditions for well-being within the gerontological social work and care, well-being must be achieved in an interrelationship between the elderly resident and the care professional, whereby both parties contribute commitment and interest in understanding each other. In this commitment, it might be fruitful to ask not only (1) how do we as care professionals get the person of old age to feel well-being but also (2) how should we feel for the elderly person as a psychological and social individual and for well-being related to that matter?

From a practical viewpoint, well-being can be supported by providing the resident with opportunities for social contact and socializing with others (social interactions) as well as opportunities to be listened to (presence). This approach highlights the elderly's perspective on well-being and the use of the knowledge that takes into account the elderly perspective on well-being as a starting point for the care work. This could, for example, be done by encouraging the elderly resident to talk about their well-being. In this context, the care professional's role is to pay good attention to the elderly person rather than ask questions. With reference to Sudbery [11], the key features of helping relationships are not the techniques used by the helper but rather the qualities of the helper—that they give unconditional positive regard, show empathy, and offer non-possessive warmth. When this happens, the relationship itself provides well-being to the resident at the RCH.

Author details

Ulla Hellström Muhli[1]* and Ann-Marie Svensson[2]

*Address all correspondence to: ulla.hellstrom_muhli@soc.uu.se

1 Department of Sociology, Uppsala University, Sweden

2 Skara Municipality, Sweden

References

[1] The Swedish National Board of Health and Welfare. *Vad tycker de äldre om äldreomsorgen? En rikstäckande undersökning av äldres uppfattning om kvaliteten i hemtjänst och äldreboenden* 2011. [What Do Elderly Persons Think About Elderly Care? A Nationwide Review of the Elderly's Opinion of the Quality of Home Services and Special Housing Accommodation 2011]. Stockholm, Sweden: The National Board of Health and Welfare; 2011

[2] Dahlberg K, & Segesten K. *Hälsa & vårdande i teori och praxis* [Health & Caring in Theory and Practice]. Lund: Studentlitteratur. p. 101102

[3] Blomgren Mannerheim A, Hellström Muhli U & Siouta E. Parents' Experiences of Caring Responsibility for Their Adult Child with Schizophrenia. Schizophrenia Research and Treatment. 2016: Volume 2016. http://dx.doi.org/10.1155/2016/1958198

[4] Eriksson K *Vårdprocessen*. [The Process of Care]. Stockholm: Almqvist & Wiksell; 1985

[5] Graham JR & Shier ML. The social work profession and subjective well-being: The impact of a profession on overall subjective well-being. British Journal of Social Work. **40**(5);15531572

[6] Muldoon MF, Barger SD, Flory JD & Manuck SB. What are quality of life measurements measuring? British Medical Journal. 1998;**316**:542-545

[7] WHO. Constitution of the World Health Organisation. Geneva:1946

[8] WHO. Quality of life. Constitutions of the World Health Organization; 1947;**1**(1-2):29-43

[9] Walker A. Ageing and Quality of Life in Europe. In: Dannefer D and Phillipson C, editors. The Sage Handbook of Social Gerontology. London: Sage Publications; 2013

[10] Sarvimäki A. Quality of life in old age described as a sense of well-being, meaning and value. Journal of Advanced Nursing. 2000;**32**(4):1025-1033

[11] Sudbery J. Human Growth and Development. An introduction for social workers. London and New York: Routledge; 2010

[12] Svensson A-M. *Äldres erfarenheter av välbefinnande inom särskilt boende för äldre* [Elderly People's Experience of Wellbeing at Nursing Homes]. (Licentiate thesis). Jönköping, Sweden: University of Jönköping, School of Health Sciences; 2014

[13] Lundin A, Hellström Muhli U & Berg L-E. Witnessing presence: Swedish care professionals' experiences of supporting resident's well-being processes within the frame of residential care homes. Journal of Aging Studies. 2016;**37**:1-9. doi:10.1016/j.jaging.2016.01.001

[14] Svensson A-M, Mårtensson L & Hellström Muhli U. Well-being dialogue: Elderly women`s subjective sense of well-being from their course of life perspective. International Journal of Qualitative Studies on Health and Well-being. 2012;**7**:19207. http://dx.doi.org/10.3402/qhw.v7i0.19207

[15] Lundin A, Berg L-E. & Hellström Muhli U. Feeling existentially touched - A phenomenological notion of the well-being of elderly living in special housing accommodation from the perspective of care professionals. International Journal of Qualitative Studies on Health and Well-being. 2013;**8**:20587. http://dx.doi.org/10.3402/qhw.v8i0.20587

[16] Ohnstad A. *Det goda samtalet*. [The Good Dialogue]. Lund: Studentlitteratur; 1995

[17] Hellström Muhli U & Delbene R. The ethical dimension of presence in elderly care talk-encounters. In: José Carlos Gonçalves (red). (2013). Presence in Healthcare Communication: Implication for Professional Education in Health Care. Rio de Janeiro: Editora da Universidade Federal Fluminense (UFF). ISBN 9788522808496

[18] Brannigan M. Precence in Suffering: Lessons from the Buddhist Four Noble Truths. NY: 2009. New York: Lexington Books.

[19] Mayeroff M. On Caring. New York: Harper Collins Publishers; 1999

[20] Wiklund L. Vårdvetenskap i klinisk praxis. Stockholm: Kultur och Natur; 2003

[21] Maslow AH & Frager R. Motivation and Personality. New York: Longman; 1954/2003

[22] Robertsson A. Narrativ analys [Narrative analysis]. In: Bergström IG & Boréus K, editors. *Textens mening och makt* [The Meaning and Power of the Text]. Lund: Studentlitteratur; 2012

[23] Hydén L-C & Hydén M. *Att studera berättelser* [Studying Narratives]. Stockholm: Liber; 1997

[24] Reissman CK. Narrative Analysis. London: Sage Publications; 1993

[25] Berger P & Luckmann T. The Social Construction of Reality. London: Penguin Books Ltd; 1991/1996

[26] Tornstam L. *Åldrandets psykologi* [The Psychology of Ageing]. Stockholm: Nordstedts;2011

[27] Öberg P. Livet som berättelse. Om biografi och åldrande [Life as a Narrative. About Biography and Aging]. Uppsala: Acta universitatis Upsaliensis; 1997

[28] Ricoeur P. Time and Narrative. (Övers: McLaughlin K & DPellauer). Chicago: The University of Chigo Press Ltd; 1984

[29] World Medical Association (WMA). World Medical Association Declaration of Helsinki, Ethical Principles for Medical Research Involving Human Subjects. 59th WMA General Assembly Seoul. 2008. Retrieved february 10, 2011 from http://www.wma.net/en/30publications/10policies/b3/17c.pdf

[30] SFS (2003:460). *Etikprövningslagen* [Ethical Review Act]. The Swedish Government directive. Stockholm: Utbildningsdepartementet. Hämtad July 8, 2014, från http://www.riksdagen.se/sv/Dokument-Lagar/Lagar/Svenskforfattningssamling/Lag-2003460-ometikprovning_sfs-2003-460/

[31] Mead GH & Morris CW. Mind, Self and Society. From the Standpoint of a Social Behaviorist. Chicago: University of Chicago Press; 1967

[32] Meinow B, Parker MG & Thorslund M. Consumers of eldercare in Sweden: The semblance of choice. Social Science & Medicine. 2011;**73**(9)12851289

[33] Neugarten B. Personality and Ageing. In: Birren JE and Schaie KW, editors. Handbook of the Psychology of Ageing. New York: Van Nostrand Reinhold; 1977

[34] Atchley RC. A continuity theory of normal aging. The Gerontologist. 1989;**29**(2):183-190

Chapter 8

Well-being and Quality of Life—Medical Perspective

Nkporbu A.K.

Additional information is available at the end of the chapter

http://dx.doi.org/10.5772/intechopen.68730

Abstract

The prevalence of chronic diseases, both communicable and non communicable, has continued to increase worldwide and its consequences including psychiatric co-morbidity with accompanying decrease in quality of life of the sufferers. The aim of this study, therefore, was to determine the quality of life of patients with HIV, Tuberculosis, Diabetes Mellitus and Hypertension and the effect of psychiatric comorbidity on the quality of life and well-being of these patients who attend the different out-patient clinics of the University of Port Harcourt Teaching Hospital(UPTH). Following ethical approval and informed consent from the participants, 120 subjects from each group of HIV, Tuberculosis, Diabetes Mellitus and Hypertension were recruited based on the study's inclusion and exclusion criteria. Subjects were further administered with the study's instruments including the socio-demographic/clinical questionnaire, GHQ-12, WHO Composite International Diagnostic Interview (WHO CIDI) and the WHO QOL-Bref. The data were analyzed using the Statistical Package for Social Sciences (SPSS) version 20 statistical package. Confidence interval was set at 95% while P- value of less than 0.05 was considered statistically significant. The study found a prevalence of psychiatric co-morbidity of 43.75% in all. Tuberculosis had the highest prevalence of psychiatric comorbidity with 63(52.5%), followed by . hypertension with 59(49.17%), and Diabetes mellitus 46 (38.3%), while HIV was the least with 42(35%). There were significant relationships between the presence of psychiatric co-morbidity and level of education(p=0.004), income class(p=0.001) in all four medical conditions and self stigma for HIV and tuberculosis. Hypertension had the highest average mean quality of life with 6.34+253 followed by HIV with 60. , diabetes with 5.8 while tuberculosis had the least with mean quality of life 5.3(p=0.001). However, when quality of life was associated with psychiatric comorbidity, the mean quality of life was 5.9, 5.6, 5.2 and 4.9 for Hypertension, diabetes, HIV and tuberculosis respectively. The management of patients with these medical conditions should therefore necessarily include adequate attention to the mental health status the sufferers.

Keywords: Chronic medical diseases, Psychiatric Co-morbidity, QOL, Well-being, UPTH

1. Introduction

The prevalence of chronic medical diseases, both communicable including HIV [1] and tuberculosis infections [2] and non-communicable including diabetes mellitus [3] and hypertension [4], has continued to increase worldwide with a corresponding increase in the awareness of their possible comorbidities. These four medical conditions are among the most common chronic conditions in Sub-Saharan Africa and indeed worldwide [3, 5]. Psychiatric comorbidities have been identified as one of the commonest groups of comorbidities associated with chronic medical conditions [6–8]. Knowledge of psychiatric comorbidities particularly in these four chronic medical conditions has increased tremendously over the past decades and their effects on chronic disease outcome particularly quality of life and well-being have remained a growing concern [9, 10]. Even in the presence of all these findings, there is still low level of awareness of the psychological effects of these medical diseases particularly among clinicians who are non-psychiatrists. This often leads to poor management of patients, high morbidity, mortality and poor prognostic outcome.

In 2010, it was reported that diabetes mellitus affected about 285 million adults (age 20–79 years) worldwide and this figure may increase to about 439 million adults by 2030 [3]. Similarly, tuberculosis is reported as one of the leading causes of mortality worldwide [12], with the World Health Organization (WHO) estimation of about two billion people, having latent tuberculosis. This number of persons makes up almost a third of the world's population [5, 11]. Every year about eight million people develop this disease, and some three million die of it, with more than 95% of these from developing countries [13]. A study [14] estimates a prevalence of between 10 and 15% of adult populations worldwide to have high blood pressure which also agrees with similar report in Africa [15]. But this is different from other studies which reported a prevalence of 15–30% in adults worldwide [16]. Five key factors have been identified to be associated with hypertension in the Western world: increasing age, obesity, elevated pressure in blood relative, environment and race [17]. Furthermore, over 36 million people have contracted HIV infection worldwide and about 16 million people are said to have died from the disease [1, 7]. In the 1960s and 1970s, HIV-related disorders and AIDS were increasingly common but unrecognized, particularly in Africa and North America. In Nigeria, the following prevalence rates have been reported for the four medical conditions: about 10–15% for diabetes mellitus [4] and about 4.6% for HIV infection [1], 10–15% for essential hypertension [14], and increasing from 11.2% in 1990 to 27.9% in 2010 in rural communities in the Niger Delta, and 44.3% in urban Lagos [18, 19] while tuberculosis is 36–45% [20].

Although these four medical conditions may appear dissimilar, within them lie some comparative similarities. Diabetes mellitus and hypertension have genetic component, adverse environmental factors, and prevalent life style as major risk factors [21–24], while HIV and tuberculosis are both chronic infectious diseases and are mainly acquired from infected objects or persons [25, 26]. In fact, a strong genetic association between diabetes mellitus and depression has been reported [27, 24]. Also, HIV and tuberculosis infections, unlike diabetes and hypertension, are associated with high level of stigmatization and social discrimination

[28, 29]. These diseases have similar complications which include among others diabetic ketoacidosis, HIV and hypertensive encephalopathies, and CNS disseminated tuberculosis which may affect all the brain cells, leading to altered sensorium, neuro-affections, neuro-deficits, cognitive impairment, organic mental syndrome and in some individuals seizures may occur. Similarly, all four illnesses could directly impair important neurotransmitter functions due to their direct toxic effects on the brain cells (neurons) either from the viral cells [30] or other opportunistic infections [31], disseminated tuberculosis to the CNS, hypertensive encephalopathy or due to ketoacidotic complication. All these are capable of disrupting the neurotransmitter system and this may affect the limbic system, the center that regulates mood in humans.

2. Chronic medical conditions and psychiatric comorbidity

Chronic medical diseases, besides being associated with high rate of mortality, are equally associated with various forms of comorbidities and psychiatric comorbidity appears to be most common and most disabling [31–37]. It has been reported that about 20% of patients with somatic diseases suffer from major depression [38]. Strong causal associations have equally been found to exist between mental disorders especially depression, anxiety and hypertension and diabetes [39, 40]. Besides the genetic components, persistent environmental stressors are also well-known triggers for hypertension and diabetes mellitus [23]. They rearrange and over intensify the sympathetic system [42] by initiating or enhancing the release of norepinephrine which stimulates the sympathetic pathway. Thus, this neurotransmitter is consequently hyperactivated and becomes oversensitized in most chronic diseases. This causes both hypertension and anxiety disorders including panic disorder, General Anxiety Disorder (GAD), acute stress disorder and posttraumatic stress disorder (PTSD) [43–45]. Also, most of the psychotropic drugs have been known to be diabetogenic either by slowing down body metabolism including carbohydrates and or causing increased appetite. Again, a genetically causal link has been established between depression and diabetes with abnormal chromosome 21 being implicated. Additionally, the presence of anxiety alone could lead to hypertension and diabetes mellitus in predisposed individuals [39, 40].

Infectious chronic medical diseases like HIV and tuberculosis equally create multiple burdens for patients, including the need to deal with pain, suffering, reduced quality of life, poor compliance to medication, premature mortality, financial costs, stigma, discrimination, isolation, rejection and familial emotional trauma [46]. The co-existence of psychiatric disorders with various communicable and non-communicable diseases (NCDs) is a relevant phenomenon, which tend to be chronic in nature and has significant implications in the patient's well-being, quality of life, cost of treatment as well as longevity of the patient. Other available evidence have clearly showed an increasing prevalence of psychiatric comorbidities among communicable and NCD patients [32–34, 47, 48]. The chronic nature of these medical illnesses, their persistent and recurrent symptoms, impairment in functioning capacity, as well as other adverse environmental psycho-social burdens and even the worried thought of these can also in turn cause anxiety and may also quickly drive the patient into depression, suicidal ideation or attempt, and ultimately suicide. It is therefore important to note that the adverse environmental factors could lead to both the medical conditions and mental illness and can equally aggravate them. Although HIV- and tuberculosis-infected patients under anti-retroviral (arbacire and niverapine) and anti-tuberculosis drugs (isoniazid, ethambutol and cycloserine) therapy infrequently suffer acute organic psychotic complications [49–52], the chronicity of the disease places them at greater risk for psychiatric comorbidity than the general population [53]. Acute episodes of mania or manic phase of bipolar affective disorder tend to occur in HIV and tuberculosis patients [54, 55].

In addition to the effect of psychiatric comorbidity on quality of life in patients with chronic diseases, depression also contributes to poor self-care, reduced immunity, poor adherence to medical treatment, higher rates of medical morbidity and mortality and increased health-care costs [56, 57]. The enduring thought of the prolong illness, and the tendency for long-term treatment including the inconvenience of routine or daily parenteral or oral medication, the functional incapacitation, the burden of care including financial, the stigma (both externalized and internalized) and the social rejection are more than enough psycho-social factors to impact mental ill health. In addition to these, the presence of depressive symptoms is also related to hypertensive and diabetes self-care, reflected by a worse adherence to medications, diet and exercise advice including reduction on alcohol use [58, 59]. The available studies suggest that psychiatric comorbidity particularly depression may indeed precede a decrease in quality of life among patients suffering from chronic medical illness [60]. However, the possibility of reversed causality, for example, that a reduced quality of life or physical function in individuals with chronic diseases preceding the development of depressive symptoms cannot be ruled out. Chronic diseases, psychiatric comorbidity particularly depression and anxiety and quality of life are closely related triad. The causality and time course and path of these relations remain largely unknown. There is therefore need for an integrated chronic disease care approach that addresses both the practical and emotional issues.

It is also worthy of note that while some medications that are used in the treatment of these medical conditions are associated with neuropsychiatric side effects like channel blockers and propranolol for hypertension, strong association has also been found between some medications used in the treatment of some mental disorders and diabetes mellitus (Selective Serotonin Reuptake Inhibitor (SSRI) and Olanzepine) and hypertension (TCA). Mental disorders can equally predispose individuals to hypertension or diabetes [39–41], and also to acquiring HIV and or tuberculosis due to poor sense of judgment, leading to sexual indiscretion and other risk bearing practices [61]. In addition, the mentally ill are often taken advantage off and subjected to sexual assaults, which is one of the major modes of transmission of HIV infection. Furthermore, some medications have been known to be immunosuppresants in their action and by so doing can increase the vulnerability of individuals to these infections. Additionally, use of alcohol and other drugs of abuse, which is common in most chronic diseases, makes them vulnerable to enter casual or coercive sexual relationship [62–64]. Mental disorders particularly depression can lower immunity thereby making the individuals vulnerable to these infectious diseases.

3. Quality of life and well-being in patients with chronic medical conditions

The terms 'quality of life' and, more specifically, 'health-related quality of life' refers to the physical, psychological and social domains of health, seen as distinct areas that are influenced by a person's experiences, beliefs, expectations and perception [65, 66], ('which is referred here collectively as perceptions of health'). The emphasis within the above definition is first, on the subjective nature of QOL and second, on the need to assess all those aspects of life considered as having significant impact on QOL and well-being of any individual. Quality of life assessment is aimed at measuring changes in physical, functional, mental and social health of the individual and then the population in order to evaluate the human and financial cost and benefits of new programs and interventions [65, 66].

'Quality of life' is characterized by an individual's awareness on his ability to function well, cope with his or her daily demands and be in good health socially, physically and mentally [67]. A self-reporting questionnaire which is reliable and has valid numbers can use to measure one's quality of life. This questionnaire is categorized in three main groups: generic, disease and domain specifications [68]. The generic specification measures patients' quality of life irrespective of the presence of any diseases. The disease specification deals with the consequences, cost and depth of a specific disease on the patients' quality of life. While the final domain specified on certain domains in patients' quality of life, for example, if there are any physical inabilities.

Quality of life therefore is affected in a complex way by the individual's physical health, psychological state, as well as their sense of independence, level of social relationships, personal beliefs and their other relationships to salient features of their environment. Quality of life also consists of fulfilling needs, meeting of social expectations and assessing opportunities by using abilities. Abilities are impaired by ill health and worse still chronic medical illnesses [68–70]. Quality of life can be altered by both the immediate and the long-term consequences of treatment especially the case of chronic illnesses [71–78]. Hence, the focus of medical practice has always been directed toward relieving physical symptoms of chronic medical diseases which often overlook the huge impact on the physical and psychological well-being, psychiatric comorbidity and the overall patient's quality of life. This has often resulted in monumental adverse health outcomes [79–81]. Also, emphasis is usually on treating the basic abnormality and most times symptomatology of the disease leaving the individuals' own views of their state of well-being.

Most chronic medical conditions weaken patients' physical functioning and impair their quality of life [82, 83]. It has become important that tuberculosis and HIV control programs at public health clinics design strategies to improve the quality of health of tuberculosis and HIV separately or co-infected patients. Tuberculosis affected all predicted domains of QOL, including general health perceptions, somatic sensation, psychological health, spiritual well-being, and physical, social and role functioning [74–76]. Social stigmatization, isolation, pill burden, long duration of therapy, sexual dysfunction, loss of income and fear are associated with chronic medical diseases and all affect patients' QOL and well-being. Despite available curative therapy, chronic diseases and their treatment still have significant short- and long-term consequences on patients' QOL [69, 73, 75].

Several factors are known to influence both depressive symptoms and quality of life, including age, sex, marital status, educational level or income [77, 81]. These factors may confound the association between depressive symptoms and quality of life. Depressive symptoms in individuals with chronic medical diseases are associated with a worse quality of life [84–88]. Generic, disease specific, as well as domain specific quality of life are poorer in the presence of depressive symptoms [89]. In addition, there is evidence that depressive symptoms can predict the development of functional limitations in the future, suggesting a causal relationship between depressive symptoms and functional disability [90].

4. Aim

The aim of this study, therefore, was to determine the quality of life of patients with chronic medical diseases: HIV, tuberculosis, diabetes mellitus and hypertension and the effect of psychiatric comorbidity on the quality of life and well-being of these patients who attend the different out-patient clinics of the University of Port Harcourt Teaching Hospital (UPTH).

5. Methodology

Following ethical approval and informed consent from the participants, 120 subjects from each group of HIV, tuberculosis, diabetes mellitus and hypertension were recruited based on the study's inclusion and exclusion criteria. Patients with additional medical condition including having two of the four illnesses under consideration, severely ill patients, for example, with evident AIDS and multi-drug resistant tuberculosis were excluded. Age range of 10–70 years was a criterion. A pilot study was carried out. Subjects were further administered with the study's instruments including the socio-demographic/clinical questionnaire, GHQ-12, WHO composite international diagnostic interview (WHO CIDI) and the WHOQOL-Bref. The data were analyzed using the SPSS version 20 statistical package. Confidence interval was set at 95% while P value of less than 0.05 was considered statistically significant.

5.1. WHO QOL-Bref

The World Health Organization quality of life assessment instrument, short version (WHOQOL-Bref, used in this study) is a 26-item questionnaire that assesses how patients feel about their life [91, 92].The WHOQOL produces a quality of life profile with four domain scores and general health facet. The four domain scores and general health facet test an individual's perception of quality of life in each particular domain. Scores on domain 1 assess physical health; domain 2 psychological health; domain 3 social relationship and domain 4 assess individual's perception of the environment. Domain scores are scaled in a positive direction (higher score denotes higher quality of life) [91, 92]. QOL represents the effect of an illness on a patient, as perceived by the patient, and yields complementary information to medical or epidemiological data, and it is often used as an outcomes measurement [93]. QOL has also been characterized as 'the ultimate goal of all health interventions'.

5.2. The general health questionnaire (GHQ-12)

It is a self-administered screening instrument, which was designed to detect short-term changes in mental health (anxiety, depression, somatic symptoms and social dysfunctions) and to screen for psychiatric morbidity [94–96]. It is aimed at distinguishing between psychological ill health and well-being. It can detect disorders of less than two-week duration. It focuses on breaks in normal functioning and is concerned with a person's inability to continue with normal healthy functions and the experience of new phenomena of a distressing nature.

6. Results

6.1. Psychiatry diagnosis

From the study, tuberculosis had the highest prevalence of psychiatric comorbidity with 63 (52.5%), followed by hypertension with 59 (49.17%), and diabetes mellitus 46 (38.3%), while HIV was the least with 42 (35%). Depression was the most common psychiatric disorder among all the medical conditions with 11.67, 15.83, 9.16 and 15% among diabetes, hypertension, HIV and tuberculosis, respectively, followed by generalized anxiety disorder. In a total of 12 patients (5 among the HIV and 7 among the tuberculosis) had psychosis, 2 and 3 cases which occurred following commencement of anti-retroviral and anti-tuberculosis medications, respectively (see **Table 1**).

S/N	Psychiatric morbidity total	Diabetes mellitus (%)	Hypertension (%)	PLWHV (%)	Tuberculosis (%)
		n = 120	n = 120	n = 120	n = 120
1	Depressive disorders (with or without psychosis)	14 (11.67)	19 (15.83)	11 (9.16)	18 (15.0)
2	GAD	8 (6.06)	10 (8.33)	7 (5.83)	11 (9.16)
3	Sexual dysfunctions	2 (1.66)	5 (4.16)	2 (1.66)	2 (1.66)
4	Mixed anxiety and depressive disorders	6 (5.0)	8 (6.66)	4 (3.33)	7 (5.83)

S/N	Psychiatric morbidity total	Diabetes mellitus (%)	Hypertension (%)	PLWHV (%)	Tuberculosis (%)
		n = 120	n = 120	n = 120	n = 120
5	Substance abuse	3 (2.5)	5 (4.16)	3 (2.5)	5 (5.83)
6	Adjustment disorder	2 (1.66)	1 (0.83)	2 (1.66)	1 (0.83)
7	PSTD	0 (0)	0 (0)	3 (2.5)	3 (2.5)
8	Panic without agoraphobia	0 (0)	0 (0)	1 (0.83)	1 (0.83)
9	Dysthymia	3 (2.5)	4 (3.33)	2 (1.66)	3 (2.5)
10	Phantom disorders	2 (1.66)	0	0 (0)	0 (0)
11	Social phobia	2 (1.66)	3 (2.5)	2 (1.66)	2 (1.66)
12	Somatization	4 (3.33)	3 (2.5)	3 (2.5)	5 (4.16)
13	Personality disorders	0 (0)	1 (0.83)	1 (0.83)	2 (1.66)
14	BAD	0 (0)	0 0	1 (0.83)	3 (2.5)
15	Nil (No psychiatric) illness)	74 (62.0)	61 (51.0)	78 (65.0)	57 (47.5)

Note: Percentages reflect proportions within each medical conditions.

Table 1. Psychiatric comorbidity among diabetes.

6.2. Socio-demographic and clinical correlates of psychiatric morbidity

In all four medical conditions, there was no significant relationship between the presence of psychiatric comorbidity and age class ($P = 0.350$), gender ($P = 0.22$), level of education ($P = 043$), income class ($P = 0.81$) and occupation. Persons who were married were significantly more likely to have a psychiatric comorbidity ($P < 0.001$). Also, those who reacted with either 'very sad' or a 'wish to die' when they received the diagnosis of the medical conditions were more likely to have psychiatric comorbidity ($P = 0.001$). There was also no significant relationship between age of onset of illness ($P = 0.60$), duration of illness ($P = 0.73$), duration of treatment ($P = 0.82$) and self-stigma ($P = 0.15$) (see **Table 2**).

6.3. Association of psychiatric comorbidity with quality of life among patients with diabetes mellitus, hypertension, HIV and tuberculosis

On domains 1 and 4, PLWHIV who had psychiatric comorbidity had better performance on quality of life, followed by hypertension, diabetes mellitus while tuberculosis had the least in similar domains. Furthermore, PLWHIV free of psychopathology also scored higher on general health facet (GHF) than similar group among the other three medical conditions (see **Table 3**).

http://dx.doi.org/10.5772/intechopen.68730

Variable	DM		Hypertension		HIV		TB		Statistical Analysis
	Psychiatric diagnosis	No psychiatric diagnosis	Psychiatric diagnosis	No psychiatric diagnosis	Psychiatric diagnosis	No psychiatric diagnosis	Psychiatric diagnosis	No psychiatric diagnosis	
Age									
<20	2 (1.67%)	5 (4.17%)	0 (0)	0 (0%)	4 (3.33%)	3 (2.5%)	3 (2.5%)	1 (0.83%)	
20–29	4 (3.33%)	13 (10.83%)	5 (4.17%)	13 (10.83%)	8 (6.67%)	18 (15%)	8 (6.67%)	15 (12.5%)	
30–39	5 (4.17%)	12 (10%)	9 (7.57%)	11 (9.12%)	13 (10.83%)	26 (21.57%)	19 (15.83%)	18 (15%)	$P = 0.35$
40–49	11 (9.12%)	18 (15%)	18 (15%)	15 (12.5%)	9 (7.5%)	17 (14.17%)	17 (14.17%)	15 (12.5%)	
50–59	16 (13.33%)	15 (12.5%)	13 (10.83%)	10 (8.33%)	5 (4.17%)	12 (10%)	9 (7.5%)	6 (5%)	
>60	8 (6.67%)	10 (8.33%)	14 (11.67%)	12 (10%)	2 (1.67%)	1 (0.83%)	6 (5%)	2 (1.67%)	
Sex									
Male	1 (11.67%)	31 (25.83%)	18 (15%)	29 (24.12%)	16 (13.33%)	31 (25.33%)	29 (24.12%)	27 922.5%)	$P = 0.22$
Female	32 (26.67%)	43 (35.83%)	41 (34.12%)	32 (26.67%)	26 (21.67%)	47 (39.12%)	34 (28.33%)	30 (25%)	
Marital status									
Married	21 (17.5%)	32 (26.67%)	23 (19.2%)	26 (21.67%)	15 (5.8%)	31 (25.3%)	25	20 (16.6%)	$P < 0.001$
Single	13 (5%)	20 (9.16%)	21 (9.2%)	24 (10%)	18 (15%)	34 (28.3%)	22 (18%)	18 (15%)	
Divorced	4 (3.33%)	8 (15%)	6 (5%)	4 (3.33%)	4 (3.33%)	6 (5%)	6 (5%)	7 (5.8%)	
Separated	5 (4.17%)	10 (6.65%)	4 (3.3%)	3 (2.5%)	5(4.17%)	5 4.17%)	5 (4.17%)	5 (4.17%)	
Widowed	10 (8.33%)	13 (10.8%)	15 (12.6%)	16 (13.3%)	1 (0.8%)	2 1.67%)	4 (3.33%)	7 (5.8%)	
Education									
None	5 (4.17%)	6 (5%)	7 (5.83%)	4 (3.33%)	5 (4.17%)	9 (7.5%)	8 (6.67%)	6 (5%)	$P = 0.43$
Primary	11 (9.17%)	21 (17.5%)	20 (16.67%)	21 (17.5%)	12 (10%)	23 (19.2%)	21 (19.2%)	14 (11.07%)	
Secondary	13 (10.83%)	19 (15.83%)	19 (15.83%)	20 (16.67%)	13 (10.08%)	21 (17.5%)	19 (15.8%)	16 (13.3%)	
Tertiary	18 (15%)	27 (22.5%)	13 (10.83%)	16 (13.3%)	12 (10%)	26 (21.57%)	15 (5.8%)	2319 (2%)	

Variable	DM		Hypertension		HIV		TB		Statistical Analysis
	Psychiatric diagnosis	No psychiatric diagnosis	Psychiatric diagnosis	No psychiatric diagnosis	Psychiatric diagnosis	No psychiatric diagnosis	Psychiatric diagnosis	No psychiatric diagnosis	
Average monthly income									
Low	14 (11.67%)	15 (12.5%)	16 (13.3%)	17 (14.17%)	14 (11.67%)	20 (16.67%)	28 (23.3%)	21 (17.5%)	$P = 0.81$
Average	22 (18.33%)	35 (29.2%)	30 (25%)	34 (28.3%)	20 (16.67%)	39 (32.5%)	27 (22.5%)	29 (24.17%)	
High	10 (8.33%)	24 (20%)	12 (10%)		16 (13.3%)	8 (6.67%)	19 (15.8%)	8 (6.67%)	
Reaction to diagnosis									
Normal	12 (10%)	37 (30.83%)	14 (11.67%)	24 (20%)	4 (3.33%)	10 (8.3%)	3 (2.5%)	5 (4.17%)	$P < 0.001$
Sad	23 (19.2%)	26 (21.67%)	23 (19.2%)	21 (17.50%)	11 (9.17%)	31 (25.83%)	23 (19.2%)	25 (20.83%)	
Very sad	11 (9.17%)	12 (10%)	22 (18.33%)	16 (13.3%)	19 (15.83%)	28 (23.33%)	26 (21.67%)	25 (20.83%)	
Wish to die	0 (0%)	0 (0%)	0 (0%)	0 (0%)	10 (8.3%)	9 (7.50%)	11 (9.17%)	2 (1.67%)	
Occupation									
Managers	1 (0.8%)	2 (1.67%)	1 (0.8%)	1 (0.83%)	0 (0%)	0 (0%)	0 (0%)	0 (0%)	
Professionals	1 (0.8%)	3 (2.5%)	1 (0.8%)	2 (1.67%)	1 (0.83%)	3 (2.5%)	0 (0%)	0 (0%)	
Technicians and Associates professionals	3 (2.5%)	6 (5%)	2 (1.67%)	3 (2.50%)	3 (2.50%)	4 (3.33%)	1 (0.83%)	2 (1.67%)	$P = 0.14$
Clerical support workers	5 (4.17%)	8 (6.67%)	5 (4.17%)	5 (4.18%)	5(4.18%)	9(7.50%)	4 (3.33%)	4 (3.33%)	
Service and sales workers									
Skilled agricultural forestry and fishery workers	7 (5.83%)	11 (9.17%)	7 (5.83%)	8 (6.67%)	6 (5%)	10 (8.3%)	5 (4.18%)	6 (5%)	
Craft and related trade workers	7 (5.83%)	13 (10.8%)	12 (10%)	13 (10.83%)	5 (4.18%)	15 (12.50%)	11 (9.17%)	13 (10.83%)	

Variable	DM		Hypertension		HIV		TB		Statistical Analysis
	Psychiatric diagnosis	No psychiatric diagnosis	Psychiatric diagnosis	No psychiatric diagnosis	Psychiatric diagnosis	No psychiatric diagnosis	Psychiatric diagnosis	No psychiatric diagnosis	
Plant and machine operators and assemblers	6 (5%)	10 (8.33%)	8 (6.67%)	11 (9.17%)	6 (5%)	9 (7.5%)	13 (10.83%)	12 (10.00%)	
Elementary occupation	10 (8.33%)	13 (10.08%)	11 (9.17%)	12 (10%)	9(7.50%)	14 (11.67%)	18 (15.00%)	13 (10.83%)	
Armed forces occupation	0 (0%)	0 (0%)	3 (2.5%)	2 (1.67%)	2 (1.67%)	4 (3.33%)	0 (0%)	0 (0%)	
Unemployed	6 (5%)	8 (6.67%)	7 (5.83%)	4 (3.33%)	5(4.18%)	10 (8.33%)	11 (9.17%)	7 (5.8%)	
Age of onset of disease									
<20	4 (3.3%)	7 (5.83%)	0 (0%)	0 (0%)	7 (7.5%)	15 (12.5%)	9 (7.50%)	5 (4.17%)	
20–29	9 (7.5%)	13 (10.83%)	10 (8.3%)	11 (9.17%)	18 (15%)	30 (25%)	13 (10.83%)	12 (10.00%)	
30–39	15 (12.5%)	20 (16.67%)	16 (13.3%)	16 (13.3%)	12 (10%)	23 (19.17%)	24 (20%)	19 (15.83%)	$P = 0.004$
40–49	11 (9.17%)	24 (20%)	24 (20%)	24 (20%)	4 (3.3%)	12 (10%)	12 (10%)	15 (12.5%)	
>50	7 (5.8%)	10 (8.3%)	9 (7.5%)	10 (8.3%)	1 (0.8%)	5 (4.16%)	5 (4.16%)	6 (5%)	
Duration of illness									
<5 years	13 (10.83%)	18 (15%)	20 (16.67%)	21 (17.5%)	16 (13.3%)	28 (23.3%)	20 (16.67%)	16.13 (3%)	
6–19 years	22 (18.3%)	44 (36.67%)	27 (22.5%)	29 (24.17%)	15 (5–8%)	30 (25%)	27 (22.5%)	27 (22.5%)	$P = 0.002$
>11 years	11 (9.17%)	12 (10%)	12 (10%)	11 (9.17%)	11 (9.16%)	10 (8.33%)	16 (13.3%)	14 (11.67%)	
Stigma									
Yes	2 (1.67%)	0 (0%)	0 (0%)	0 (0%)	35 (29.17%)	48 (40%)	49 (40.8%)	40 (33.3)	$P = 0.002$
No	44 (67%)	74 (61.67%)	59 (49.17%)	61 (50.8%)	7 (5.8%)	30 (25%)	14 (11.67%)	17 (14.17%)	

Table 2. Socio-demographic variables and psychiatric comorbidity among patients with diabetes, hypertension, HIV and tuberculosis.

Medical condition	Status of psychiatric comorbidity	Domains of quality of life				
		Domain 1 (Physical)	Domain 2 (Psychological)	Domain 3 (Social Relationship)	Domain 4 Environment)	GHF
DM	All patients	51.97 ± 14.77	56.20 ± 22.19	57.51 + 26.13	52.01 + 16.91	48.34 + 22.44
	Psychiatric comorbidity	45.98 + 13.064	56.60 + 24.914	48.06 + 26.114	44.95 + 14.831	47.98 + 21.896
	No psychiatric comorbidity	60.46 ± 12.788	61.05 + 13.362	66.80 + 21.378	59.62 + 16.503	51.91 + 23.319
HPT	All patients	50.97 + 14.671	54.20 + 22.186	54.51 + 26.13	50.01 + 16.91	49.34 + 22.44
	Psychiatric comorbidity	45.98.+ 13.064	56.60 + 24.914	48.06 + 26.114	44.95 + 14.831	47.98 +21.896
	No psychiatric comorbidity	60.46 + 12.788	61.05 + 13.362	66.80 + 21.378	59.62 + 16.503	51.91 + 23.319
HIV	All patients	60.71 + 15.57	62.34 + 26.32	61.57 + 25.04	55.15 + 14.00	65.81 + 21.84
	Psychiatric comorbidity	46.70 + 10.103	48.67 + 15.016	46.84 + 21.032	50.33 + 10.456	63.83 + 20.349
	No psychiatric comorbidity	66.36 + 13.698	67.85 + 27.870	57.09 + 14.888	67.50 + 24.102	66.61 + 22.418
TB	All patients	50.97 + 14.671	54.20 + 22.186	54.51 + 26.13	50.01 + 16.91	49.34 + 22.44
	Psychiatric comorbidity	45.98 + 13.064	56.60 + 24.914	48.06 + 26.114	44.95 + 14.831	47.98 + 21.896
	No psychiatric comorbidity	60.46 + 12.788	61.05 + 13.362	66.80 + 21.378	59..62 + 16.503	51.91 + 23.319
Statistical analysis		*P* = 0.001	*P* = 0.001	*P* = 0002	*P* = 0.004	*P* = 0.24

DM = diabetes mellitus, HPT = hypertension, HIV = human immunodeficiency virus, TB = tuberculosis, GHF = general health facet.

Table 3. Association of psychiatric comorbidity with quality of life among patients with diabetes mellitus, hypertension, HIV and tuberculosis.

7. Discussion

Young adults from 35 years and above predominated the diabetic mellitus, hypertension and the tuberculosis groups. This is in agreement with earlier studies [4, 14, 18, 19, 35, 97, 98]. Apart from the influence of genetic predisposition, diabetic mellitus and hypertension are largely diseases of lifestyles including unhealthy eating habits, alcohol intake and sedentary lifestyles, which would manifest most among adults [4, 14, 18, 19]. Similarly, tuberculosis is an infective disease and exposure to it may be more in adults who work in health-care facilities, are mostly close care givers to already infected persons, and also in individuals whose immunity may have waned due to poor nutrition, alcohol use, stress and other emotional illnesses, most times occasioned by the medical illness, as well as other infective diseases. HIV had the most of youngest population with age group as young as 10–19 years taking 16%, while the peak age was 30–39 years. This is in line with earlier finding that HIV is more prevalent among young persons [25].

The prevalence of essential hypertension was noted to be increasing with age and was about twice higher in the age groups 40–49 and 50–59 and above, compared to age group of 30–39, and about six times higher compared with age group 20–29. This result is consistent with earlier studies which reported that about 4.3 million Nigerians above the age of 15 years are classified as being hypertensive [15]. Furthermore, the prevalence has been said to be related to age, particularly in females, with a substantial increase occurring after the age of 50 [15]. Africans usually seek medical attention mostly when illness has presented with disabling symptoms, and in most cases late, in spite of awareness of the diagnosis [99–101]. This is particularly more so for essential hypertension and diabetes mellitus. Voluntary testing which is a key aspect of the HIV prevention and control program is yet to achieve 100% success rate.

A preponderance of females was observed among all disease conditions, especially PLWHIV and hypertension which recorded more proportions of females compared to diabetes and tuberculosis. HIV is equally well-known to be commoner among females. Several reasons have been postulated for this observation. Females, by reason of their reproductive anatomy, have larger surface area for the transmission of the virus. It has also been reported that the sperm contains heavy viral concentrations per/ml compared to vaginal fluid. Unfortunately, cultural factors further heighten this adversity among the female gender. Although essential hypertension is more common in males, females may have been over-represented in this study due to two reasons. First, African females tend to have lower blood pressure than males early in life with a reversal of the trend after the age 45–50 years [15]. This may be due to hormonal changes associated with the preparation for or actual menopause occurring at this age, couple with the increasing family and domestic (stressors) responsibilities shouldered by females in this age group. Males predominated among those with psychiatric comorbidity in all medical conditions except among PLWHIV. The tendency and willingness for females to present early to care facility may equally translate to early understanding of the nature and extent of their illness, early reduction in symptomatology and early reduction of emotional burden, which may all impact better mental health, good quality of life and well-being.

Also from the study, it was found that more of each of the categories (married, separated and widowed) had more psychiatric comorbidity among the hypertensives and diabetes than their corresponding groups in PLWHIV and tuberculosis. A possible explanation could be that in hypertension and diabetes, marital difficulties, separation and even widowhood may serve as baseline psycho-social factors that may act either singly or in synergy with the medical stressor to cause psychiatric comorbidity. The presence of these psycho-social stressors alone can equally predispose to hypertension and diabetes due mainly to stress. In PLWHIV and tuberculosis, however, they are likely to influence the outcome of the disease, so there is more likelihood of marital difficulty, separation and even divorce in the chronic infective diseases, adding to the medical stressor. Thus, in hypertension and diabetes, they are both causal and effectual while in HIV and tuberculosis infections, they are more of effects.

Most of the subjects in this study had attained various levels of formal education especially secondary and tertiary. Perhaps the influence of westernization and urbanization in Rivers State, Niger Delta and Nigeria might have played an important role. Furthermore, the cosmopolitan nature of Port Harcourt, domiciling majority of ethnic groups in Nigeria, with over 50% of Nigeria's oil and gas businesses, makes education a priority. A good number (39%) of the subjects with hypertension and diabetes had tertiary education. The fact that they were educated may have increased their chances of employment and possibly ability to seek quality health care and timely too [102]. It is equally important to note that perhaps the older you become, and probably with more education, the more your socio-economic and family responsibilities, with their accompanying stressors. A number of studies have implicated environmental stressors as important aetiological factors in high blood pressure and diabetes mellitus, particularly in already genetically predisposed individuals [21–24]. Occupational environments in Nigeria had remained stressful due to lack of job security and poor wages, confronting countless demands from members of the family in a poverty ravaged economy such as ours. In all groups, the incidence of psychiatric comorbidity was lowest among those with tertiary education. Hence, education tended to have some protective influence on the psyche of patients [102]. Expectedly, those with higher level of education were more likely to secure better employment, earn better income and have better access to quality health care, hence better quality of life and well-being.

In this study, there was significantly reduced rate of stigma among patients with hypertension and diabetes mellitus. Expectedly, therefore, the negative effect of stigmatization was felt more among patients with the infective conditions. Several studies have reported high rate of stigmatization among PLWHIV and tuberculosis [28, 29, 46, 103]. Due to fear of stigmatization with its accompanying negative psychological impact on the individual [104], such as to experience a second psychological trauma, many people living with HIV rather choose to avoid disclosure of their seropositive status, to even close relative, while such concealment is difficult in tuberculosis because of the natural course and eventual symptomatology of the disease. They have learned to absorb their shock alone because of a possible rebound negative effects of disclosure.

Majority of the respondents reacted 'sadly' or 'very sadly' to the diagnosis of their medical conditions. Of particular note was among subjects with HIV and tuberculosis infections where those who reacted 'very sadly' and even with a 'wish to die' were significantly higher compared to subjects with essential hypertension and diabetes. Death wish is the forerunner of suicidal ideation that may ultimately end in suicide. Thus, it is not surprising that 4 and 2 cases of suicide attempt were seen among PLWHIV and tuberculosis, respectively. The possible explanation to this is the degree of stigma and social rejection usually associated with the chronic infective conditions, mainly born out of ignorance and socio-cultural factors [103, 104]. From the results, initial reaction to the diagnosis of chronic medical conditions seems to assume some importance in this study with regards to psychiatric comorbidity. Also of note here is that more people reacted very emotionally to diagnosis of communicable diseases than the non-communicable diseases. This is expected considering the level of stigma, discrimination, social rejection and isolation, need for adjustment, lowered self-esteem, fear of spreading the illness to others, helplessness brought about by incapacitation, associated with the former. Theses ultimately generate enormous psychological trauma, increase morbidity and mortality [29, 56, 104].

The study found a prevalence of psychiatric comorbidity of 43.75% in all. Tuberculosis had the highest prevalence of psychiatric comorbidity with 63 (52.5%), followed by hypertension with 59 (49.17%) and diabetes mellitus 46 (38.3%), while HIV was the least with 42 (35%). Although, previous studies found variable prevalence rates of 82.6 [105], 46 [106] and 5–30% [107] for HIV, 72 [108] and 46–47.2% [109] for tuberculosis, 35% for hypertension [15] and 37–48% for diabetes [36, 37], the lower prevalence rates of psychiatric comorbidity found in some of the medical conditions in this study particularly for HIV could be a reflection of increased awareness, following concerted efforts and aggressive campaigns by both government and voluntary organizations, more access to existing medical care, articulated interventional measures by government, such as the anti-retroviral and anti-tuberculosis schemes, and sustained efforts aimed at reducing stigma and discrimination. Furthermore, variable rates reported by previous studies may be due to the setting as well as methodology of the various studies.

Depressive illness was the most common in all four medical conditions, followed by generalized anxiety disorder. In a total of 17 patients (6 among the HIV and 11 among the tuberculosis) had psychosis which occurred following commencement of anti-retroviral and anti-tuberculosis medications, respectively.

For HIV, depressive illness was … and this figure is similar to the finding of Olisah of 14.2% in a study in Nigeria [110] and also consistent with several other studies which put the prevalence of depression (among PLWHIV) at 5–25 [111] and 10–40% [112]. Females recorded higher percentage of depressive illness in all medical conditions, consistent with the male to female ratio of 1:2 in depression [110, 111]. This might also be due to reasons earlier given for overall psychiatric disorders. Anxiety and depression are frequently and highly occurring mental disorders in patients with chronic medical diseases [110–112]. When depression and anxiety comorbid with chronic medical diseases, it leads to poor adherence to medication, which is an important barrier to global control of chronic disease particularly tuberculosis and HIV and increases the risk of morbidity and mortality [113–116].

The possible aetiological mechanisms of depression in all four illnesses have been explained from the biological (physical) as well as the psychological points of view. For HIV, the viral cells and even some opportunistic infections and ketoacidosis and glucose deposit in the cerebral cortex may either directly destroy the brain cells responsible for emotions in the limbic system or in both cases, cause direct affectation of the neurotransmitter system altering their release or uptake, and in effect causing depression. From the psychological point of view, the burden of the illness, the thought of almost life-time use of medications, fear of and uncertainty about the future, chronic sense of rejection and feeling of loss, possible loss of functional capability with reduced quality of life, associated stigma in the case of HIV, and chronic leg ulcer and possible amputation in cases of complicated diabetic foot ulcers, regimented lifestyle particularly food intake for diabetes and fear of impending death, are all depressogenic factors and may have had additive effects for depression. Several studies have also established similar lines of thought [113–116]. The threat to life and perceived loss or difficulty having a life partner may be responsible for the anxiety among the PLWHIV and tuberculosis. The predominant feeling of loss involves functional incapability, loss of job, relationship and even difficulty to secure a life partner. Truly, these are both depressogenic and anxiogenic. Another reason for the frequent comorbidity is that there are commonly shared risk factors for the development of a variety of psychiatric and medical disorders including smoking and low socio-economic status [117, 118].

Depression, generalized anxiety disorder, somatization, posttraumatic stress disorder (PTSD), acute stress disorder, panic disorder, adjustment disorder, substance abuse, sexual disfunction, psychosis and bipolar affective disorder are the most common stress-related conditions of chronic medical disorders in this study [45, 50, 119–121]. HIV and tuberculosis, with all the psycho-social difficulties including stigma and social isolation, and associated economic burden due to its chronic nature with very little hope of long-term survival, could be weighed as catastrophic to many sufferers. Chronic medical conditions are commonly associated with stressful conditions like loss of losses and unemployment, prolonged difficulties and people at war front. It has been found that stress, which potentially causes anxiety, also increases the level of cortisol which in turn causes increased deposition of arterosclerotic deposits in the intima of blood vessels. These deposits gradually narrow the lumen of the vessels. This in turn increases arterial pressure, resulting in hypertension. Excessive cortisol predisposes to diabetes mellitus.

Dependence on alcohol and other drugs could be the response to anxiety and depression as many of the sufferers may use them to self-medicate these disabling psychological symptoms and disorders [62–64]. There is a bidirectional relationship between substance abuse and chronic diseases, that is, substance abuse particularly alcohol can cause hypertension while hypertension, on the other hand, can precipitate substance abuse due to frustration [91–93]. However, the true relationship still remains difficult to establish in this study. People with chronic medical condition tend to abuse substance mainly to self-medicate their depression or to abate the many anxiety or anxiety-like symptoms that characterized hypertension, hence the use of propranolol and diazepam which have anxiolytic effects.

It is important to note that the viral cells alone or acting together with opportunistic infections, disseminating tuberculosis to the CNS or most substances of abuse, particularly alcohol, can directly impair or damage brain cells. Thus, these will synergistically hasten the deterioration of the health of the individual, thereby negatively affecting his/her psychological well-being and quality of life. Second, interactions between substances of abuse, particular alcohol, and anti-retroviral and anti-tuberculosis medications have also been associated with poor drug adherence, as well as reduced effectiveness of medications, which may result in unbearable side effects, often contributing to poor drug adherence.

Majority of the patients in all medical conditions fared well on most domain 65, 66, 76, 122–124]. The possible reasons being their focus on physical strength (e.g. evident physical health, absence of symptoms, ability to work around, available family support and a strong religious belief) than on their weaknesses (e.g. social discrimination, difficulty in having intimate relationships and reduced job opportunity). On domains 1 and 4, PLWHIV who had psychiatric comorbidity had better performance on quality of life, followed by hypertension, diabetes mellitus while tuberculosis had the least in similar domains. Furthermore, PLWHIV free of psychopathology also scored higher on GHF than similar group among the other three medical conditions.

It is true that the stress, depression and lack of adequate social support which often may synergistically hasten the progression of HIV to AIDS, and by so doing, cause rapid deterioration in their psychological well-being and quality of life. However, in this study, it appeared that the combined effects of Government intervention programs and education for the general public as well as the supportive involvement by significant family members may have helped to increase the level of social support, reduced economic burden of care and associated stress which ultimately reduced the prevalence and severity of depression and other psychiatric disorders. This often results in improved quality of life. At present, more concern and care are given to PLWHIV, unlike diabetic mellitus, hypertensive and tuberculosis patients, from both governmental and non-governmental organizations, other social groups as well as the family.

Some authors have opined that quality of life will be poor in the developing countries like Nigeria where factors of finances, social relationships, health and personal safety are considered to be poor [124, 125]. Chronic medical conditions have high impact on the economy and with consequent low quality of life of individuals. The results in this study are consistent with many studies elsewhere [35, 125]. All domains of quality of life in those whose illness was complicated with psychiatric disorders among the medical conditions were affected, significantly worse in domains 1–4, compared to those without psychiatric comorbidity. This finding varied a little from previous studies where all domains of QOL except domain 4 (environment) and health satisfaction were affected by psychiatric comorbidity [125]. The result of this study is also consistent with the finding of Oliseh in his study done in Nigeria in which depression was the most prevalent psychiatric comorbidity, found that 63.6% subjects with depression had poor overall QOL while only 11.3% of subjects with no depressive disorder had poor overall quality of life and asserted that this could be representative, as depression has been identified as the most prevalent psychiatric illness in PLWHIV/AIDS [64].

In the presence of psychiatric comorbidity, the quality of life in PLWHIV and tuberculosis patients was significantly affected, particularly on the psychological and social domains compared with diabetes mellitus and hypertension. This suggests that the stigma and social rejection associated with the communicable disease may play a significant role in the development of psychological illness. This also implies that even though psychiatric comorbidity is equally common in the diabetes mellitus and hypertension and affecting quality of life generally, the presence of psychiatric comorbidity in PLWHIV and tuberculosis tended to have more severe negative impact on quality of life.

The presence of symptoms of tuberculosis, hypertension and diabetes mellitus alone appear to be more disabling than those in PLWHIV, bearing in mind that acute cases were excluded. Moderate to severe cases of diabetes mellitus, hypertension and tuberculosis cause more symptoms and they are more disabling. This may account for the better quality of life among PLWHIV on the physical domain and the lower quality of life scores on both the physical and environment domains among them compared with PLWHIV.

In this study, it was observed that socio-demographic and clinical characteristics of respondents in all medical conditions had some significant relationships with quality of life. Factors like increased age, marital status (married), later age of onset of illness, education, employment, average to high monthly income, shorter duration of illness, longer duration of treatment and emotional stability positively affected both psychiatric comorbidity and quality of life in all the groups. These results are consistent with several studies [124]. These suggested that a good number of psycho-social and clinical factors affected the outcome of the medical conditions. The implication of this is that these factors have to be addressed in the holistic management of these and indeed other chronic medical conditions.

The effect of gender was variable. Although, females were more affected in most of the psychiatric illnesses which is consistent with existing literature, this did not translate into lower quality of life for females on most domains of quality of life. This suggests that psychiatric comorbidity in females tended to have better prognosis compared with males. From the study, females performed better on the overall quality of life, domains 2–4, while males fared better on health satisfaction and domain 1. This might be due to their better health-seeking behavior. Second, the obvious fact that most of the psychiatric disorders found in this study are usually associated with better prognosis in females, might have contributed to their better quality of life.

8. Recommendations

1. Based on the findings of this study, that chronic medical conditions carry the risk of psychiatric comorbidity with consequent affectation of quality of life and well-being, it becomes imperative that renewed efforts by Government, aimed at both primary and secondary prevention, be intensified for these chronic medical conditions.
2. That the intervention of Government in terms of increasing awareness campaign as well as free anti-retroviral and anti-tuberculosis scheme of the federal government for PLWHIV and tuberculosis should be sustained and further extended to non-communicable diseases.
3. Attention of clinicians should be drawn to diagnosing early psychiatric complications of chronic medical diseases and adequate management instituted or referrals made.

9. Conclusion

From the study, presence of psychiatric comorbidity and other psycho-social factors also influence the outcome of these medical conditions and to some extent significantly determined the psychological well-being and QOL. On account of the above, advocacy for a formal integration of functional Liaison-psychiatric practice has become imperative in the management of most chronic medical conditions for optimal benefits of patients and physicians, as early identification and prompt attention to the psychological components will, no doubt, go a long way in improving the clinical outcomes of the sufferers.

Furthermore, efforts of both governmental and non-governmental organizations in the management of chronic health conditions, including communicable and non-communicable conditions should have clear objectives backed by legislation and not on *ad hoc* basis, as this would guarantee its sustenance.

Author details

Nkporbu A.K.

Address all correspondence to: nakpigi2008@yahoo.com

Department of Neuropsychiatry, University of Port Harcourt Teaching Hospital, Port Harcourt, Nigeria

References

[1] Centres for Disease Control and Prevention HIV/AIDS Surveillance Report. 2004;**12** (No.1):1-42

[2] Kochi A. The global tuberculosis situation and the new control strategy of the World Health Organization. Tubercle. 1991;**72**(1):1-6

[3] Shaw JE, Sicree RA, Zimmet PZ. Global estimates of the prevalence of diabetes for 2010 and 2030. Diabetes Research and Clinical Practice. 2010;**87**(1):4-14

[4] Unachukwu CN, Agomoh DI, Alasia DD. Pattern of non-communicable diseases among medical admissions in Port Harcourt, Nig. Nig. Journal of Clinical Practice. 2008;**11**(1):14-17

[5] World Health Organization. Global Tuberculosis Control 2010. Geneva, Switzerland: World Health Organization; 2010. http://www.who.int/tb/publications/global_report/en/index.html

[6] Aydin IO, Ulusahin A. Depression, anxiety comorbidity, and disability in tuberculosis and chronic obstructive pulmonary disease patients: applicability of GHQ-12. General Hospital Psychiatry. 2001;**23**(2):77-83

[7] Cook J, Grey D, Burke J, et al. Depressive symptoms and AIDS-related mortality among a multiside cohort of HIV-positive women. American Journal of Public Health. 2004;**94**:1 133-1140

[8] Carroll D, Phillips AC, Gale CR, Batty GD. Generalized anxiety and major depressive disorders, their comorbidity and hypertension in middle-aged men. Psychosomatic Medicine. 2010;**72**:16-19

[9] Gwatkin D, Guillot M, Heuveline P. The burden of disease among the global poor. Lancet, 2000;**354**:586-589

[10] Murray CJ, Lopez AD. Mortality by cause for eight regions of the world. Global burden of disease. Lancet. 1997;**349**:1269-1276

[11] Rajeswari R, Balasubramanian R, Muniyandi M, Geetharamani S, Thresa X, Venkatesan P. Socio-economic impact of tuberculosis on patients and family in India. International Journal of Tuberculosis and Lung Disease. 1999;**3**(10):869-877

[12] Dye C, Scheele S, Dolin P, Pathania V, Raviglione MC. Global burden of tuberculosis: Estimated incidence, prevalence, and mortality by country. Journal of the American Medical Association. 1999;**282**(7):677-686

[13] Waisbord S. Behavioral Barriers in Tuberculosis Control: A Literature Review. Washington, DC, USA: The CHANGE Project/Academy for Educational Development; 2004

[14] Kearney PM, Whelton M, Renoids K, Muntner P, Whenton KP. Global burden of hypertension: Analysis of worldwide data. Lancet. 2005;**365**:217-223

[15] Erhun WO, Olayiwola, G, Agbani EO, Omotoso NS. Prevalence of hypertension in a University Community in South Western Nigeria. African Journal of Biomedical Research Ibadan Biomedical Communication Group. 2005;**8**:15-19

[16] Van de Sande MAB, Bailey R, Faal H, Banya WA, Dolin P, Nyan OA, Ceesay SM, Walraven GE, Johnson GJ, McAdamKP. Nationwide prevalence study of hypertension and related non-communicable diseases. Tropical Medicine & International Health. 1997;**2**:1039-1048. CrossRefMedline

[17] Briganti EM, Shaw JE, Chadban SJ, Zimmet PZ, Welborn TA, McNeil JJ, Atkins RC. Untreated hypertension among Australian adults: The 1999-2000 Australian diabetes, obesity and lifestyle study. Medical Journal of Australia. 2003;**179**:135-139

[18] Adefuye BO, Adefuye PO, Oladepo TO, Familoni OB, Olurunga TO. Prevalence of hypertension and other cardiovascular risk factors in an African Urban, sub-urban religious community. Nigerian Medical Practitioner. 2009;**55**(1-2):4-8

[19] Wokoma FS, Alasia DD. Blood pressure pattern in Barako: A rural community in Rivers State, Nigeria. The Nigeria Journal. 2011;**11**:813

[20] Ota M., Issiki M. An Outbreak of tuberculosis in a long-term care unit of a mental hospital. Kekkaku. 2004;**79**(10):579-586

[21] Lifton RP, Gharavi AG, Geller DS. Molecular Mechanisms of Human Hypertension. ScienceDirect. 2001;**104**(4):545-556

[22] Wilson FH, Disse-Nieodme S, Choate KA, et al. Human hypertension caused by mutations in WNK kinases. Science. 2001;**293**(5532):1107-1112

[23] Pickering TG. Devereux RB, James GD, et al. Environmental influences on blood pressure and the role of job strain. Journal of Hypertension Supplement. 1999;**341**:120-124

[24] Bharat S, Aggleton P. Facing the challenge: household responses to AIDS in Mumbai, India. AIDS Care. 1999;**11**:31-44

[25] Windle M. The trading of sex for money or drugs, sexually Transmitted Diseses (STDs) and HIV related risk behaviors among polysubstance use and alcoholic inpatients. Drug and Alcohol Dependence. 1997;**49**:33-38

[26] Susser E, Valencia E, Miller M, Tsai WY, Meyer-Bahlburg H, Conover S. Sexual behavior of homeless mentally ill men at risk for HIV. American Journal of Psychiatry. 1995;**1552**:583-587

[27] Bird ST, Bogart LM, Delahanty DL. Health-related correlates of perceived discrimination in HIV care. AIDS Patient Care STDs. 2004;**18**:19-26

[28] Courtwright A, Turner AN. Tuberculosis and stigmatization: pathways and interventions. Public Health Reports. 2010;**125**(supplement 4):34-42

[29] Johnson RT. HIV. In: Johnson A, editor. Viral Infections of the Nervous System. 2nd ed. PA. USA: Lippincott-Raven; 1998. pp. 287-231

[30] Center for Disease Control and Prevention. Prevention and treatment of tuberculosis patients infected with human immunodeficiency virus: Principles of therapy and revised recommendations. MMWR. Morbidity and Mortality Weekly Report. 1998;**47**(RR-20): 1-51

[31] World Health Organization. Burden: Mortality, Morbidity and Risk Factors, Global Status Report on Non-communicable Diseases. 2010. Available from: http://www.who.int/nmh/publications/ncd_report_chapter1pdf [Accessed: 1 October 2012]

[32] Sulehri MA, Dogar IA, Sohail H, et al. Prevalence of depression among tuberculosis patients. Agricultural Produce Market Committee. 2010;**4**:133-137

[33] Issa BA, Yussuf AD, Kuranga SI. Depression comorbidity among patients with tuberculosis in a university teaching hospital outpatient clinic in Nigeria. Mental Health and Family Medicine. 2009;**6**:133-138

[34] Deribew A, Tesfaye M, Hailmichael Y, et al. Common mental disorders in TB/HIV co-infected patients in Ethiopia. BMC Infectious Diseases. 2010;**10**:201

[35] Chaudhry R, Mishra P, Mishra J, Parminder S, Mishra BP. Psychiatric morbidity among diabetic patients: a hospital-based study. Indian Journal of Psychiatry. 2011;**19**(1):47-49

[36] De Ornelas Maia AC, Braga Ade A, Brouwers A, Nardi AE, De Oliveira e Silva AC. Prevalence of psychiatric disorders in patients with diabetes types 1 and 2. Comprehensive Psychiatry. 2012;**53**(8):1169-1173

[37] Campbell LC, Clauw DJ, Keefe FJ. Persistent pain and depression: a biopsychosocial perspective. Biological Psychiatry. 2003;**54**(3):399-409

[38] Jonas BS, Franks P, Ingram DD. Are symptoms of anxiety and depression risk factors for hypertension? Longitudinal evidence from the National Health and Nutrition Examination Survey I Epidemiologic Follow-up Study. Archives of Family Medicine. 1997;**6**:43-49

[39] Markovitz JH, Matthews KA, Kannel WB, Cobb JL, O'Agostino RB. Psychological predictors of hypertension in the Frarningham Study. Journal of the American Medical Association. 1993;**270**:2439- 2443

[40] Engum A. The role of depression and anxiety in onset of diabetes in a large population-based study. Journal of Psychosomatic Research. 2007;**62**(1):31-38

[41] Takahashi H. Sympathetic hyperactivity in hypertension (in Japanese). Nippon Rinsho. Japanese Journal of Clinical Medicine. 2008;**66**(8):1495-1502

[42] Katze S, Nevid JS. Risk factors associated with posttraumatic stress disorder symptomatology in HIV-infected women. AIDS Patients Care and STDs. 2005;**19**(2):110-120

[43] Olley BO, Zeier MD, Seedat S, Stein DJ. Posttraumatic stress disorder among recently diagnosed patients with HIV/AIDS in South Africa. AIDS Acre. 2005;**17**(5):550-557

[44] Peltzer K, Naidoo P, Matseke G, Louw J, McHunu G, Tutshana B. Prevalence of post-traumatic stress symptoms and associated factors in tuberculosis (TB), TB retreatment and/or TB-HIV co-infected primary public health-care patients in three districts in South Africa. Psychology Health & Medicine. In press

[45] Jaggarajamma K, Ramachandran R, Charles N, Chandrasekaran V, Muniyandi M, Ganapathy S. Psycho-social dysfunction: Perceived and enacted stigma among tuberculosis patients registered under revised national tuberculosis control programme. The Indian Journal of Tuberculosis. 2008;**55**(4):179-187

[46] Guruprasad KG, Niranjan MR, Ashwin S. A study of association of depressive symptoms among the type 2 diabetic outpatients presenting to a tertiary care hospital. Indian Journal of Psychological Medicine. 2012;**34**:30-33

[47] Poongothai S, Anjana RM, Pradeepa R, Ganesan A, Umapathy N, Mohan V. Prevalence of depression in relation to glucose intolerance in urban south Indians: The Chennai Urban Rural Epidemiology Study (CURES-76). Diabetes Technology & Therapeutics. 2010;**12**:989-994

[48] Prasad R, Garg R, Verma SK. Isoniazid- and ethambutol-induced psychosis. Annals of Thoracic Medicine. 2008;**3**(4):149-151

[49] Martin SJ, Bowden FJ. Ethambutol toxicity manifesting as acute onset psychosis. International Journal of STD and AIDS. 2007;**18**(4):287-288

[50] Kennedy NA, Oluwaseun A, Denis AA, Chukwuemeka SP. Cycloserine induced-psychosis in a 22-year old male pharmacy student: A case report. American Journal of Psychiatry and Neuroscience. 2016;**1**:124-126

[51] Rabkin JG, Ferrando S, Lin SH, Sewell M, McElhiney M. Psychological effects of HAART: A 2-year study. Psychosoatic Medicine. 2000;**62**:413-422

[52] Wells K, Golding J, Burman M. Psychiatric disordesr In a sample of the general population with and without chronic medical conditions. American Journal of Psychiatry. 1988;**145**:976-981

[53] Ellen SR, Judd FK, Mijch AM, Cockram A. Secondary mania in patients with HIV infection. Australian & New Zealand Journal of Psychiatry. 1999;**33**:353-360

[54] Venugopal D, Patil TB, Gupta D, Murl M, Kar N, Sharam PSVN. AIDS related mania. Indian Journal of Psychiatry. 2001;**43**:242-245

[55] Benatar SR. Prospects for global health: Lessons from tuberculosis. Thorax. 1995;**50**(5): 487-489

[56] Grange JM, Festenstein F. The human dimension of tuberculosis control. Tubercle and Lung Disease. 1993;**74**(4):219-222

[57] Lerman I, Lozano L, Villa AR, et al. Psychosocial factors associated with poor diabetes self-care management in a specialized center in Mexico City. Biomedicine & Pharmacotherapy. 2004;**58**:566-570

[58] Lin EH, Katon W, Von Korff M, et al. Relationship of depression and diabetes self-care, medication adherence, and preventive care. Diabetes Care. 2004;**27**:2154-2160

[59] Bruce DG, Davis WA, Davis TM. Longitudinal predictors of reduced mobility and physical disability in patients with type 2 diabetes: The Fremantle diabetes study. Diabetes Care. 2005;**28**:2441-2447. [PubMed]

[60] Watters JK, Estilio MJ, Clark GL, Lorvick, J. Syringe and needle exchange as HIV/AIDS prevention for injection drug users. Journal of the American Medical Association. 1994;**271**:115-120

[61] Baum N, Rafle C, Lal S, Sales S., Page J, Campa A. Alcohol use accelerates HIV disease progression. AIDS Research and Human Retroviruses. 2010;**26**:511-518

[62] Kresina T, Fexner C, Sinclair J, et al. Alcohol use and HIV pharmacotherapy. AIDS Research and Human Retroviruses. 2002;**18**:757-770

[63] Peltzer K, Louw J, Mchunu G, Naidoo P, Matseke G, Tutshana B. Hazardous and harmful alcohol use and associated factors in tuberculosis public primary care patients in South Africa. International Journal of Environmental Research and Public Health. 2012;**9**(9): 3245-3257

[64] Bergner M. Quality of life, health status, and clinical research. Concepts of Health-related quality of life. Medical Care. 1989;**27**(Suppl):S148-S156

[65] Levine S, Croog S. What constitutes quality of life? A conceptualization of the dimensions of life quality in healthy Population and patients with cardiovascular disease. In: Wenger N, Mattson ME, Furgerg CD, Elinson J, editors. Assessment of Quality of Life in Clinical Trials of Cardiovascular Therapies. New York: Le Jacq; 1998. pp. 46-58

[66] Robinson S, Young T, Roos L. Estimating the burden of disease. Comparing administrative data and self-reports. Medical Care. 1997;**35**(9):932-947

[67] Patrick DL, Erickson P, editors. Concepts of health-related quality of life. In: Health Status and Health Policy: Quality of Life in Health Care Evaluation and Resource Allocation. New York: Oxford University Press; 1993. pp. 76-112

[68] Testa MA, Hollenberg Anderson RA, Williams GH. Assessment of quality of life by patient and spouse during antihypertensive therapy with atenolol and nifedipine gastrointestinal therapeutic system. American Journal of Hypertension. 1991;**4**:363-373

[69] Oliva J, Fernández-Bolaños A, Hidalgo A. Health-related quality of life in diabetic people with different vascular risk. BMC Public Health. 2012;**12**:812

[70] Ahari SS, Arshi S, Iranparvar M, Amani F, Siahpush H. The effects of type II diabetes on quality of life. Journal of Ardebil University of Medical Sciences. 2008;**8**(4):394-402

[71] Ahmadi A, Hasanzadeh J, Mediseh MR, Lashkari L: Factors affecting quality of life in patients with type 2 diabetes in Chaharmahal & Bakhtiari province. Journal of North Khorasan University of Medical Sciences. 2011;**3**(1):7-13

[72] Borzou R, Salavati M, Safari M, Hadadinejad SH, Zandieh M, Torkman B: Quality of life in type II diabetic patients referred to Sina Hospital, Hamadan. Zahedan Journal of Research in Medical Sciences. 2010;**13**(4):43-46

[73] Haririan H, Moghadasian S, Aghajanlou A. Quality of life and its dimensions in diabetic patients referred to diabetes center of Tabriz medical university. Iranian Journal of Diabetes and Lipid Disorders. 2009;**9**(2):152-160

[74] Brown J, Capocci S, Smith C, Morris S, Abubakar I, Lipman M. Health status and quality of life in tuberculosis. International Journal of Infectious Diseases. 2015;**32**:68-75

[75] Rajeswari R, Muniyandi M, Balasubramanian R, Narayanan PR. Perceptions of tuberculosis patients about their physical, mental and social well-being: a field report from south India. Social Science & Medicine. 2005;**60**(8):1845-1853. DOI: 10.1016/j.socscimed.2004.08.024

[76] Adeyeye OO, Ogunleye OO, Coker A, Kuyinu Y, Bamisile RT, Ekrikpo U, et al. Factors influencing quality of life and predictors of low quality of life scores in patients on treatment for pulmonary tuberculosis: A cross sectional study. Journal of Public Health in Africa. 2014;**5**(2). DOI: 10.4081/jphia.2014.366

[77] Balgude A, Sontakke S. Study of impact of antitubercular therapy on quality of life. Indian Journal of Medical Sciences. 2012;**66**(3-4):71-77

[78] Aggarwal AN. Health-related quality of life: A neglected aspect of pulmonary tuberculosis. Lung India. 2010;**27**(1):1-3

[79] Chamla D. The assessment of patients' health-related quality of life during tuberculosis treatment in Wuhan, China. International Journal of Tuberculosis and Lung Disease. 2004;**8**(9):1100-1106

[80] Marra CA, Marra F, Cox VC, Palepu A, Fitzgerald JM. Factors influencing quality of life in patients with active tuberculosis. Health and Quality of Life Outcomes. 2004;**2**:58. DOI: 10.1186/1477-7525-2-58

[81] Marra CA, Marra F, Colley L, Moadebi S, Elwood RK, Fitzgerald JM. Health-related quality of life trajectories among adults with tuberculosis: Differences between latent and active infection. Chest. 2008;**133**(2):396-403

[82] Louw J, Peltzer K, Naidoo P, Matseke G, McHunu G, Tutshana B. Quality of life among tuberculosis (TB), TB retreatment and/or TB-HIV co-infected primary public health care patients in three districts in South Africa. Health and Quality of Life Outcomes. 2012;**10**:77. DOI: 10.1186/1477-7525-10-77

[83] KatonWJ. Epidemiology and treatment of depression in patients with chronic medical illness. Dialogues in Clinical Neuroscience. 2011;**13**:7-23

[84] Katon W, Lin EH, Kroenke K. The association of depression and anxiety with medical symptom burden in patients with chronic medical illness. General Hospital Psychiatry. 2007;**29**:147-155

[85] Chandra PS, Deepthi V, Jairam KR, Thomas T. Relationship of Psychological morbidity and quality of life to illness-related disclosure among HIV-infected persons. Journal of Psychosomatic Research. 2003;**54**:199-203

[86] Ii W, Liu L, Purnte JG, Li Y, Jiang X, Jin S, Ma. H, Kong L, Ma L, He X, Ma S, Chen C. Hypertension and Health related quality of life: An epidemiological study in Patients attending hospital clinic in China. Journal of Hypertension. 2005;**23**(9):1667-1676

[87] Olusina AK, Ohaeri JU, Subjective quality of life of recently discharged Nigerian Psychiatric patients. Social Psychiatry and Psychiatric Epidemiology. 2003;**38**(12):707-704

[88] Ustun TB, Ayuso-Mateos JL, Chatterji S, et al. Global burden of co-morbid depression in adults with Type 1 diabetes; systematic literature review. Diabetic Medicine. 2006;**23**(4); 445-448

[89] Black SA, Markides KS, Ray LA. Depression predicts increased incidence of adverse health outcomes in older Mexican Americans with type 2 diabetes. Diabetes Care. 2003;**26**:28222828

[90] Pierdomenico SD, Di Nicola M, Esposito AL, et al. Prognostic value of different indices of blood pressure variability in hypertensive patients. American Journal of Hypertension. 2009;**22**(8):842-847

[91] Crosby G, Stall R, Paul J, Barrett D. Substances use and HIV risk profile of gay/bisexual males who drop substance abuse treatment. AIDS Education and Prevention. 2000;**12**:38-48

[92] WHOQOL-HIV Group. Preliminary development of the World Health Organization's Quality of Life HIV instrument (WHOQOL-HIV): Analysis of pilot version. Social Science & Medicine. 2003;**57**:1259-1275

[93] Hiyasu Z, Kabir M, Abubarkar IS, Babashani M, Zubair ZA. Compliance to antiretroviral therapy among AIDSs patients in Aminu Kano Teaching Hospital, Kano, Nigeria. Nigerian Journal of Medicine. 2005;**14**:290-295

[94] Adamson TA, and Sijuwola, OA. Ilewo-Orile. Nigerian Epidemiological Study: Psychiatric morbidity in a rural community using the 12-item GHO. Nigerian Journal of Psychiatry. 2001;**1**(5):315-321

[95] Peyrot M, Rubin RR. Levels and risks of depression and anxiety symptomatology among diabetic adults. Diabetes Care. 1997;**20**:585-590

[96] Kaholokula JK, Haynes SN, Grandinetti A, Chang HK. Biological, psychosocial, and sociodemographic variables associated with depressive symptoms in persons with type 2 diabetes. Journal of Behavioral Medicine. 2003;**26**:435-458

[97] Owoaje EE, Rotimi CN, Kaufman JS, Tracy J, Cooper RS. Prevalence of adult diabetes in Ibadan, Nigeria. East African Medical Journal. 1997;**74**:299-302

[98] Ogunrin AO, Odiase FE, Ogunniyi A. Reaction time in patients with HIV/AIDS and correlation with CD4 count: A case-control study. Transactions of the Royal Society of Tropical Medicine and Hygiene. 2007;**101**(5):517-522

[99] Levy A, Laska F, Abelhauser A, Delfraissy JE, Goujard C, Bowe F, et al. Disclosure of HIV seropositivity. Journal of Clinical Psychology. 1999;**55**:1041-1049

[100] Dressler WW. Social and cultural influences in cardiovascular disease: A review trans-cultural. Psychiatric Research Review. 1984;**21**:1

[101] Taghidisi MH, Borhani M, Solhi M, Afkari ME, Hosseini F. The effect of an education program utilizing PRECEDE model on the quality of life in patients with type 2 diabetes. Health Education Journal. 2012;**71**(2):229-238

[102] Chandra PS, Deepthivarma S, Manjula V. Disclosure of HIV infection in South India: Patterns, reasons and reactions. AIDS Care. 2003;**15**:207-215

[103] Boardman J, Petrusthkin H, Ovuga E. Psychiatric disorders in HIV Positive Individuals in Urban Uganda. British Journal of Psychiatry. 2005;**29**:455-460

[104] Jacob KS, Eapen V, John JK. Psychiatric morbidity in HIV infected individuals. Indian Journal of Medical Research. 1991;**93**:62-66

[105] Johnson J, Williams J, Rabkin J, et al. Axis 1 psychiatric symptomatology associated with HIV infection and personality disorder. American Journal of Psychiatry. 1995;**152**:551-554

[106] Aamir S, AishaComorbid anxiety and depression among TB patients. Journal of College of Physicians and Surgeon Pakistan. 2010;**20**(10):703-704

[107] Lopez AG. Tuberculosis and the severely mentally ill. American Journal of Psychiatry. 1994;**151**(1):151-152

[108] Olisah VO, Biayewu O, Sheikh TL. Depression Underdiagnosis and the effects on quality of life in outpatients with HIV at a Nigerian University Teaching Hospital. African Journal of AIDS Research. 2011;**10**(3):247-254

[109] Morrison MF, Prtitto JM, Ten Hanc T, Grttes DR, Chiappini MS, Weber AL, et al. Depressive and anxiety disorders in women with HIV infection. American Journal of Psychiatry. 2002;**159**:789-796

[110] Chandra PS, Ravi V, Desai A. Anxiety and depression among HIV-infected heterosexuals- A report from India. Journal of Psychosomatic Research. 1998;**45**:401-409

[111] Barllett J. Addressing the challenges of adherence. Journal of Acquired Immune Deficiency Syndromes. 2002;**29**(Suppl 1)

[112] Bagchi S, Ambe G, Sathiakumar N. Determinants of poor adherence to anti-tuberculosis treatment in Mumbai, India. International Journal of Preventive Medicine. 2010;**1**(4):223-232

[113] Manoharam E, John KR, Joseph A, Jacob KS. Psychiatric morbidity, patients' perspectives of illness and factors associated with poor medication compliance among the tuberculous in Vellore, south India. Indian Journal of Tuberculosis. 2001;**48**:77-80

[114] Lustman PJ, Anderson RJ, Freeedland KE, et al. Depression and poor glycemic control: A meta-analytic review of the literature. Diabetes Care. 2000;**23**:934-942

[115] Patel V. Mental health in low- and middle-income countries. British Medical Bulletin. 2007;**81-82**(1):81-96

[116] Slama K, Chiang CY, Enarson DA, et al. Tobacco and tuberculosis: a qualitative systematic review and meta-analysis. International Journal of Tuberculosis and Lung Disease. 2007;**11**(10):1049-1061

[117] Chandra PS, Ravi V. Somatization disorder among HIV seronegative men. Indian Journal of Psychiatry. 1994;**37**:41-42

[118] Nakimuli-Mpungu E, MusIsl S, Mpungu S, Katabira E. Clinical presentation of bipolar mania in HIV-positive patients. Uganda Psychosomatics. 2009;**50**:325-330

[119] Martinez A, Israleski D, Walker C. Posttraumatic stress disorders in women attending human immunodeficiency virus outpatient clinics. AIDS Patient Care STDS. 2002;**16**:283-289

[120] Liu C, Johnson l, Ostrow D, Silvestre A, Visscher B, et al. Predictors of lower quality of life in the HAART era among HIV infected men. Journal of Acquired Immune Deficiency Syndromes. 2006;**24**(4):470-477

[121] Goldney RD, Phillips PJ, Fisher LJ, Wilson DH. Diabetes, depression, and quality of life—a population study. Diabetes Care. 2004;**27**:1066-1070

[122] Larsson D, Lager I, Nilsson PM. Socio-economic characteristics and quality of life in diabetes mellitus / relation to metabolic control. Scandinavian Journal of Public Health. 1999;**27**:101-105

[123] Okunoye OC, Asekomeh GE, Iyagba MA, Owuchekwa AC. Qualiti of life of people living with HIV/AIDS in Port Harcourt. The Nigerian Health Journal. 2014;**14**:8-12

[124] Nor K, Tlou S, Norr J. The threat of AIDS for women in developing countries. In: Cohen F, Durham JD, editors. Women Children and HIV. New York: Springer Publishing Co; 1993. p. 263

[125] Demi A, Bakeman R, Richard S, Linda M, Brenda S. Suicidal thoughts of women with HIV infection: Effect of stressors and moderating effects of family cohesion. Journal of Family Psychology. 1998;**12**:344-353

Leisure Time Activities as Cognitive Training: Can Cognitive Fitness Improve Well-Being of Older Adults?

Andrea Vranic

Additional information is available at the end of the chapter

http://dx.doi.org/10.5772/67929

Abstract

Economic reports show that traditional economic variables are becoming insufficient in indicating nation's progress and that self-reported well-being is emerging as an important indicator of nation's prosperity and quality of life of its inhabitants. Since average life expectancy is consistently increasing all over the world, practitioners and scholars are increasingly confronted with the question of what can be done to improve the well-being of growing elderly population. An often highlighted characteristic of older adults is cognitive decline, and interventions aimed at improving or maintaining cognitive abilities of elderly are being extensively designed. In this chapter we will review studies indicating a link between cognition and well-being in aging and advocate (long-term) involvement in leisure activities as a form of cognitive training. Apart from its availability to a much wider audience than those participating in cognitive interventions, a key argument in favor of leisure activities is its unambiguous positive association with well-being.

Keywords: well-being, cognition, aging, leisure activity, cognitive training

1. Introduction

All over the world the population is aging. Essentially every country in the world is faced with the expanding number and growing portion of its older inhabitants. Within the next 30 years, the number of persons aged 60 years or over is expected to more than double, reaching to more than 2 billion in 2050. Moreover, average life expectancy is consistently increasing, i.e., older population is itself aging—the share of people over 80 (the "oldest old") within the older population was 14% in 2013 and is projected to reach 19% in 2050 [1]. This so-called graying of the population will most certainly be the number one trademark of the twenty-first century's societal changes. Furthermore, due to the specificities of older population, worldwide aging will affect probably all sectors of society, from family structure to labor market, as well as goods and services needed. A growing field of gerontology has thus a main aim of providing societal decision-makers with advices on creating retirement and social security policies adequate for the growing size of aging population and on ensuring conditions, which support and promote successful aging, i.e., appropriate housing, family relations, and free time of the elderly. What we usually refer to as successful aging is the set of individual and societal conditions under which older people get a maximum of satisfaction and happiness, while the society maintains balance among satisfaction of all age and gender groups [2].

In psychology, life satisfaction and happiness are studied under the umbrella term of well-being [3]. Other componential constructs include optimal experience and functioning [4]. Although seemingly a complex construct, well-being has been repeatedly associated with a number of health, family, work, and economic benefits, for example, decreased risk of disease [5, 6], better immune functioning [7], better coping and speedier recovery [8], and increased longevity [9]. Well-being is also associated with more job-related success and higher incomes, and individuals with higher well-being are more likely to engage in community projects [10]. Globally speaking, countries with higher average well-being are wealthier and have better civil liberties, more equality, and more governmental stability and political engagement [11]. It is of no surprise that well-being of the population has become almost a societal imperative and public policy makers are getting more and more concerned with this topic. Economic reports show that traditional economic variables, such as gross domestic product, average income, or housing conditions, are becoming insufficient in indicating nation's progress and that self-reported well-being is emerging as a variable that must be taken into account [12].

What certainly comes to mind when mentioning old population is the decline of their cognitive capabilities. Until recently, findings of cognitive decline have been relatively unified and disheartening. Beginning of the 1990s, efforts to improve cognitive functioning of healthy older adult have begun. Evidence is accumulating that cognitive training—a therapeutic intervention aimed at empowering cognition—might actually improve cognitive performance and slow down the inevitable cognitive decline. There are now quite robust findings of variability and trainability of aspects of cognition in the old age [13, 14]. Consequently, the topic of understanding and promoting successful cognitive aging is granted high scholarly and practical importance. Since older adults are usually faced with more spare time in the years following their retirement, an important direction in promoting overall well-being of elderly is to identify activities in which they might participate and that might improve their functional status and overall quality of life.

In sum, well-being of elderly is obviously becoming a major societal concern, and upsurge of interest in interventions and activities that could contribute to the life satisfaction and happiness of older people is of no surprise. In this chapter we will (1) give a theoretical overview of well-being research and recapitulate findings on the sociodemographic correlates of well-being, most specifically age; (2) offer accounts on "successful aging," with a focus on cognitive aging; (3) present evidence linking cognition and well-being of elderly; (4) recapitulate benefits of cognitive training programs for elderly; and, finally, (5) advocate leisure activities as a widely available platform to train cognition.

2. Well-being in psychological research

National surveys measuring well-being are becoming increasingly popular and are being carried out worldwide [15, 16]. Measurement of well-being is a focus of debate itself, and the issue has raised considerable controversies. This controversy stems partly from how "optimal functioning" and "good life" are defined at the level of an individual and is exponentially gaining complexity as we go beyond individual level toward various, cultures, economies, and personalities. Since the 1960s, when the movement of human potentials has spread throughout the Western countries, in which psychological advancements of the time were mostly nested, research on well-being has generated two distinct philosophies, labeled hedonism and eudaimonism.

Hedonism mirrors the notion of well-being, which is built around subjectively pleasant and positive everyday experiences. Hedonic well-being is measured by self-reports in which respondents rate their experiences of positive and negative affect operationalized through adjectives, such as happy, sad, and worried. It is not reduced to physical pleasure and preferences, but it can also be achieved by realization of valued and imported personal goals and aspirations. Despite numerous ways by which this pleasure/pain continuum can be assessed, most research within the hedonic approach has used the construct subjective well-being (SWB). SWB deals with how and why people experience their lives positively. It includes affect (positive and negative) and cognitive judgments of life satisfaction, and it is considered a valuable indicator of optimal functioning [17, 18]. During the past two and a half decades, SWB has become a primary index of well-being and is often employed as a major outcome variable in public health studies and medical and epidemiological testing of diseases and risk factor prevention [4, 19].

Eudaimonism expresses the belief that well-being requires actualization of one's potential. Being well is about fulfilling one's true nature—daimon [20]—and it is distinct from mere happiness. From eudaimonic standpoint, happiness cannot be equaled with well-being because, although maybe pleasure producing, not all outcomes yield achievement of well-being [4]. Realization of eudemonic happiness, and/or well-being, happens when our values are intertwined with our activities and when one is living an authentic life. In operationalizing eudemonic well-being, researchers focus on judgments about meaning and purpose of one's life, called psychological well-being (PWB), and view it as distinct from SWB. Since factors promoting SWB do not necessarily overlap with does yielding PWB, assessment of PWB is cognitively more engaging and demands considerable reflection about self-actualization, vitality, and mental health. PWB is a more diverse construct, and a multimodal approach to its measurement has been proposed. Ryff and Singer (2000), for example, specify six aspects of human actualization: autonomy, personal growth, self-acceptance, life purpose, mastery, and positive relatedness. Also, they report evidence that eudemonic well-being might influence physiological systems and consequently promote immunological functioning and overall health [21].

Within these two approaches, a number of measurement instruments have been proposed and are widely used. Kobau et al. [19] conducted an extensive study to examine descriptive and psychometric properties of widely used scales and to examine the level of well-being in a representative sample of community-dwelling adults. The study covered (1) satisfaction with life [22]; (2) meaning in life [23]; (3) basic psychological needs of autonomy, competence, and relatedness [24]; (4) domain-specific life satisfaction [25]; and (5) select positive and negative affect items [26]. The scales demonstrated acceptable psychometric properties, and the results confirmed previous findings of preponderance of mildly positive level of SWB in adults [11, 27]. Except for the autonomy, competence, and relatedness scales, all other scales showed good variability and expected differences across sociodemographic subgroups.

2.1. Sociodemographic correlates of well-being

Sociodemographic picture of well-being, as seen through the lenses of psychological research, is more or less clear. In a stratified sample of almost 5500 respondents, Kobau et al. [19] tested for the well-being differences in gender, race/ethnicity, educational level, and household income. Men and women generally reported similar degree of SWB or PWB, i.e., exhibited similar levels of meaning in life, positive or negative affect, global happiness, autonomy, and competence. Women exhibit higher satisfaction with spiritual, religious, and philosophical beliefs, as well as with housing, family and social life, and their ability to help others. Men, however, reported of being more satisfied with their energy level than women did. Also, only slight differences in well-being domains were found across race/ethnicity. Higher educational level is found related to higher well-being across life domains. Respondents of low-level education experienced similar level of positive affect as those highly educated, yet they reported of experiencing significantly more negative affect and lower levels of happiness. In terms of household income, households with lower incomes are generally associated with lower levels of well-being. It seems that income provides clear advantages, with high-income respondents reporting more satisfaction with life, while lower-income respondents generally report lower levels of satisfaction across all life domains.

What does lifespan perspective say about well-being? The U-shaped association between age and well-being is usually taken as standard finding with the most salient finding being the so-called paradox of aging [4]. Not only does well-being not decline, but it usually increases, despite many challenges and losses experienced in the old age. Still, depending on the approach and instruments used, the findings from various studies reveal a relatively dynamic age-related trajectory of well-being. When a more descriptive approach to lifespan changes in well-being is taken, interesting theoretical and methodological issues arise. Namely, when a multidimensional way of measuring well-being is applied, studies show that age is related to (1) people's conception of well-being and (2) content of life aspects contributing to well-being. Young adults draw their well-being from the perceived self-knowledge and competence and are more involved in gaining new experiences. Older adults draw their satisfaction from experiences of positive coping with change, as well as depth and concentration directed toward tasks at hand. Carstensen et al. [28] proposed a socio-emotional selectivity theory to explain these findings. They argue that age is associated with increasing motivation to extract emotional meaning from life and decreasing motivation to expand one's horizons. As they age, people accumulate emotional wisdom, which leads them to select more emotionally gratifying events, relationship, and experiences. By limiting their

http://dx.doi.org/10.5772/67929

set of social contacts and experiences, despite of deteriorating health, lowered income, and losses related to deaths and retirement, older people maintain and even increase their self-reported well-being.

3. Aging research: is age just a number?

When speaking of aging, we usually refer to the physiological and behavioral changes leading to senescence. Disciplines dealing with senescence, such as sociology, biology, and psychology, are not consistent in their definitions and criteria of old age, yet ages 60–65 are most often marked as the beginning of the old age. As the population is "graying," suggested phases of aging are being more and more refined. The rule of thumb in categorization of age cohorts is three life stages: the young-old (approximately 65–74), the middle-old (ages 75–84), and the old-old (over age 85). Yet criteria of old age are very broad and subjective—it can be the loss of reproductive ability, attainment of wisdom, or retirement. Although highly correlated with age, at an individual level, these indicators of old age may arrive at very different ages. Such a broad range of interindividual differences in older adults proves that chronological age is not a reliable guide in understanding the aging process.

Therefore, today age is held as one of many variables accounting for interindividual differences among elderly. As people move through the lifespan, adult development is less and less under control of physical ontogeny. In adulthood, the information of one's age tells relatively little besides the fact that one has lived for a certain number of years. Knowing solely one's age is informative neither of one's health condition, cognitive status, physiological state, socioeconomic status, nor their lifestyle. Other contributing variables include various physical factors, such as exercise or nutrition [29], psychological health (e.g., Ref. [30]), social factors [31], and various lifestyle factors [32].

Probably the best-documented changes related to aging come from physiology. Aging heart becomes more vulnerable to disease; vision and hearing undergo qualitative changes and so do the skeletal and endocrine systems. Advancements in medicine, throughout the last century, have cushioned much of age-related changes and bolstered longevity. Psychological aspects of aging have not captured researchers' attention until much later—beginnings of organized research in psychology of aging are set in the mid-1950s. Psychology defines aging as a "result of ecological relationships – a particular genetic background is expressed in particular social and physical environments and modified by the strategic capacities of the individual" [33, pg. 3]. In terms of psychological characteristics of aging, the most studied ones come from the domain of cognition.

3.1. Cognitive aging: cognition in older age

What is known about everyday cognitive abilities along adult lifespan? Over the last decades, the magnitude of cognition-related age differences reported in the literature has shrunk. This is likely due to a more systematic investigation of the influences of other concomitant variables that could account for a significant share of variance that was previously attributed to chronological age. Human mind, as suggested by lifespan psychology, is viewed as a complex system composed of many intellectual abilities that develop with different rates and trajectories, and lifespan changes in cognitive ability should be considered in a differential manner [34].

The most influential and certainly still the most resilient perspective on cognitive abilities in aging is Cattell-Horn's legacy of fluid and crystallized intelligence [35, 36]. Fluid and crystallized intelligence represent a categorization of numerous discrete abilities into two different sets of abilities with different trajectories over the course of development—while crystallized abilities incrementally improve throughout life, fluid abilities peak at around 20 years of age and start declining after 40, with decline progressing after the age of 65. And indeed, some abilities, mostly those broadly termed fluid, decrease with age, such as processing speed [37], working memory capacity [38], associative memory [39], executive functioning [40, 41], fluid intelligence, and reasoning (e.g., Ref. [42]). On the other hand, abilities associated with experience, cultural and social processes, and measures of crystallized intelligence, for example, vocabulary, remain preserved even in very old age [43].

However, if a more differential approach is taken and age differences in various tasks are decomposed depending on the contribution of content, strategy use, or other more reality-related criteria to the efficacy of executing the task, findings show a slightly different perspective. For example, problem solving is often found to show marked age differences with young excelling the old. Yet when problems are analyzed in terms of strategies used to solve them or social and emotional impact a solution of the problem, performance is stable throughout lifespan and sometimes even improved in late adulthood [44]. It seems that young adults excel older ones in problem solving only when tasks are based on fluency or involve single solution [45]. In other words, human development can be seen as a continuous and dynamic interplay of cognitive gains and losses. To underline this thesis, Baltes et al. [46] showed that older adults can profoundly profit from guided practice in tasks or when they are shown and taught strategies for problem solving. In a sample of healthy older adults, they found improvement in fluid ability of healthy older participants when tutor-guided training was provided. A rearview mirror view would probably reveal this study as the beginning of the end of Cattell-Horn's era of innate abilities and the dawning of utility hypothesis of cognitive abilities which is slowly, but steadily, getting recognition under the auspices of *use-it-or-lose-it* hypothesis (e.g., Ref. [47]).

4. The relatedness of cognition and well-being: hot or not?

In recent years, a body of evidence suggesting that well-being might be a potential resource for healthier aging is growing. Older people with higher well-being are less likely to develop mobility problems or other activities of daily life [48]. High positive affect seems to lower the risk of frailty [49]. However, it is unclear whether positive well-being might act protectively with regard to cognitive aging, another important component of healthy and successful aging [50]. Some recent cross-sectional analyses of several older age cohorts have found small to

http://dx.doi.org/10.5772/67929

moderate positive association between well-being and cognition [51, 52]. Studies have found links of stronger sense of perceived control, a component of eudaimonic well-being, with higher scores in memory performance [53, 54].

Enkvist et al. [55] tested six cognitive domains (executive function, processing speed, episodic memory, semantic memory, spatial ability, and working memory) in a sample of oldest old (aged 78–98), and after the adjustment conducted for potential confounding factors including depression, processing speed and spatial ability were positively associated with life satisfaction 3 years later. Somewhat stronger evidence that poorer fluid abilities might have a detrimental effect on well-being is found in the study investigating whether fluid cognitive ability predicts exposure and emotional reactivity to daily stressors [56]. Findings showed that higher levels of fluid ability were associated with smaller stressor-related increases in negative mood and smaller stressor-related decreases in positive mood, suggesting that cognitively better-off individuals may be more emotionally resilient in the face of daily stress. Allerhand et al. [42] used multilevel modeling to estimate association between cognitive function and positive well-being in four waves of data, collected on a sample of over 10,000 participants, aged 50–90. They found that, although most variation in cognitive function was explained by age and most variation in well-being was explained by depression, small but significant associations between cognition and well-being remained after variation in age and depression were controlled.

It remains unclear whether these significant cross-sectional associations reflect the effect of well-being on cognition or vice versa. It might also be plausible that the relationship between positive well-being and cognitive function is bidirectional. For example, cognitive success in younger age might lead to higher self-efficacy and feeling of mastery, which in turn leads to an increase in cognitive appraisal of satisfaction with life, i.e., increase in well-being. On the other hand, at older ages impaired cognition may constrain managing of daily activities and hence cause detrimental well-being. Studies suggest that cognitive decline leads to diminished well-being, more in terms of its eudaimonic than hedonic aspects [57]. A more rapid cognitive decline in a 5-year period preceding well-being assessment, as measured by Scales of Psychological Well-being [58], was associated with lower level of nearly all aspects of well-being. The extent of the association varied across well-being dimensions. Also, rate of decline of episodic memory, semantic memory, and perceptual speed was associated with rate of decline in purpose in life, and the association between rates of decline in working memory and purpose in life was of borderline significance. A prospective epidemiological study of community-dwelling elderly has also found that greater purpose in life is associated with a reduced risk of incident Alzheimer's disease and mild cognitive impairment [59].

People's cognitive responses to various life events are also associated with different levels of well-being. Happier and more satisfied individuals are more characterized by optimistic strategies and biases when facing different outcomes. They tend to perceive life's circumstances positively [60], expect favorable future [61], experience more internality in terms of control [62], and are confident about their strengths and skills [63]. Inverse relation to well-being is found in inclination to encode negative aspects of events and ruminate about one's problems [64].

Further elaboration associating cognition and well-being comes from the research on self-efficacy. Significant associations between well-being and self-efficacy have been reported [62], and self-efficacy is referred to as a strong predictor of subjective well-being and a mediator of the relation between personality and SWB [65]. Twin and family studies have found that genetic factors may account for 30–40% of the variance in SWB, leaving a fair share of variance of SWB under environmental influences [66]. Environmental factors can obviously play an important role in individual differences in SWB. Therefore, it is reasonable to assume that, partially at least, interventions aimed at various cognitive correlates of well-being could change the self-reported perception of well-being. In sum, an association between cognition and well-being is a sensible assumption, although it is still escaping firm explanations and mechanisms enabling it. In the following paragraphs, we will review types of cognitive interventions offered to elderly and then offer an account of arguments for leisure time activities as a form of cognitive training.

5. Cognitive training: how can cognition be trained?

Although nowadays research argues that age is one of many variables accounting for individual differences in cognition, it is still often taken as a cause of decline in cognitive abilities. Therefore, it is not surprising that practitioners and scholars are increasingly confronted with the question of what can be done to maintain cognitive abilities by postponing degenerative (yet normative) aging processes, thus improving the quality of life in elderly. Pushing the limits of what is currently known about sustaining cognitive functioning in the old age and about the ways of improving cognitive status of the elderly has become one of the most intriguing endeavors of psychologists, gerontologists, and cognitive scientists alike. Also, it is a pursuit which brings together not only scientists but also clinicians and entrepreneurs. In 2012, the brain-fitness market had total revenues of more than $1 billion, and forecasts for 2020 are settling around $6 billion [67]. The metaphors of the brain gym and workout for the mind have never been more appealing. Although cognitive training programs could be tailored to suit various age cohorts, aging baby boomers seem to be their number one consumers.

Similar to children, adults do not often function at the limits of their capacities; when placed in conditions that support and foster their higher-level functioning, resources are activated, but they are not spontaneously used [68]. In many areas of life, performance of older adults is below their actual competence level. This idea of an unused reserve, which is not activated because the situation is not requiring it, is partly described in the concept of cognitive reserve. Cognitive reserve refers to the brain's capacity to cope actively with neuropathological damage through the implementation of cognitive processes [69]. It is the ability to adapt to neural damage by employing alternative cognitive processes and/or strategies in compensation. In essence cognitive training procedures are based on this idea that the brain, even in old age, can change for the better. Findings of brain's plasticity [70], i.e., brain's ability to change and keep its vitality, suggest that it resembles muscles. As physical training can improve physical abilities and overall physical fitness, cognitive training can improve trained activities and cognitive status. Such training can take many shapes. The use of instruction and guided practice

http://dx.doi.org/10.5772/67929

on tasks related to specific cognitive abilities or cognitive appraisals of one's own ability has proven to be a key to success of such programs intended for the elderly. In the consecutive order of their development, these types of training are listed: strategy training, multimodal training, cardiovascular exercise, and process training [71].

Strategy training involves training strategies instrumental in increasing performance on trained tasks. These strategies most often used mnemonics and reported memory improvements for different mnemonics taught, ranging from imagery (d = .14), and peg-word (d = .62), to method of loci (d = .80), and organization (d = .85) [72]. Strategy trainings are featured by large and lasting effects on the trained task but are scarce in transfer effects, i.e., acquired mnemonics has limited generalizability to untrained tasks. As a response to these limited transfer abilities of strategy training, *multimodal trainings* were developed. These more complex interventions include cognitive and social component and sometimes even lifestyle changes. Multimodal training, as the name suggests, includes multiple modes of meaningful and joyful activities, for example, learning new skills which is cognitively demanding (e.g., board and card games, Sudoku, quilting). This approach offers wider transfer to different cognitive and other psychological domains (e.g., Ref. [73]), yet from a methodological point of view, it is difficult to determine which mode, or activity, was crucial or most beneficial for the improvement. *Cardiovascular training* features improvement in cognitive function of elderly via designated aerobic exercise. Greatest benefits of cardiovascular training are witnessed in executive function (g = .68 for flanker task) and are moderate for performance in speeded tasks (g = .43 for spatial or g = .27 for reaction time tasks) [74]. These trainings can have large and widespread effects, but are not necessarily accessible to persons with physical disabilities. Finally, *process training* trains specific processes, such as working memory [75] by exposing the participants to a "cognitive drill" on tasks tapping specific process. These trainings are promising, in terms of transfer effects, yet require rigorous task analysis to determine exactly which process is being trained. In applied sense, process trainings are known to be tiring and, consequently, demotivating, which in turn can have adverse effects on adherence.

Cognitive interventions are best delivered in a group format, although individual sessions and even self-help trainings [76] have been successfully implemented. Similar to real-life experiences, adaptive tasks are instrumental for training success, i.e., tasks which grow more challenging as the performance is getting better [72]. There is evidence that even relatively simple cognitive exercises can help to improve aspects of cognition. For example, older people can benefit from cognitive interventions in forms of computerized board games of chance [76]. Authors argue that the reason for this improvement lies in the newly acquired skill (i.e., computer use) and draw attention to higher feelings of mastery and self-efficacy as a potential reason behind the improvement. Other computer-based procedures come to similar conclusions. Whitlock et al. [77] confirm enhancement of well-being and life satisfaction after a multimodal game-based intervention and, in line with other studies on cognitive exercise and well-being, suggest a link between gaming and better emotional functioning and higher well-being. In sum, programs for the elderly which are now needed are those targeting multiple cognitive and physical functions because age-related decline is experienced in both of these areas. This decline is often accompanied by emotional difficulties, which might affect cognition. Thus, interventions which incorporate multiple components, such as regular physical and cognitive activity and stimulation, and are depression protective, are needed—most likely in midlife and earlier—to maximize their potential of slowing down cognitive decline in later age. Finally, cognitive training procedures may have practical values and benefits even if they do not change the rate of cognitive decline, in terms of statistical significance. Even if the rate of decline returns to normal after temporary improvement in the level of function due to the intervention, it may delay the point at which an individual reaches sufficient degrees of impairment to impact daily life in terms of dependency, housing, or constant medical care [71].

6. Leisure time activities: engagement that counts

Two contrasting, although not mutually exclusive, views have been proposed to account for the contribution of activities in successful aging. According to the *activity theory* [78], maintenance of activities and attitudes of middle age as far and as long as possible, in terms of one's age, is the key to successful aging. On the other hand, *disengagement theory* [79] suggests acceptance and willingness to disengage from active life as factors holding the key to aging successfully.

Ever since the 1960s, both of these theories have been a benchmark for testing various hypotheses explaining why some people age more adaptively, in terms of successfully standing up to the challenges which aging brings. They both hold an interdisciplinary approach on aging, taking into account not just medical/physiological data but also social, psychological, and interpersonal factors. Most importantly, they have both called scholars' attention to the positive and healthy aspects of aging instead of the usual studies on frailty and decline in relation to age. Furthermore, both theories apply a developmental perspective to late adulthood, a standpoint in which aging involves a progression through consecutive stages rather than a decline from middle adulthood toward the end of one's life [2]. Activity and disengagement theories have provided a fruitful theoretical platform for the study of aging and have extensively advanced our knowledge and appreciation of aging.

Studies identify three categories of leisure time activity as valuable constituents of "successful aging"—social engagement, physical exercise, and mental stimulation [50]. Many recent studies have found that social network size (number of people we see or meet in certain periods) is inversely related to the risk of cognitive decrement [80, 81]. Large social networks provide cognitive and physical engagement, which mediates cognitive decline, while preserved cognitive abilities are mostly a condition sine qua non for lasting social network. As for the physical component of leisure time activities, studies show that long-standing aerobic fitness leads to positive cognitive outcomes, even in old age, and cardiovascular trainings can improve cognitive performance in otherwise mostly sedentary older adults (e.g., Refs. [13, 82]). Cognitively demanding leisure activities, such as reading books and high-level journals and visiting plays and exhibitions, intuitively perpetrate higher cognitive abilities [83]. However, there are methodological issues involved in this intuition—it might be that highly functioning individuals seek complex, more cognitively demanding, leisure activities.

http://dx.doi.org/10.5772/67929

Interestingly, there are three lifestyle factors important in slowing the rate of cognitive decline: social network, regular physical activity, and cognitive leisure activities [84]. The integration of these aspects into neurorehabilitative methods is strongly recommended [71]. Combined interventions are often designed to be enjoyable or socially meaningful for older adults, increasing the chances that they will maintain the activities and skills even after the formal training period has ended. Compared with just physical or cognitive training programs, combined trainings provide significantly greater benefits in various cognitive functions of older adults [85, 86]. Given the similarity of training components contributing to training success and lifestyle factors important in slowing the rate of cognitive decline, it is justified to ask whether leisure activities could serve as a form of cognitive training.

An important aspect to bear in mind, in the discussion of leisure time activities as a "substitute" for cognitive trainings, is their relation to well-being. Overall, training and cognitive aging studies, although not extensively, provide evidence of a positive association between cognitive status and well-being in the old age. Literature on leisure activities almost unambiguously demonstrates that leisure activities can improve well-being. Well-being is positively associated with frequency of participation in enjoyable activities (e.g., different hobbies, reading, socializing) [87, 88]. The level of participation of older adults (age 60+) in leisure activity was found to predict life satisfaction 7 years later. A similar longitudinal finding is the one of older adults (age 72+) who volunteered in community work and other helping behaviors and were found to have greater life satisfaction 3 years later [89], while more hours of volunteering resulted in even greater benefit in well-being in older (age 60+) adults [90]. Overall, both cross-sectional and longitudinal evidences speak in favor of leisure time activities in terms of well-being. Long-term participation in leisure activities is not just protective of cognition in old age but is also regarded as a factor contributing to its improvement.

It seems that leisure time activities have some striking similarities to cognitive training program. These similarities certainly lie in factors proven to slow down the rate of cognitive decline and components of frequent leisure time activities in which elderly are involved are striking, namely, social network, physical activity, and cognitively demanding tasks. Also, just as cognitive trainings provide users with the sense of mastery and self-efficacy, especially if new skills are required through the course of the training, so do leisure time activities. Based on their facial validity, i.e., instantly seen pleasurable purpose of these activities, development of mastery and self-efficacy in leisure time activities is probably even more pronounced. Furthermore, leisure activities circumvent crucial shortcoming of cognitive trainings—they are mostly free and available to almost everyone. Cognitive training procedures are usually not offered to older population on a larger scale. They are usually based in validity studies and are at disposition to volunteers in research institution—university centers, geriatric units, or aging institutes. Web-based programs are becoming an economically available alternative, yet they still require resources and, after all, a home computer with Internet access. Finally, traditional leisure time activities are perceived as much more enjoyable than strategy trainings and extensive practice tasks, and this enjoyment might ensure for higher motivation and adherence of older participants in pursing them.

7. Concluding remarks

A successful example in which an initially leisure time activity, volunteering, was tested as a cognitive protection intervention [91]. Older adults (ages 60–86) worked with elementary school students in supportive interactive roles for 15 hours weekly over the school year. Physical activity, strength, reported social support networks, and cognitive activity significantly increased for these elders. This study is an example of targeting social engagement interventions to enhance cognitive aging for older adults and a further support of the theory that active involvement with society and engagement in meaningful activity are critical to "successful aging." More studies like this are needed to have a sound, psychometrically proven, argument of leisure activity being the No. 1 choice of cognitive training. However, such studies are usually not commercially or grant-wise interesting because the final products—a participation in activity—cannot be sold and there are no financial incentives; it is most often open to community dwellers in forms of various hobby centers, book clubs, or volunteering opportunities.

At the end, which is then the right way to healthy aging and cognitive vitality in the old age? Is it the activities, is it training programs either web-based or offered by research in others, or might it be both? Aging is a dynamic progress in which we win some and we lose some. Losses are mostly there by default; wins mostly require hard work. Studies on cognitive plasticity show that behavioral effects of experience are quite narrow and the effects of participation in activities show because engaged lifestyle pulls attentional resources to support abilities and maintain skills needed to pursue such lifestyle [92]. In other words, abilities are maintained through consistent use of exercising them. Exercise and engagement can be operationalized in various ways, but they usually involve challenging levels of activity situated in real-life contexts. So the question of activity or disengagement as a key to successful aging can best be rephrased by emphasizing choices we make about our activities and involvements in them. Such a claim is advanced in the Dumbledore hypothesis of cognitive aging—suggested by Elizabeth Stine-Morrow and inspired by the wise words of Dumbledore, a kind wizard in the Harry Potter saga—it is not our abilities that define us; it is our choices that show what we truly are [92]. What seems crucial for successful aging and higher well-being of elderly is to find the right balance between one's ability, choice of activities, and the level of engagement in these activities. Choosing to engage in leisure activities and adjusting this engagement to own aging abilities, yet making it inspiring enough to provide a mastery experience, which would lead to higher self-efficacy, seem to be the path to maximize our own potentials and be well.

Author details

Andrea Vranic

Address all correspondence to: avranic@ffzg.hr

Department of Psychology, University of Zagreb, Croatia

References

[1] United Nations Department of Economic and Social Affairs, Population Division. [Internet]. 2015. Available from: http://www.un.org/en/development/desa/population/ [Accessed: 2017-01-15]

[2] DeLiema, M, Bengston, VL. The history of disengagement, activity, and successful aging theories. In: Pachana NA, editor. Encyclopedia of Geropsychology. New York, NY: Springer; 2015.

[3] Diener E. Subjective well-being. Psychological Bulletin. 1984; 95:542-575. DOI: 10.1037/0033-2909.95.3.542

[4] Ryan RM, Deci EL. On happiness and human potentials: A review of research on hedonic and eudaimonic well-being. Annual Review of Psychology. 2001; 52:141-166. DOI: 10.1146/annurev.psych.52.1.141

[5] Ostir GV, Markides KS, Peek MK, Goodwin JS. The association between emotional well-being and the incidence of stroke in older adults. Psychosomatic Medicine. 2001; 63:210-215. DOI: 10.1097/00006842-200103000-00003

[6] Pressman SD, Cohen S. Does positive affect influence health. Psychological Bulletin. 2005; 131:925-971. DOI: 10.1037/0033-2909.131.6.925

[7] Lyubomirsky S, King L, Diener E. The benefits of frequent positive affect: Does happiness lead to success?. Psychological Bulletin. 2005; 131:803-855. DOI: 10.1037/0033-2909.131.6.803

[8] Tugade MM, Fredrickson BL, Feldman Barrett L. Psychological resilience and positive emotional granularity: Examining the benefits of positive emotions on coping and health. Journal of Personality. 2004; 72:1161-1190. DOI: 10.1111/j.1467-6494.2004.00294.x

[9] Kirkcaldy B, Furnham A. Positive affectivity, psychological well-being, accident- and traffic-deaths and suicide: An international comparison. Studia Psychologica. 2000; 42:97-104.

[10] Tov W, Diener E. The well-being of nations: Linking together trust, cooperation, and democracy. In: Sullivan BA, Snyder M, Sullivan JL, editors. Cooperation: The Political Psychology of Effective Human Interaction. Malden, MA: Blackwell Publishing; 2008. pp. 323-342. DOI: 10.1002/acp.1509

[11] Diener E, Suh EM. National differences in subjective well-being. In: Kahneman DE. Diener E, Schwarz N, editors. Well-being: The Foundations of Hedonic Psychology. New York: Russell Sage Foundation; 1999. pp. 434-452. DOI: 10.1017/s0953820806231972

[12] Steptoe A, Deaton A, Stone AA. Subjective wellbeing, health, and ageing. Lancet. 2015; 385:640-648. DOI: 10.1016/S0140-6736(13)61489-0

[13] Simone PM, Haas A. Cognition and leisure time activities of older adults. LLI Review. 2009; 4:22-28.

[14] Hedden T, Gabrieli JD. Insights into the ageing mind: A view from cognitive neuroscience. Nature Reviews Neuroscience. 2004; 5:87-96. DOI: 10.1038/nrn1323

[15] Stiglitz J. Report by the Commission on the Measurement of Economic Performance and Social Progress. [Internet]. 2009. Available from: https://www.ofce.sciences-po.fr/pdf/dtravail/WP2009-33.pdf [Accessed: 2017-01-15]

[16] Seaford C. Policy: Time to legislate for the good life. Nature. 2011; 477:532-533. DOI: 10.1038/477532a.

[17] Simone PM, Haas AL. Frailty, leisure activity and functional status in older adults: Relationship with subjective well-being. Clinical Gerontologist. 2013; 36:275-293. DOI: 10.1080/07317115.2013.788114

[18] Keyes CLM. Mental illness and/or mental health? Investigating axioms of the complete state model of health. Journal of Consulting and Clinical Psychology. 2005; 73:539-548. DOI: 10.1037/0022-006X.73.3.539

[19] Kobau R, Sniezek J, Zack MM, Lucas R, Burns A. Well-being assessment: An evaluation of well-being scales for public health and population estimates of well-being among US adults. Applied Psychology: Health and Well-Being. 2010; 2:272-297. DOI:10.111 1/j.1758-0854.2010.01035.

[20] Waterman AS. Two conceptions of happiness: Contrasts of personal expressiveness (eudaimonia) and hedonic enjoyment. Journal of Personality and Social Psychology. 1993; 64:678-691. DOI: 10.1037/0022-3514.64.4.678

[21] Ryff CD, Singer BH, Dienberg Love G. Positive health: Connecting well-being with biology. Philosophical Transactions of the Royal Society of London. Series B, Biological Sciences. 2004; 359:1383-1394. DOI: 10.1098/rstb.2004.1521

[22] Diener E, Emmons RA, Larsen, RJ, Griffin S. The satisfaction with life scale. Journal of Personality Assessment. 1985; 49:71-75. DOI: 10.1207/s15327752jpa 4901_13

[23] Steger MF, Frazier P, Oishi S, Kaler M. The meaning in life questionnaire: Assessing the presence of and search for meaning in life. Journal of Counseling Psychology. 2006; 53:80-93. DOI: 10.1037/0022-0167.53.1.80

[24] Deci EL, Ryan RM. The "what" and "why" of goal pursuits: Human needs and the self-determination of behavior. Psychological Inquiry. 2000; 11:227-268. DOI: 10.1207/S15327965PLI1104_01

[25] Diener E, Suh, EM, Lucas RE, Smith HE. Subjective well-being: Three decades of progress. Psychological Bulletin. 1999; 125:276-302. DOI: 10.1037/0033-2909.125.2.276

[26] Watson D, Clark LA, Tellegen A. Development and validation of brief measures of positive and negative affect: The PANAS scales. Journal of Personality and Social Psychology. 1988; 54:1063-1070. DOI: 10.1037/0022-3514.54.6.1063

[27] Frey BS, Stutzer A. Happiness and Economics. Princeton, NJ: Princeton University Press. 2002. 232 p. DOI: 10.1515/9781400829262

[28] Carstensen LL, Fung HH, Charles ST. Socioemotional selectivity theory and the regulation of emotion in the second half of life. Motivation and Emotion. 2003; 27:103-123. DOI: 10.1023/A:1024569803230

[29] Vaillant GE, Mukamal K. Successful aging. American Journal of Psychiatry. 2001; 158:839-847. DOI: 10.1176/appi.ajp.158.6.839

[30] Depp CA, Jeste, DV. Definitions and predictors of successful aging: A comprehensive review of larger quantitative studies. The American Journal of Geriatric Psychiatry. 2006; 14:6-20. DOI: 10.1097/01.JGP.0000192501.03069.bc

[31] Meng X, D'Arcy C. The projected effect of risk factor reduction on major depression incidence: A 16-year longitudinal Canadian cohort of the National Population Health Survey. Journal of Affective Disorders. 2014; 158:56-61. DOI: 10.1016/j.jad.2014.02.007

[32] Arenaza-Urquijo EM, Wirth M, Chételat G. Cognitive reserve and lifestyle: Moving towards preclinical Alzheimer's disease. Frontiers in Aging Neuroscience. 2015; 7:134. DOI: 10.3389/fnagi.2015.00134.

[33] Birren JE, Schroots JF. Autobiographical memory and the narrative self over the life span. In: Birren JE, Schaie KW, editors. Handbook of the Psychology of Aging. 6th ed. San Diego, Ca: Academic Press; 2006. pp. 477-498. DOI: 10.1016/b978-0-12-101264-9.x5000-9

[34] Schaie KW. Developmental Influences on Adult Intelligence: The Seattle Longitudinal Study. 2nd ed. New York, NY: Oxford University Press. 2013. 600 p. DOI: 10.1093/acprof:osobl/9780195386134.001.0001

[35] Cattell RB. Theory of fluid and crystallized intelligence: A critical experiment. Journal of Educational Psychology. 1963; 54:1-22. DOI: 10.1037/h0046743

[36] Horn JL. Fluid and Crystallized Intelligence: A Factor Analytic Study of the Structure Among Primary Mental Abilities [thesis]. Urbana-Champaign: University of Illinois; 1965.

[37] Kerchner GA, Racine CA, Hale S, Wilheim R, Laluz V, Miller BL, Kramer JH. Cognitive processing speed in older adults: Relationship with white matter integrity. PLoS One. 2012; 7: e50425. DOI: 10.1371/journal.pone.0050425

[38] Bopp KL, Verhaeghen P. Working memory and aging: Separating the effects of content and context. Psychology and Aging. 2009; 24:968-980. DOI: 10.1037/a0017731

[39] Naveh-Benjamin M, Brav TK, Levy O. The associative memory deficit of older adults: The role of strategy utilization. Psychology and Aging. 2007; 22:202-208. DOI: 10.1037/0882-7974.22.1.202

[40] Rhoades MG. Age-related differences in performance on the Wisconsin Card Sorting Test: A meta-analytic review. Psychology and Aging. 2004; 19:482-494. DOI: 10.1037/0882-7974.19.3.482

[41] Wasylyshyn C, Verhaeghan P, Sliwinski MJ. Aging and task switching: A meta-analysis. Psychology and Aging. 2011; 26:15-20. DOI: 10.1037/a0020912

[42] Allerhand M, Gale CR, Deary IJ. The dynamic relationship between cognitive function and positive well-being in older people: A prospective study using the English longitudinal study of aging. Psychology and Aging. 2014; 29:306-318. DOI: 10.1037/a0036551

[43] Craik FIM, Salthouse TA, editors. Handbook of Aging and Cognition. 3rd ed. Hillsdale, NJ: Lawrence Erlbaum Associates. 2000. 656 p.

[44] Blanchard-Fields F, Mienaltowski A, Seay RB. Age differences in everyday problem-solving effectiveness: Older adults select more effective strategies for interpersonal problems. Journal of Gerontology Series B: Psychological Sciences and Social Sciences. 2007; 62B:61-64. DOI: 10.1093/geronb/62.1.P61

[45] Mienaltowski A. Everyday problem solving across the adult life span: Solution diversity and efficacy. Annals of the New York Academy of Sciences. 2011; 1235:75-85. DOI:10.1111/j.1749-6632.2011.06207.x

[46] Baltes PB, Sowarka D, Kliegl R. Cognitive training research on fluid intelligence in old age: What can older adults achieve by themselves?. Psychology and Aging. 1989; 4:217-221. DOI: 10.1037/0882-7974.4.2.217

[47] Hultsch DF, Hertzog C, Small BJ, Dixon RA. Use it or lose it: Engaged lifestyle as a buffer of cognitive decline in aging?. Psychology and Aging. 1999; 14:245-263. DOI: 10.1037/0882-7974.14.2.245

[48] Love Collins A, Goldman N, Rodríguez G. Is positive well-being protective of mobility limitations among older adults?. Journal of Gerontology (Series B): Psychological and Social Sciences. 2008; 63:321-327. DOI: 10.1093/geronb/63.6.P321

[49] Ostir GV, Ottenbacher KJ, Markides KS. Onset of frailty in older adults and the protective role of positive affect. Psychology and Aging. 2004; 19:402-408. DOI: 10.1037/0882-7974.19.3.402

[50] Rowe JW, Kahn RL. Successful aging. Gerontologist. 1997; 37:433-440. DOI: 10.1093/geront/37.4.433

[51] Isaacowitz DM, Smith J. Positive and negative affect in very old age. Journal of Gerontology (Series B): Psychological and Social Sciences. 2003; 58:143-152. DOI: 10.1093/geronb/63.6.P321

[52] Gale CR, Cooper R, Craig L, Elliott J, Kuh D, Richards M, Deary IJ, on behalf of the HALCyon Study Team. Cognitive function in childhood and lifetime cognitive change in relation to mental wellbeing in four cohorts of older people. PLOS One. 2012; 7(9):e44860. DOI: 10.1371/journal.pone.0044860

[53] Welch DC, West RL. Self-efficacy and mastery: Its application to issues of environmental control, cognition, and aging. Developmental Review. 1995; 15:150-171. DOI: 10.1006/drev.1995.1007

[54] West RL, Bagwell DK, Dark-Freudeman A. Self-efficacy and memory aging: The impact of a memory intervention based on self-efficacy. Aging, Neuropsychology, and Cognition. 2008; 15:302-329. DOI: 10.1080/13825580701440510

[55] Enkvist A, Ekström H, Elmståhl S. Associations between cognitive abilities and life satisfaction in the oldest-old. Results from the longitudinal population study Good Aging in Skåne. Clinical Interventions in Aging. 2013; 8:845-853. DOI: 10.2147/CIA.S45382

[56] Stawski RS, Almeida DA, Lachman ME, Tun PA. Fluid cognitive ability is associated with greater exposure and smaller emotional reactions to daily stressors. Psychology and Aging. 2010; 25:330-342. DOI: 10.1037/a0018246

[57] Wilson RS, Boyle PA, Segawa E, Yu L, Begeny CT, Anagnos SE, Bennett DA. The influence of cognitive decline on well-being in old age. Psychology and Aging. 2013; 28:304-313. DOI: 10.1037/a0031196

[58] Ryff CD. In the eye of the beholder-Views of psychological wellbeing among middle-aged and older adults. Psychology and Aging. 1989; 4:195-210. DOI: 10.1037//0882-7974.4.2.195

[59] Boyle PA, Buchman AS, Bennett DA. Purpose in life is associated with a reduced risk of incident disability among community-dwelling older people. American Journal of Geriatric Psychiatry. 2010. 18:1093-1102. DOI: 10.1097/JGP.0b013e3181d6c259

[60] DeNeve KM, Cooper H. The happy personality: A meta-analysis of 137 personality traits and subjective well-being. Psychological Bulletin. 1998; 124:197-229. DOI: 10.1037/0033-2909.124.2.197

[61] Seligman MEP. Learned Optimism: How to Change Your Mind and Your Life. New York, NY: Pocket Books; 1991. 319 p.

[62] Bandura A. Self-efficacy: The exercise of control. New York, NY: Freeman; 1997. 604 p.

[63] Lyubomirsky S, Lepper H. A measure of subjective happiness: Preliminary reliability and construct validation. Social Indicators Research. 1999; 46:137-155.

[64] Lyubomirsky S, Caldwell ND, Nolen-Hoeksema S. Effects of ruminative and distracting responses to depressed mood on retrieval of autobiographical memories. Journal of Personality and Social Psychology. 1998; 75:166-177. DOI: 10.1037/0022-3514.75.1.166

[65] Strobel M, Tumasjan A, Spörrle M. Be yourself, believe in yourself, and be happy: Self-efficacy as a mediator between personality factors and subjective well-being. Scandinavian Journal of Psychology. 2011; 52:43-48. DOI: 10.1111/j.1467-9450.2010.00826.x

[66] Rietveld CA, Cesarini D, Benjamin, DJ, Koellinger PD, DeNeve J-E, Tiemeier H, Johannesson M, Magnusson, PKE, Pedersen NL, Krueger RF, Bartels M. Molecular genetics and subjective well-being. PNAS. 2013; 110:9692-9697. DOI: 10.1073/pnas.1222171110

[67] The Economist. [Internet]. 2013. Commercialising Neuroscience: Brain Sells. Available from: http://www.economist.com/news/business/21583260-cognitive-training-may-be-moneyspinner-despite-scientists-doubts-brain-sells [Accessed: 2017-01-12]

[68] Flavell JH. Cognitive Development. 2nd ed. Upper Saddle River, NJ: Prentice-Hall; 1995. 338 p.

[69] Stern Y. Cognitive reserve. Neuropsychologia. 2009; 47:2015-2028. DOI: 10.1016/j.neuropsychologia.2009.03.004

[70] Kramer AF, Bherer L, Colcombe SJ, Dong W, Greenough WT. Environmental influences on cognitive and brain plasticity during aging. Journal of Gerontology (series A): Biological Sciences and Medical Sciences. 2004; 59:940-957. DOI: 10.1093/gerona/59.9.M940

[71] Lustig C, Shah P, Seidler R, Reuter-Lorenz PA. Aging, training, and the brain: A review and future directions. Neuropsychological Review. 2009; 19:504-522. DOI: 10.1007/s11065-009-9119-9.

[72] Verhaeghen P, Marcoen A, Goossens L. Improving memory performance in the aged through mnemonic training: A meta-analytic study. Psychology and Aging. 1992; 7:242-251. DOI: 10.1037/0882-7974.7.2.242

[73] Vranic A, Španić AM, Carretti B, Borella E. The efficacy of a multifactorial memory training in older adults living in residential care settings. International Psychogeriatrics. 2013; 25:1885-1897. DOI: 10.1017/S1041610213001154.

[74] Colcombe S, Kramer AF. Fitness effects on the cognitive function of older adults: A meta-analytic study. Psychological Science. 2003; 14:125-30. DOI: 10.1111/1467-9280.t01-1-01430.

[75] Borella E, Carretti B, Riboldi F, De Beni R. Working memory training in older adults: Evidence of transfer and maintenance effects. Psychology and Aging. 2010; 25:767-778. DOI: 10.1037/a0020683.

[76] Cujzek M, Vranic A. Computerized tabletop games as a form of a video game training for old-old. Aging, Neuropsychology, and Cognition. DOI: 10.1080/13825585.2016.1246649.

[77] Whitlock LA, McLaughlin AC, Allaire JC. Individual differences in response to cognitive training: Using a multi-modal, attentionally demanding, game-based intervention for older adults. Computers in Human Behavior. 2012; 28:1091-1096. DOI: 10.1016/j.chb.2012.01.012.

[78] Havinghurst RJ. Successful Aging. Gerontologist. 1961; 1:8-13. DOI: 10.1093/geront/1.1.8.

[79] Cummings E, Henry W. Growing Old: The Process of Disengagement. New York, NY: Basic books; 1961. 293 p. DOI: 10.1093/sw/7.3.122.

[80] Bennett DA, Schneider JA, Arvanitakis Z, Kelly JF, Aggarwal NT, Shah RC, Wilson RS. Neuropathology of older persons without cognitive impairment from two community-based studies. Neurology. 2006; 66:1837-1844. DOI: 10.1212/01.wnl.0000219668.47116.e6.

[81] Gow AJ, Whiteman MC, Pattie A, Whalley L, Starr J, & Deary IJ. Lifetime intellectual function and satisfaction with life. British Medical Journal. 2005; 331:141-142. DOI: 10.1136/bmj.38531.675660.F7.

[82] Kosmat H, Vranic A. The efficacy of a dance intervention as cognitive training for the old-old. Journal of Aging and Physical Activity. 2017; 25:32-40. DOI: 10.1123/japa.2015-0264.

[83] Schooler C, Mulatu MS. The reciprocal effects of leisure time activities and intellectual functioning in older people: A longitudinal analysis. Psychology and Aging. 2001; 16:466-482.

[84] Fratiglioni L, Paillard-Borg S, Winblad B. An active and socially integrated lifestyle in late life might protect against dementia. Lancet Neurology. 2004; 3:343-353. DOI: 10.1016/S1474-4422(04)00767-7.

[85] Fabre C, Chamari K, Mucci P, Massé-Biron J, Préfaut C. Improvement of cognitive function by mental and/or individualized aerobic training in healthy elderly subjects. International Journal of Sport Medicine. 2002; 23:415-421. DOI: 10.1055/s-2002-33735.

[86] Oswald WD, Gunzelmann T, Rupprecht R, Hagen B. Differential effects of single versus combined cognitive and physical training with older adults: The SimA study in a 5-year perspective. European Journal of Aging. 2006; 3:179-192. DOI: 10.1007/s10433-006-0035-z.

[87] Pressman SD, Matthews KA, Cohen S, Martire LM, Scheier M, Baum A, Schulz R. Association of enjoyable leisure activities with psychological and physical well-being. Psychosomatic Medicine. 2009; 71:725-732. DOI: 10.1097/PSY.0b013e3181ad7978.

[88] Zimmer Z, Lin H-S. Leisure activity and well-being among the elderly in Taiwan: Testing hypotheses in an Asian setting. Journal of Cross-Cultural Gerontology. 1996; 11:167-186. DOI: 10.1007/BF00114859.

[89] Kahana E, Bhatta T, Lovegreen LD, Kahana B, Midlarsky E. Altruism, helping, and volunteering: Pathways to well-being in late life. Journal of Aging and Health. 2013; 25:159-187. DOI: 10.1177/0898264312469665.

[90] Van Willigen M. Differential benefits of volunteering across the life course. Journal of Gerontology (series B): Psychological Science and Social Science. 2000. 55:S308-S318. DOI: 10.1093/geronb/55.5.S308.

[91] Fried LP, Carlson MC, Freedman M, Frick KD, Glass TA, Hill J, McGill S, Rebok GW, Seeman T, Tielsch J, Wasik BA, Zeger S. A social model for health promotion for an aging population: Initial evidence on the Experience Corps Model. Journal of Urban Health. 2004; 81:64-78. DOI: 10.1093/jurban/jth094.

[92] Stine-Morrow EAL. The dumbledore hypothesis of cognitive aging. Current Directions in Psychological Science. 2007; 16:295-299. DOI: 10.1111/j.1467-8721.2007.00524.x

Chapter 10

How Air Pollution Affects Subjective Well-Being

Murat Darçın

Additional information is available at the end of the chapter

http://dx.doi.org/10.5772/67742

Abstract

Clean air is considered as one of the basic requirements for human being. Pollution-related diseases due to air pollution continue to rise at an alarming rate and affect people's quality of life. Air pollution also adversely affects welfare inequality. Air pollution as a significant risk factor affects health and sense of subjective well-being (SWB). In this study, the effect of air pollution on subjective well-being (life satisfaction, happiness, and optimism) is investigated. Relationship between well-being and air quality is a fundamental issue to design public policies. Hence, the studies about the link are of grooving interest in the literature. The results show that air pollutants have an effect on subjective well-being. The link between life satisfaction and happiness is very strong. Optimism is also positively correlated with life satisfaction and happiness. Good air quality and optimism make people happier. Air quality is likely to have an effect on people's sense of life satisfaction, happiness, and optimistic view. The results indicate that the PM 2.5 level is significantly and negatively related to on optimism. It is obvious that there is a very strong relationship between air quality and subjective wellbeing. Relationship between wellbeing and air quality is a fundamental issue to design public policies.

Keywords: air pollution, happiness, optimism, life satisfaction, subjective wellbeing

1. Introduction

Air is the most important natural resource that forms the basis of life on Earth. The air in the atmosphere provides oxygen and other gases that are essential elements for survival of life for living beings. Clean air is vital to sustain the delicate balance of life on the Earth. However, the quality of air can be affected by air pollution. To maintain human life in a healthy environment is becoming increasingly difficult because of pollution caused by power plants, transportation, industry, agriculture, and naturally occurring sources [1].

The European Environment Agency defines air pollution as "the presence of contaminant or pollutant substances in the air at a concentration that interferes with human health or welfare, or produces other harmful environmental effects" [2]. Air pollution may also be defined as any atmospheric condition in which chemicals, particulate matter, or biological materials are present in the air at concentrations high enough above their normal atmospheric levels, causing diseases, allergies, death, harm or discomfort to humans, damage to other living organisms such as animals and food crops, or the natural or built environment [3]. Air pollution also affects the quality of life and subjective well-being (SWB) [4, 5].

Air pollution occurs when certain gases and dust particles are introduced into the atmosphere in a way to such levels that they can cause harm to our health, causing breathing and respiratory problems, and even resulting in premature death, as well as damaging the environment, animals, and plant around us [6]. Air pollutants can originate from manmade sources, including emissions from internal combustion engines or the burning of fossil fuels such as coal, oil, petrol, or diesel, but can also come from natural sources such as forest fires, wind erosion, and volcanic eruptions.

A new World Health Organization (WHO) air quality model reveals that 92% of the world's population lives in places where air quality levels exceed the WHO limits. Exposure to outdoor air pollution is the actual cause of death for as many as three million people worldwide each year [7]. According to the WHO, it is known that air pollution is even the most dangerous environmental risk affecting everyone [8]. As an important public health threat of this century, the level of air pollution increases the risk of the global burden of disease from respiratory infections, heart disease, and lung cancer [9]. Since people might adapt to poor environmental quality, it has to be assumed that negative effects of pollution are even underestimated [8].

Pollution-related diseases due to air pollution continue to rise at an alarming rate, and affects people's quality of life [5]. Air pollution also adversely affects welfare inequality [10]. Association between well-being and air quality is one of the fundamental issues in designing public policies about health of society as a whole. Therefore, the studies about the relation are of grooving interest in the literature [11–18]. These literatures propose that air quality is likely to have an effect on life satisfaction, happiness, and optimism. This chapter is concerned with the effects of air pollution on human well-being. The relationship between air pollution and subjective well-being is analyzed by using data about air pollutants and quality of life. The chapter is organized as follows:

- Section 2 describes major air pollutants.
- Section 3 reveals effect of pollutants on human health.
- Section 4 responds to the question "What is subjective well-being?"
- Section 5 analyzes the link between air pollution and subjective well-being.
- Section 6 evaluates the link between air pollution and subjective well-being.
- Section 7 gives conclusion remarks.

http://dx.doi.org/10.5772/67742

2. Major air pollutants

Major primary air pollutants, which can have effects as both directly and precursors of secondary air pollutants (chemicals are formed as a result of reactions between primary pollutants and other elements in the atmosphere), include the following [19–21].

- sulfur oxides (SOx)
- nitrogen oxides (NOx; NO, and NO_2 referred together as NOx)
- carbon monoxide (CO)/carbon dioxide (CO_2)
- volatile organic compounds (VOCs)
- particulate matter (PM)
- chlorofluorocarbons (CFCs)
- ammonia (NH_3)

Primary pollutants are emitted directly into the air from its man-made or natural sources. Sources of air pollution, primary and secondary pollutants are shown in **Figure 1**.

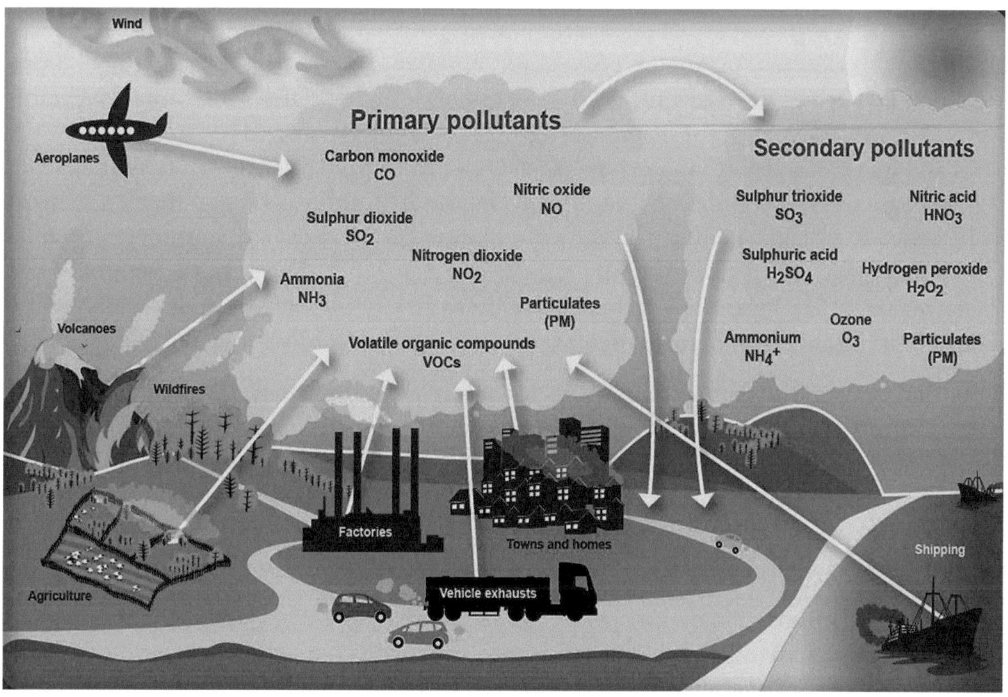

Figure 1. Sources of air pollution [22].

3. Effect of pollutants on human health

Clean air is considered as one of the basic requirements for human health and subjective well-being. Air pollutants, such as nitrogen dioxide (NOx), sulfur dioxide (SOx), particulate matter (PM), carbon monoxide (CO), ozone (O_3), can cause serious health problems [23]. The associations between pollutants and health have usually been reported in previous epidemiological studies that air pollution is an important risk to public health. Air pollution leads to the deterioration of people's health conditions and damage to human health [24–27]. People's exposure to air pollutants is also more likely to experience anxiety and depression [28, 29].

Air pollution is regarded as one of the most serious public health threats facing by countries throughout the world [9]. Results from the studies about health and air pollution have evidenced that exposure to pollution is associated with a very wide range of adverse health or health-related outcomes such as cardiovascular diseases, heart attacks, decreased lung function, chronic non-cancer lung disease, chronic bronchitis, increased respiratory symptoms, asthma exacerbations, emphysema, earlier death from cardiovascular as well as respiratory causes, and cancer, especially lung cancer [3, 25, 30].

The World Health Organization reports that in 2012 alone, around seven million people died as a result of air pollution exposure [30]. A 2016 comparative risk assessment of the Global Burden of Disease 2015 study found similar estimates. It was found that air pollution exposure was linked with almost 6.5 million deaths worldwide [31]. In 2012, air pollution was linked with one out of eight deaths across the world [30]. Particularly, vulnerable groups to air pollution include elderly people, children, people with a preexisting chronic condition [3] such as heart or lung disease, asthmatics, and socially disadvantaged groups. Air pollution is also one of the leading dangers to children's health [31]. It is linked with diseases and infections, which kills around 600,000 children under 5 years old per year, globally [32]. Risk assessment of the Global Burden of Disease 2015 study estimates that a nearly one in 10 under-five deaths is linked to the air pollution [31]. Almost one million children die from pneumonia each year. Air pollution is directly linked with pneumonia and other respiratory diseases [31]. Pneumonia accounts for up to 16% of all under-five deaths [33]; more than half of childhood pneumonia deaths are directly associated with air pollution [31, 34].

These findings confirm that air pollution is the single most deadly environmental health risk. About 3.3 million people a year are killed prematurely by outdoor air pollution, meaning that only outdoor air pollution is responsible for more deaths than both of terrible scourges, HIV/AIDS and malaria combined, each year [30]. Around 18,000 people die each day as a result of air pollution. In fact, the number of deaths due to air pollution each year is more than the number from HIV/AIDS, tuberculosis, and road injuries combined [32, 35].

World Health Organization defines health as "a state of complete physical, mental, and social well-being and not merely the absence of disease or infirmity." It also recognizes health as all physical condition, mental, and social well-being, and does not realize it as merely the absence of disease or infirmity. Thus, health is both absence of disease and presence of well-being [36]. The 1947 World Health Organization definition of health includes not just physical health but complete social well-being [37]. The quality of air and current potential risk to our health play

http://dx.doi.org/10.5772/67742

a crucial role in state of people's physical, mental, and social well-being. It is obvious that reducing air pollution not only saves millions of lives but also prevents the damage resulting from air pollution to people's well-being [36]. Many aspects of the physical environmental quality can affect people's lives and well-being as a key factor in people's well-being [38–40]. There is a close relationship between quality of life and the environment, which has a direct impact on human health and well-being [38, 41, 42]. However, much less is known about how air pollution damages well-being, quality of life, and mental health of people. Some major pollutants, their sources, and health effects are listed in **Table 1**.

Pollutant	Sources	Health effects
Nitrogen dioxide (NO_2)		
• A brownish highly reactive gas that forms quickly when fuels burned at high temperatures. • In the atmosphere can be converted to nitric acid (HNO_3). • Contributes to the formation of ground-level ozone and fine particle pollution. • A main component of acid deposition.	• Motor vehicles. • Power plants. • Industrial plants.	• Reduction in overall lung functions. • Increases symptoms of chronic lung disease. • Lung inflammation, irritation and damage. • Respiratory infections. • Symptoms of bronchitis and asthmatic diseases in children increase with long term exposure. • Wheezing and exacerbation of pneumonia, asthma, bronchial symptoms. • High risk factor of emphysema. • Premature death. • Aggravate existing hearth disease exacerbation of lung.
Particulate matter (PM) (fine dust)		
• Mixture of visible or microscopic solid particles and liquid droplets in the air.	• Burning coal and solid fuel in power and industrial plants. • Combustion – including emission from vehicles, ships, power generation and households. • Natural sources, such as sea salt, wind-blown soil and sand. • Road dust. • Sea spray. • Construction.	• Wheezing and exacerbation of asthma. • Respiratory infections. • Chronic bronchitis and chronic obstructive pulmonary disease. • Exacerbation of chronic obstructive pulmonary disease. • Decreased lung function. • Eye, nose, and throat irritation. • Irregular heartbeat/Nonfatal heart attacks. • Premature death • Mutations, reproductive problems or even cancer.

Pollutant	Sources	Health effects
Sulfur dioxide (SO_2)		
• Highly reactive colorless and irritating gas. • Odorless at low concentrations. • Pungent at very high concentrations. • Related with the formation of acid rain and aerosols.	• Produced by volcanoes, power plants, and other industrial facilities. • Mainly generated from combustion of sulfur-containing fossil fuels, such as coal and petroleum.	• Wheezing and aggravation of asthma and chronic bronchitis/Breathing problems/Respiratory illness. • Exacerbation of chronic obstructive pulmonary disease. • Cardiovascular disease. • Eye, nose and throat irritation. • Increases symptoms of chronic lung disease. • Causing coughing. • Mucus secretion.
Ammonia (NH_3)		
• The most abundant alkaline gas in the atmosphere. • Major component of total reactive nitrogen. • A gas with a pungent odor characteristic. • Colorless, pungent-smelling, caustic (corrosive) gas.	• Decaying organic matters. • The excreta of humans and animals. • Agricultural processes, including animal husbandry and NH_3-based fertilizer applications. • Industrial processes. • Vehicular emissions. • Volatilization from soils and oceans.	• Eye, nose, and throat irritation. • Burning the skin eyes, throat, or lungs might be cause permanent blindness, lung disease, or death. • Can cause life-threatening accumulation of fluid in the lungs (pulmonary edema). • Long-term exposure may harm the respiratory system.
Volatile organic compounds (VOCs)		
• They are often divided into separate categories of methane (CH4) and nonmethane (NMVOCs). • Contain carbon. • Along with carbon, they contain elements such as oxygen, hydrogen, bromine, chlorine, fluorine, sulfur, or nitrogen. • Organic chemicals that have a high vapor pressure.	• Fossil fuel combustion. • Industrial activities. • Solvents, paints, glues, and other products that are used and stored at home and at work. • Oil and gas fields. • Road vehicles. • Household heating. • Power generation. • Natural emissions from vegetation and fires.	• Eye, nose, and throat irritation. • Headaches. • Dizziness. • Fatigue. • Nausea. • Emesis. • Epistaxis. • Loss of co-ordination. • Damage to liver. • Damage to kidney. • Damage to central nervous system. • Allergic skin reaction dyspnea. • Cancer in humans.

http://dx.doi.org/10.5772/67742

Pollutant	Sources	Health effects
Ozone (O_3)		
• Secondary pollutant. • Colorless gas with a strong odor. • Forms as a result of chemical reactions between NOx, VOCs, and oxygen. • Highly active oxidization gas.	Is not emitted directly from any source but is formed in sunlight when certain chemicals react. These chemicals from: • Motor vehicles. • Electric utilities. • Refineries. • Factories. • Petrochemicals. • Vegetation. • Landfills. • Miscellaneous small sources such as gas stations.	• Coughing, chest tightness, chest pain, and wheezing. • Throat irritation and congestion. • Worsen bronchitis, emphysema, and asthma. • Asthma attacks. • Decreased lung function. • Permanent lung damage. • Damage to lung tissue. • Aggravates chronic lung disease. • Irritates respiratory system. • Headaches and weariness. • Increase premature death.

Table 1. Main air pollutants, sources, and effects [22, 43–46].

4. What is subjective well-being (SWB)

Interest in SWB has been increasing rapidly in recent years. SWB that refers to how people experience and evaluate their lives and specific domains and activities in their lives [47, 48] encompasses both cognitive judgments of satisfaction and effective appraisals of moods and emotions [49]. There is no generally-accepted single definition of SWB. Diener et al. [41] define SWB as judging life positively and feeling good. "Thus a person is said to have high [SWB] if she or he experiences life satisfaction and frequent joy, and only infrequently experiences unpleasant emotions such as sadness or anger. Contrariwise, a person is said to have low [SWB] if she or he is dissatisfied with life, experiences little joy and affection and frequently feels negative emotions such as anger or anxiety" [50].

"SWB is a broad category of phenomena that includes people's emotional responses, domain satisfactions, and global judgments of life satisfaction" [51]. SWB is defined as a person's cognitive and affective evaluations of his or her life [52].

Snyder and Lopez [53] define SWB as "a broad concept that includes experiencing pleasant emotions, low levels of negative moods, and high life satisfaction." Eid and Diener [54] proposed that SWB "refers to one's multidimensional evaluation of their lives, including cognitive judgments of life satisfaction as well as affective evaluations of moods and emotions." "Well-being, which we define as people's positive evaluations of their lives, includes positive emotions, engagement, satisfaction, and meaning" [55]. Kahnemann and Riis [56] consider SWB as being a hybrid concept with two components, which can be labeled "experienced well-being" and "evaluated well-being." Both components are subjective and refer to a time of Ref. [56].

In the terms of Diener [57], "SWB is an umbrella term for the different valuations people make regarding their lives, the events happening to them, their bodies and minds, and the circumstances in which they live." Lyubomirsky [58] defines SWB as "the experience of joy, contentment, or positive well-being, combined with a sense that one's life is good, meaningful, and worthwhile." According to Friedman, SWB refers to "the psychological wellbeing of a person and how satisfying a person believes his or her life is" [59].

Reconciling these various definitions, OECD [60] builds an inclusive definition of SWB as: "good mental states, including all of the various evaluations, positive and negative, that people make of their lives and the affective reactions of people to their experiences." This definition of subjective wellbeing includes three elements:

- Life evaluation—a reflective assessment on a person's life or some specific aspects of it.
- Affect—a person's feelings or emotional states, typically measured with reference to a particular point in time.
- Eudaimonia—a sense of meaning and purpose in life, or good psychological functioning.

The three primary components of SWB include positive/negative affect, happiness, and life satisfaction [45, 61–63]. Affect represents the emotional side of SWB. The balance of positive and negative affects refers to the emotions, moods, and feelings a person has. These can be all negative, positive, or a mix of both positive and negative [64]. Subjective well-being is at the heart of happiness. In fact, SWB is considered to be a much broader concept than just happiness. In other words, happiness is only one element of SWB. Veenhoven [65] defines happiness as "the degree to which an individual judges the overall quality of his/her own life-as-a-whole favorably." Life satisfaction global judgments of one's life and satisfaction with specific life domains (e.g. work satisfaction) are considered cognitive components of SWB [63] because they are based on evaluative beliefs (attitudes) about one's life. Positive and negative affects assess the affective component of SWB [51]. Life satisfaction represents one's assessment of one's own life. It is described by the OECD as "measures how people evaluate their life as a whole rather than their current feelings" [66].

Figure 2, summarizing the various items of a simple measurement frame for subjective well-being, shows the different dimensions of the three measurement concepts and how they relate to the subjective well-being determinants [60].

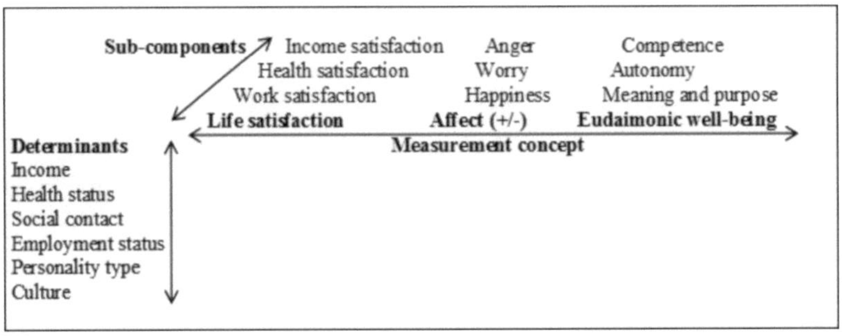

Figure 2. A simple model of subjective well-being.

SWB is strongly correlated to personality characters [67]. Higher incomes, unemployment, poor health, poor work/life balance and higher time spent commuting, social connections, democracy and levels of generalized trust in a country, higher quality environment, and lower crime are associated with subjective well-being [61, 68–72].

5. Analysis of the link between air pollution and subjective well-being

5.1. Methodology

The data used in this research is derived from the European Environment Agency (EEA), the European Commission—Joint Research Centre (JRC)/Netherlands, the Environmental Assessment Agency (PBL), OECD, and the 3rd European Quality of Life Survey (3EQLS). The countries included in this study are Austria, Belgium, Bulgaria, Czech Republic, Denmark, Estonia, Finland, France, Germany, Ireland, Italy, Latvia, Lithuania, Luxembourg, Malta, the Netherlands, Romania, Spain, Sweden, and the United Kingdom. For this study, the countries were selected due to the data quality and data availability of (EEA), the European Commission—(JRC)/(PBL), OECD, and (3EQLS).

Air pollutants data set (*Y* variables) is defined as, *y*1: NH_3; *y*2: NOx; *y*3: NMVOCs; *y*4: SOx; *y*5: CO_2; and *y*6: PM 2.5. Emissions of NH_3, NOx, NMVOCs, and SOx are provided from EEA statistics in 2014. CO_2 emission from fossil fuel use and cement production in 2014 is taken from the European Commission—(JRC)/(PBL) statistics. Mean population exposure to PM 2.5 $\mu g/m^3$ in 2013 is extracted from OECD statistics [73–75]. The 3rd European Quality of Life Survey is used as the basic source of subjective well-being data set (*X* variables set).

The 3rd European Quality of Life Survey (3EQLS) is a conventional tool for recording and examining quality of life in the EU. Research for the (3EQLS) in 27 member states took place from September 2011 to February 2012. This is a survey of people aged 18 and above, residents in EU for at least 6 months. Depending on the size of population 1000–3000 interviews were finished in each member states [76]. There are a large number of possible indicators of SWB in the (3EQLS) data, such as satisfaction with life, happiness, optimism about future, liking one's life, and perceived social exclusion [76]. The most widely used indicators of subjective well-being have chosen for this study as, *x*1: life satisfaction; *x*2: happiness; and *x*3: optimism.

In this study, one of the most common multivariate techniques—canonical correlation analysis (CCA) is used to explore the link between air pollutant variables set and subjective well-being variables set. CCA is a multivariate statistical analysis method that describes the associations between two sets of variables. For more information about CCA, see the study of Darcin and Darcin [77].

5.2. Results

Descriptive statistics (the mean values and standard deviation) of each variable measured in both sets are presented in **Table 2**.

Variable	Observations	Minimum	Maximum	Mean	Std. deviation
NH_3	20.00	1.59	739.79	170.07	223.10
NMVOC	20.00	3.00	1041.36	275.83	324.05
NOx	20.00	6.46	1224.29	311.70	372.18
SO_2	20.00	1.56	388.03	100.24	113.62
CO_2	20.00	2.41	767.15	143.89	191.01
PM 2.5	20.00	7.07	18.58	13.41	3.56
life satisfaction	20.00	5.50	8.40	7.18	0.72
happiness	20.00	6.30	8.20	7.40	0.49
optimism	20.00	33.00	85.00	59.20	12.94

Table 2. Descriptive statistics.

The Pearson's correlations between air pollutant and subjective well-being variables are presented in **Table 3**. According to the result of the present study, the relationship between life satisfaction and happiness is very strong. Optimism is also positively correlated with life satisfaction and happiness. Similar results are found in a study from Turkey [5]. The present study also suggests that there is a negative association between optimism and PM 2.5.

The CC (0.913) between the first pair was found significant ($p < 0.01$) from the likelihood ratio test. The remaining canonical correlation is not statistically significant ($p > 0.05$). By construing the first canonical variable, it is possible to find relationship between air pollution and subjective well-being as rate of 83.3%. For the variable, there is very strong negative significant correlation between air pollution and subjective well-being.

Variables	NH_3	NMVOC	NOx	SO_2	CO_2	PM 2.5	life satisfaction	happiness	optimism
NH_3	**1.00**	0.89	0.93	0.76	0.89	0.20	0.05	0.07	−0.33
NMVOC	0.89	**1.00**	0.98	0.86	0.93	0.20	0.02	0.05	−0.36
NOx	0.93	0.98	**1.00**	0.87	0.95	0.18	0.05	0.09	−0.34
SO_2	0.76	0.86	0.87	**1.00**	0.86	0.21	−0.23	−0.16	−0.33
CO_2	0.89	0.93	0.95	0.86	**1.00**	0.23	0.04	0.07	−0.25
PM 2.5	0.20	0.20	0.18	0.21	0.23	**1.00**	−0.31	−0.34	−0.65
Life satisfaction	0.05	0.02	0.05	−0.23	0.04	−0.31	**1.00**	0.97	0.58
Happiness	0.07	0.05	0.09	−0.16	0.07	−0.34	0.97	**1.00**	0.55
Optimism	−0.33	−0.36	−0.34	−0.33	−0.25	−0.65	0.58	0.55	**1.00**

Table 3. Correlation matrix.

6. Evaluating relationship between air pollution and SWB

Studies about the effect of air quality on public welfare are of vital importance for policy development and evaluation [45]. However, there is relatively little empirical evidence available on the relationship between air pollution and well-being [3]. Because it is difficult to make a clear relationship between pollution and well-being when air pollution tends to be reported at a country level and well-being is an individual measure [45]. Air quality is an important determinant of subjective wellbeing and an important policy issue [12]. Environmental policies and regulations are realized to improve air quality and thus subjective well-being [78].

There are a number of papers analyzing the relationship between air pollution and subjective well-being. The significant effects of air pollution on people's subjective well-being have been explored by using measures of self-reported well-being and cross-sectional and panel data measured for air quality and for several pollutants [11, 12, 79].

The relationship between subjective measures of well-being and individual environmental attitudes, using data from the British Household Panel Survey, is examined by Ferrer-i-Carbonell and Gowdy [80]. They find that people's attitudes and experiences toward effects of environmental issues have an effect on individual's well-being. Rehdanz and Maddison [15], using data drawn from the German socio-economic panel, make an attempt to explain differences in selfreported levels of well-being in terms of environmental quality. They find that air pollution and noise levels significantly reduce subjective well-being [15]. A paper with a focus on Spanish regions recommends that environmental variables have a significant impact on individual reported subjective well-being [81]. The results of present study are also compatible with the results of work done in different countries that suggest a relationship between air pollution and subjective well-being.

In the literature, different air pollutants, such as PM10 [12, 16, 83], SO_2 [18, 82], and NO_2 [12, 17], were evidenced a negative impact on individual well-being in different studies for different countries or areas. Being exposed to local air pollution in terms of mass concentration of PM10 significantly reduces individual's well-being [45, 83]. A significant negative and robust relationship between sulfur dioxide (SO_2) emissions at the country level and subjective well-being data in several European countries is found [82]. Smyth et al. [18], using pollution data in 30 cities in urban China, also find a clear negative impact of SO_2 emission on subjective well-being. Luechinger [84] finds a significant negative impact of SO_2 pollution on well-being by using annual mean concentrations of SO_2 at 533 monitoring stations in Germany over a 19-year period. Welsch [85] suggests that air pollution plays a statistically significant role as a predictor of differences in subjective well-being between countries and between time periods. The relationship was stronger for NO_2 than for total suspended particulate (TSP) concentration [85].

Several studies show the impact of air pollution on life satisfaction that has been considered as one of the fundamental indicators of subjective well-being. Air pollution, which has objective indicators, such as air pollutants including SO_2, NO_2, and PM10, is significantly and negatively associated with the life satisfaction [8, 12, 17, 82, 84, 86]. The SO_2 concentration negatively affects selfreported life satisfaction [45, 84]. Ferreira et al. [45], using detailed regional data, report a negative and significant relationship between air pollution and individual selfreported life satisfaction. An increase in SO_2 concentrations by 1 μg/m³ is associated with a reduction in life satisfaction of between 0.016 and 0.030 points on the 11-point life satisfaction scale [45]. Furthermore, local NO_2 concentrations significantly reduce the life satisfaction. MacKerron and Mourato [17], in a study in London, find that an increase in air pollution has a negative effect on the life satisfaction. They estimate that an increase of 10 μg/m³ in annual means nitrogen dioxide concentration is associated with a drop of nearly half a point of life satisfaction on the 11-point scale [17].

Menz and Welsch [87], using data on life satisfaction for 25 OECD countries from the World Database of Happiness, report that the link between air pollution (PM10 concentration) and life satisfaction is significantly negative and stronger for young and old people than for middle-aged individuals.

A number of studies have examined the effect of air pollution on happiness [12, 14, 15, 17, 79–82, 84, 85, 87, 88]. Environmental concern can affect happiness either positively or negatively [80]. Objective measures of air pollution have significantly negative impacts on the national happiness level [12, 81]. Air pollution significantly reduces shorter-term hedonic happiness and increases the rate of depressive symptoms [88]. The association between perceived levels of noise and air pollution and selfreported happiness, using individual level data from the German socio-economic panel (SOEP) surveys which contains a total number of about 23,000 observations, is considered by Rehdanz and Maddison [15]. It is found that higher perceived air pollution significantly diminishes happiness [15]. It is found that happiness responses of around 350,000 people living in the OECD between 1975 and 1997 are negatively correlated with environmental degradation (measured by SOx emissions) [14]. Regional air pollution (measured by CO_2 emissions and number of days that PM10 exceeds a certain limit, 50 g/m³) has a negative effect on happiness. In other words, air quality affects individual happiness positively [81]. Giovanis examines the association between happiness and air pollution using the life satisfaction approach (LSA). The results of the study suggest that the O_3 and SO_2 present the strongest negative effects on happiness followed by CO and NOx [89]. Air pollution can significantly affect both human health and subjective well-being. There is evidence that health and SWB may also equally affect each other [90]. For instance, positive emotions and optimism as factors of SWB can have a positive influence on health [91].

7. Conclusion

It is obvious that air pollution has a negative significant impact on subjective well-being. This chapter also explores the association between air pollution and subjective well-being (life satisfaction-happiness-optimism). The present results also give reason for policy-maker to create clean environment. The findings propose that air pollutants, such as NH_3, NOx, NMVOCs, SOx, CO_2, and PM 2.5, may affect life satisfaction, happiness, or optimism, suggesting that

http://dx.doi.org/10.5772/67742

environmental policies focused on reducing air pollution will not only have a positive impact on the health of future generations, but at the same time will increase present individual's quality of life. The clean air can be expected to make people who live elsewhere in the world happier. There should be stronger and more stable approaches to general environmental problems and air pollution. Fight against air pollution is the responsibility of all individuals. The sensitivity of people to the environmental issue should be increased for healthier generation. The most important step to acquire awareness and sensitivity toward the environment is education. Teaching correctly and consistent information about environmental problems in schools are very important factors to raise awareness about air pollutants [92–95].

Author details

Murat Darçın

Address all correspondence to: mrtdrc@yahoo.com

Gendarmerie and Coast Guard Academy, Ankara, Turkey

References

[1] Shahzad A, Zaman K. The relationship between air quality and quality of life in Pakistan. Journal of Economic Information. 2014;**2**(1):1-12.

[2] EEA. Environmental Terminology and Discovery Service (ETDS) [Internet]. 2016. Available from: http://glossary.eea.europa.eu/terminology/concepthtml?term=air%20 pollution [Accessed: 20 December 2016]

[3] Cowie H, Crawford J, Davis A, Steinle S, Reis S, Dixon K, Morris G, Hurley F. Air Quality, Health, Wellbeing and Behaviour [Internet]. 2015. Available from: http://www.environment.scotland.gov.uk/media/126007/iom-seweb-aq-health-behaviour-review.pdf [Accessed: 10 January 2017]

[4] Darçın M. Association between air quality and quality of life. Environmental Science and Pollution Research. 2014; **21**:1954-1959.

[5] Darçın M. The link between air quality and subjective well-being. Fresenius Environmental Bulletin. 2014;**23**(9a):2235-2238.

[6] Scottish Environment Protection Agency (SEPA). The chemistry of air pollution [Internet]. 2016. Available from: https://www.sepa.org.uk/media/120465/mtc_chem_of_ air pollution.pdf [Accessed: 20 December 2016]

[7] WHO. WHO releases country estimates on air pollution exposure and health impact [Internet]. 2016. Available from: http://www.who.int/mediacentre/news/releases/2016/air-pollution-estimates/en/ [Accessed: 19 December 2016]

[8] Schmitt, M. Subjective well-being and air quality in Germany [Internet]. 2013. Available from: https://www.diw.de/documents/publikationen/73/diw_01.c.416307.de/diw_sp0541.pdf [Accessed: 19 December 2016]

[9] WHO. Air quality guidelines. Global update 2005. Particulate matter, ozone, nitrogen dioxide and sulfur dioxide [Internet]. 2006. Available from: http://www.euro.who.int/data/assets/pdf_file/0005/78638/E90038.pdf?ua=1 [Accessed: 19 December 2016]

[10] Brajer V, Mead RW, Xiao F. Adjusting Chinese income inequality for environmental equity. Environment and Development Economics. 2010;**15**(3):341-362).

[11] Welsch H. Preferences over prosperity and pollution: Environmental valuation based on happiness surveys. Kyklos. 2002;**55**:473-494.

[12] Welsch H. Environment and happiness: Valuation of air pollution using life satisfaction data. Ecological Economics. 2006;**58**:801-813.

[13] Brereton F, Clinch P, Ferreira S. Happiness, geography and the environment. Ecological Economics. 2008;**65**:386-396.

[14] Di Tella R, MacCulloch R. Gross national happiness as an answer to the Easterlin Paradox? Journal of Development Economics. 2008;**86**:22-42.

[15] Rehdanz K, Maddison D. Local environmental quality and life satisfaction in Germany. Ecological Economics. 2008;**64**:787-797.

[16] Levinson A. Valuing public goods using happiness data: The case of air quality. Journal of Public Economics. 2012;**96**(9-10), 869-880.

[17] MacKerron G, Mourato S. Life satisfaction and air quality in London. Ecological Economics. 2009;**68**:1441-1453.

[18] Smyth R, Mishra V, Qian X. The environment and well-being in Urban China. Ecological Economics. 2008;**68**(1-2):547-555.

[19] Mathew TV. Transportation systems engineering. Fuel consumption and emission studies. [Internet]. 2014. Available from: http://www.civil.iitb.ac.in/tvm/1111_nptel/583_FuelEmi/fancy.pdf [Accessed 19 December 2016]

[20] EEA. Emissions of the main air pollutants in Europe [Internet]. 2016. Available from: http://www.eea.europa.eu/data-and-maps/indicators/main-anthropogenic-air-pollutant-emissions/assessment-3 [Accessed: 12 December 2016]

[21] United States Environmental Protection Agency. Criteria Air Pollutants [Internet]. 2016. Available from: https://www.epa.gov/criteria-air-pollutants [Accessed: 20 December 2016]

[22] Scotland's Environment. Air Quality [Internet]. Available from: http://www.environment.scotland.gov.uk/get-informed/air/air-quality/ [Accessed: 20 December 2016]

[23] EEA. Air quality in Europe-2016 report [Internet]. 2016. Available from: http://www.eea.europa.eu/publications/air-quality-in-europe-2016 [Accessed: 3 Jan 2017]

[24] Currie J, Neidell M. Air pollution and infant health: What can we learn from California's recent experience? Quarterly Journal of Economics. 2005;**120**:1003-1030.

[25] Field BC, Field MK. Environmental Economics. 3rd ed. New York: McGraw-Hill; 2002.

[26] Graves PE. Environmental Economics: An Integrated Approach. New York: CRC; 2013.

[27] Anderson DA. Environmental Economics and Natural Resource Management. 4th ed. New York, NY: Routledge; 2014.

[28] Guxens M, Sunyer J. A review of epidemiological studies on neuropsychological effects of air pollution. Swiss Medical Weekly. 2012;**141**:w13322.

[29] Marques S, Lima ML. Living in grey areas: Industrial activity and psychological health. Journal of Environmental Psychology. 2011;**31**(4):314-322.

[30] WHO. 7 million premature deaths annually linked to air pollution [Internet]. 2014. Available from: http://www.who.int/mediacentre/news/releases/2014/airpollution/en/ [Accessed: 20 December 2016]

[31] UNICEF. Clear the air for children. The impact of air pollution on children. New York, NY: UNICEF Division of Data, Research and Policy; 2016

[32] WHO. Burden of Disease from Household Air Pollution for 2012: Summary of Results [Internet]. 2014. Available from: http://www.who.int/phe/healthtopics/outdoorair/databases/FINAL_HAP_AAP_BoD_24March2014.pdf [Accessed: 20 December 2016]

[33] WHO. Global Health Observatory (GHO) data [Internet]. 2016. Available from: http://www.who.int/gho/en/[Accessed: 10 December 2016]

[34] Darrow LA, Klein M, Flanders WD, Mulholland JA, Tolbert PE, Strickland MJ. Air pollution and acute respiratory infections among children 0-4 years: An 18-year time-series study. American Journal of Epidemiology. 2014;**180**(10):968-977.

[35] International Energy Agency (IEA)/OECD. Energy and Air Pollution. World Energy Outlook Special Report [Internet]. 2016. Available from: http://www.iea.org/publications/freepublications/publication/WorldEnergyOutlookSpecialReport2016EnergyandAirPollution.pdf [Accessed: 19 December 2016]

[36] WHO. Constitution of the World Health Organisation [Internet]. 1948. Available from:http://www.who.int/about/definition/en/print.html [Accessed: 20 December 2016]

[37] Callahan D. The Roots of Bioethics: Health, Progress, Technology, Death. Oxford: Oxford University Press; 2012.

[38] Streimikiene D. Environmental indicators for the assessment of quality of life. Intellectual Economics. 2015;**9**:67-79.

[39] Holman MR, Coan TG. Voting green. Social Science Quarterly. 2008;**89**:1121-1135.

[40] Kahn ME. Demographic change and the demand for environmental regulation. Journal of Policy Analysis and Management. 2002;**21**(1):45-62.

[41] Diener E, Suh E. Measuring quality of life: Economic, social, and subjective indicators. Social Indicators Research. 1997;**40**(1-2):189-216.

[42] United Nation Economic Commission for Europe (UNECE). Measuring Sustainable Development. New York/Geneva: United Nations Publication; 2009.

[43] Boamah PO, Onumah J, Takase M, Bonsu PO, Salifu T. Air pollution control techniques. Global Journal of Bio-Science and Biotechnology. 2012;**1**(2):124-131.

[44] de Kluizenaar Y, Aherne J, Farrell EP. Modelling the spatial distribution of SO2 and NOx emissions in Ireland. Environmental Pollution. 2001;**112**(2):171-182.

[45] Ferreira S, Akay A, Brereton F, Cunado J, Martinsson P, Moro M, Ningal TF. Life satisfaction and air quality in Europe. Ecological Economics. 2013;**88**:1-10.

[46] Liao P-S, Shaw D, Lin, Y-M. Environmental quality and life satisfaction: Subjective versus objective measures of air quality. Social Indicators Research. 2014;**124**:599-616.

[47] Stone AA, Mackie C (Eds.). Subjective Well-being: Measuring Happiness, Suffering, and Other Dimensions of Experience. Washington, DC: The National Academies Press; 2014.

[48] National Research Council. Subjective Well-being: Measuring Happiness, Suffering, and Other Dimensions of Experience. Washington, DC: The National Academies Press; 2013.

[49] Kesebir P, Diener E. In pursuit of happiness: Empirical answers to philosophical questions. Perspectives on Psychological Science. 2008;**3**(2):117-125.

[50] Diener E, Suh E, Oishi S. Recent findings on subjective well-being. Indian Journal of Clinical Psychology. 1997;**24**(1):25-41.

[51] Diener E, Suh EM, Lucas RE, Smith HL. Subjective well-being: three decades of progress. Psychological Bulletin. 1999;**125**(2):276-302.

[52] Diener E, Oishi S, Lucas RE. Subjective well-being: The science of happiness and life satisfaction. In: Snyder CR, Lopez SJ, editors. Handbook of Positive Psychology. Oxford and New York: Oxford University Press; 2002.

[53] Snyder CR, Lopez SJ. Handbook of Positive Psychology. Oxford University Press; 2002.

[54] Eid M, Diener E. Global judgments of subjective well-being: Situational variability and long-term stability. Social Indicators Research, 2004;**65**(3):245-277.

[55] Diener E. Seligman MEP. Toward an economy of well-being. Psychological Science in the Public Interest. 2004;**5**(1):1-31.

[56] Kahneman D, Riis J. Living and thinking about it: Two perspectives on life. In: Huppert FA, Baylis N, Keverne B, editors. The Science of Well-being. Oxford: Oxford University; 2005.

[57] Diener E. Guidelines for national indicators of subjective well-being and ill-being. Journal of Happiness Studies. 2006;**7**:397-404.

[58] Lyubomirsky S. The How of Happiness: A New Approach to Getting the Life You Want. New York, NY: Penguin Books; 2007.

[59] Friedman HS, Kern ML, Reynolds CA. Personality and health, subjective well-being, and longevity. Journal of Personality. 2010;**78**(1):179-216.

[60] OECD. OECD Guidelines on Measuring Subjective Well-being [Internet]. 2013. Available from: http://dx.doi.org/10.1787/9789264191655-en [Accessed: 20 December 2016]

[61] Dolan P, Peasgood T, White MP. Do we really know what makes us happy? A review of the economic literature on the factors associated with subjective wellbeing. Journal of Economic Psychology. 2008;**29**(1):94-122.

[62] Van Praag BMS, Ferrer-i-Carbonell A. Happiness Quantified: A Satisfaction Calculus Approach. Oxford: Oxford University Press; 2008.

[63] Diener E. Subjective well-being: The science of happiness and a proposal for a national index. American Psychologist. 2000;**55**(1):34-43.

[64] Diener E, Biswas-Diener R. Happiness: Unlocking the Mysteries of Psychological Wealth. Malden, MA: Blackwell Publishing; 2008.

[65] Veenhoven R. Greater happiness for a greater number: Is that possible? If so, how? In Sheldon KN, Kashdan TB, Steger MF, editors. Designing Positive Psychology. Oxford: Oxford University, 2011. pp. 392-409.

[66] OECD. Compendium of OECD Well-being Indicators. Paris: OECD Publishing; 2011.

[67] Steel P, Schmidt J, Shultz J. Refining the relationship between personality and subjective well-being. Psychological Bulletin. 2008;**134**(1):138-161.

[68] Boarini R, Comola M, Smith C, Manchin R, de Keulenaer F. What makes for a better life? The determinants of subjective well-being in OECD countries—evidence from the Gallup World Poll. OECD Statistics Working Papers, 2012/03. OECD Publishing; 2012.

[69] Dolan P, Metcalfe R. Measuring subjective wellbeing: Recommendations on measures for use by National Governments. Journal of Social Policy. 2012;**41**(2):409-427.

[70] Lim C, Putnam R. Religion, social networks, and life satisfaction. American Sociological Review. 2010;**75**(6):914-933.

[71] Eichhorn J. Happiness for believers contextualizing the effects of religiosity on life-satisfaction. European Sociological Review. 2012;**28**:583-593.

[72] Graham C. Happiness Around the World: The Paradox of Happy Peasants and Miserable Millionaires. Oxford: Oxford University Press; 2009.

[73] EEA. Air pollutant emissions data viewer (NEC Directive) [Internet]. 2014. Available from: http://www.eea.europa.eu/data-and-maps/data/data-viewers/emissions-nec-directive-viewer [Accessed: 10 December 2016]

[74] JRC/PBL. Emission database for global atmospheric research (EDGAR) [Internet]. 2015. Available from: http://edgar.jrc.ec.europa.eu [Accessed: 10 January 2017]

[75] OECD. Exposure to PM2.5 in countries and regions [Internet]. 2013. Available from: http://stats.oecd.org/[Accessed: 10 January 2017]

[76] Eurofound. Third European Quality of Life Survey – Quality of Life in Europe: Impacts of the Crisis. Luxembourg: Publications Office of the European Union; 2012.

[77] Darcin M, Darcin ES. Relationship between quality of life and child traffic fatalities. Accident Analysis and Prevention. 2007;**39**(4):826-832.

[78] Frey BS, Luechinger S, Stutzer A. The life satisfaction approach to environmental valuation. Annual Review of Resource Economics. 2010;**2**(1):139-160.

[79] Welsch H. Environmental welfare analysis: A life satisfaction approach. Ecological Economics. 2007;62:544-551.

[80] Ferrer-i-Carbonell A, Gowdy JM. Environmental awareness and happiness. Ecological Economics. 2007;**60**:509-516.

[81] Cunado J, Perez de Gracia F. Environment and happiness: New evidence for Spain. Social Indicators Research. 2013;**112**:549-567.

[82] Luechinger S. Life satisfaction and transboundary air pollution. Economics Letter. 2010;**107**:4-6.

[83] Ferreira S, Moro M. On the use of subjective well-being data for environmental valuation. Environmental and Resource Economics. 2010;**46**(3):249-273.

[84] Luechinger S. Valuing air quality using the life satisfaction approach. Economic Journal. 2009;**119**:482-515.

[85] Welsch H. Environment and happiness: Valuation of air pollution in ten European countries [Internet]. 2003. Available from: http://www.diw.de/documents/publikationen/73/diw_01.c. 40635.de/dp356.pdf [Accessed: 10 January 2017]

[86] Orru K, Orru H, Maasikmets M, et al. Well-being and environmental quality: Does pollution affect life satisfaction? Quality of Life Research. 2016;**25**(3):699-705

[87] Menz T, Welsch H. Population aging and environmental preferences in OECD countries: The case of air pollution. Ecological Economics. 2010;**69**:2582-2589.

[88] Zhang X, Zhang X, Chen X. Happiness in the air: How does dirty sky affect subjective well-being? Social Science Research Network Working Paper Series. Technical report, Institute for the Study of Labor (IZA); 2015.

[89] Giovanis E. Essays on Environmental Economics. A Dissertation the Degree of Doctor of Philosophy, Economics department at Royal Holloway, University of London; 2015.

[90] Okun MA, Stock WA, Haring MJ, Witter RA. Health and subjective well-being: A meta-analysis. The International Journal of Aging & Human Development. 1984;**19**(2):111-132.

[91] Diener E, Chan MY. Happy people live longer: Subjective well-being contributes to health and longevity. Applied Psychology: Health and Well-Being. 2011;**3**(1):1-43.

[92] Darçın ES, Darçın M. Secondary school students' knowledge level about environmental problems caused by vehicle emissions. Gazi University Journal of Gazi Educational Faculty. 2009;**29**(2):485-512.

[93] Cibik, AS, Darçın ES. Pre-service science teachers' knowledge level about some basic air pollutants. Journal of Baltic Science Education. 2009;**8**(1): 22-34.

[94] Erduran Avcı D, Darçın ES. Investigation of eight grade students' knowledge level about global environmental problems. Eurasian Journal of Physics and Chemistry Education. 2009;**1**(2): 93-98.

[95] Darçın ES, Bozkurt O, Hamalosmanoğlu M, Köse S. Determination of elementary students' level of knowledge and misconceptions about greenhouse effect. International Journal of Environmental and Science Education. 2006;**1**(2):104-115.

Chapter 11

What Works to Improve the Well-Being of Homeless Girls?

Susana Castaños-Cervantes

Additional information is available at the end of the chapter

http://dx.doi.org/10.5772/68000

Abstract

Homeless girls are victims of physical, sexual, and psychological abuse more frequently than the rest of the population. Consequently, their well-being is severely affected. Nonetheless, there is little information about this social group which leads to lack of proper care that, in turn, reduces their quality of life. This research was conducted to develop and test a path model of well-being in a group of 240 Mexican homeless girls aged 6–23 years. Anxiety, depression, assertiveness, and emotion regulation strategies were used as predictors of well-being. Findings reveal significant direct effects of depression and functional emotion regulation strategies on well-being. Results also show significant indirect effects of anxiety, assertiveness, and dysfunctional emotion regulation strategies through depression. The fit indices achieved, $x^2(3)=1.116$, $p=.773$; NFI=.997, RFI=.985, IFI=1.005, TLI=1.026, CFI=1.000, $p=.000$; RMSEA(90% CI)=.000(.000–.073), $p=.889$, demonstrate that the model reliably predicts well-being. Thus, the model fit is acceptable. The current study provides unique findings in terms of a path model that highlights anxiety, depression, emotion regulation strategies, and assertiveness as critical indicators for well-being in homeless girls. Hence, it is essential to consider such factors in order to promote well-being in this group, thereby improving their health and quality of life.

Keywords: well-being, homeless girls, anxiety, depression, emotion regulation

1. Introduction

The promotion of child and youth well-being requires a conscious society convinced of the importance of children and youth's development as an essential component for its future sociocultural and economic progress. However, in the present times, the children and youth population is immersed in conditions of great vulnerability.

Vulnerability is the degree to which a sociocultural collective, individual, or organization is unable to anticipate, cope, and recover from the impact of adverse natural or human-induced circumstances [1]. That is, the stressful or deficient situations that families have to face and the resources they have to solve these problems or, in the case of the child or adolescent, any condition that forces them to live an experience that violates their physical and psychological integrity, and that annuls their decisions and their capabilities [2]. So, vulnerable groups are all those who, by virtue of their age, race, sex, economic, social, physical, cultural and political circumstances or sexual orientation, may face greater obstacles in the exercise of their citizenship rights [3]. As part of vulnerable people, are homeless children and youth who are the extreme manifestation of social deterioration and exclusion because they have below the minimum needed to access services or resources [4, 5], and lack admittance to decision-making power over their own social and economic destiny [6, 7]. Among this collective are homeless girls, who are the most vulnerable group whatsoever as a result of being women, underage, and homeless. But, who are the homeless? Specifically, who are the homeless girls?

According to a previous study [8], Mexican homeless girls are a social group which in turn comprises two subgroups, those at risk of homelessness and those who have abandoned their home (sheltered, unsheltered, or emergency sheltered), characterized by:

- Having been born and living in Mexico City with less than 30% coming from different states of the Mexican Republic.
- Having an average age of ($M[SD]=12.07[3.754]$) years with an age range of 6–23 years; however, there could be younger girls (in this study, there were approximately 20 little girls between ages 3 and 5 years), and a grade level of middle school. It is important to highlight that only if they remain in an organization, they continue to study, because in their family, they do not receive any kind of support for their studies, and they are not self-sufficient. Their academic achievement and performance are low. Although few ($n<5$), some are illiterate and the vast majority ($n>50\%$) have severe problems with reading and writing.
- Participating in the formal economy system by having jobs with low salaries such as receptionist, assistant cook, saleswoman, and waitress. Similarly, although not legally considered an activity work, they contribute to domestic labor and, with increasing age, they are responsible for the care of their younger siblings. Also, they work in the informal economy system doing jobs such as street vendor, housemaid, and begging. They allocate their resources for their own support and/or assist the subsistence of their family. They can carry out criminal acts such as stealing, selling drugs, and prostitution, exposing themselves to risks that diminish their bio-psychosocial well-being.
- Consuming drugs, alcohol, and tobacco since approximately 12 years of age. The most frequently used drugs are inhalants, for their low cost and easy access, and marijuana. Substances are obtained within the family where usually there are members who use and abuse them, in their home community, at school, and in their peer group. Occasionally, in

http://dx.doi.org/10.5772/68000

the same institution they have access to these (inhalants[1]) or they themselves are able to introduce drugs. When using drugs, they can have sex under the influence of these.

- Having diverse skin, respiratory, and gastrointestinal diseases, malnutrition, and mental disorders. They usually enter the organization with a deplorable physical and mental health status, with oral and visual health problems, and a lack of hygiene and self-care behaviors. Commonly, they go to health centers for proper care. But in two or three cases, it was found that they have been victims of sexual violence by the physician who offers his services in exchange for sex or fondling. As part of their physical health, sometimes they have suffered accidents such as falls, bruises, and burns. Regarding mental health, only when they remain institutionalized, they receive psychological care. The most common mental disorders are anxiety and depression. Likewise, to a lesser extent, other typical disorders are attention deficit disorder with or without hyperactivity, personality disorders, and psychosis. Drugs commonly prescribed for psychiatric problems are sertraline, carbamazepine, valproic acid, and risperidone. They lack information concerning how to take care of their health and body, and they do not know what a psychologist does. They have suicidal thoughts, make suicide attempts, and have self-mutilation behaviors.

- Initiating sexual life usually as a consequence of having been sexually abused. As age increases, the probability of having sexual intercourse is higher. They may engage in sexual risk behaviors such as having unprotected sex, having sexual relationships with multiple partners, and/or under the influence of alcohol and drugs. They practically have very less information on contraception, sexually transmitted infections, how to take care of themselves, and their anatomy and physiology. More than 80–90% have heterosexual relationships, and approximately 10–20% involve in homosexual or bisexual relationships. They typically get pregnant and abort at ages 13–14 years. They can have children with several different men. When their partner leaves them or when the relationship is over, because they do not have any means of support, they seek aid of institutions where they can receive proper care and attendance. Their child-rearing practices are ambivalent: both violent and loving and caring.

- Engaging in risk behaviors such as drug use and abuse, self-mutilation, suicide ideation and attempts, delinquent activities (*e.g.*, theft, drug selling, and prostitution) to obtain resources necessary to survive, and consume psychoactive substances.

- Having unstable and informal couple relationships at younger ages, and more lasting and formal at older ages. Their couples provide security, support, affection, and protection. They can also become a source of violence and of risk behaviors. Their couple relationships are mainly heterosexual (n>80–85%), and, in fewer cases, homosexual and bisexual (n<15–20%).

[1]That is, at the organization they can have access to inhalants such as detergents, liquid cleaners, glue, and other chemicals used without the knowledge and consent of the institution's staff in order to get high. Additionally, when they inhale these products they do so secretly hidden from the organization's authorities.

- Establishing at a younger age broad social networks, and at older ages the quality and size of these diminish. Their peer group provides affection, support, security, and protection. Also, if they abandon their family, their group becomes their new family and their means to satisfy their basic and emotional needs. Nonetheless, their peer group can induce them to risk behaviors and can turn into a source of violence.

- Developing in dysfunctional environments characterized by physical, sexual, and psychological abuse, emotional and physical abandonment, educational, physical, and psychological neglect, lack of resources necessary to survive, drug use and abuse, and extended and reconstituted families with lousy jobs or underemployment.

- Living in communities immersed in social problems like lack of access to basic services (*e.g.*, water, paving, electricity, sewage, garbage collection, and public transportation, among others), substance use and abuse, theft and, although less frequently (n<25–30%), drug trafficking, prostitution, gangs, kidnapping, and homicides. They live in overcrowded settlements, unsuitable for housing, and usually established on the outskirts of the city; and

- Entering a non-governmental or governmental organization due to domestic violence, lack of resources to satisfy their basic needs, and to obtain better life conditions. Over time, their institutional fellows become their new family. However, with age they become more reluctant to accept rules and regulations, and to follow orders. Likewise, among their same companions, there are several problems of coexistence, acceptance, and theft. It is worth mentioning that they could have been previously staying in another institution, but because of age or behavioral problems they were channeled to the actual organization.

Girls at risk of homelessness are a social subgroup that has the following features:

- Having been born and living in Mexico City (n>80%).

- Having an average age of ($M[SD]$=10.16[2.964]) years with an age range of 6–23 years; however, there could be younger girls (in this study there were approximately 20 little girls between ages 3 and 5 years), and a grade level of middle school. It is important to highlight that because they stay in an organization they are able to obtain a college degree. Their academic achievement and performance can be low, and they have problems with reading and writing.

- Not having a job as a result of remaining in an institution since their education and biopsychosocial well-being is prioritized.

- Not using drugs. Nonetheless, as they grow older they can begin to drink alcohol and/or smoke. They have access to such substances within their peer group and family, and at school. It is worth noting that their families usually have addiction and alcohol problems.

- Having diverse skin, respiratory, and gastrointestinal diseases, oral and visual health problems, malnutrition, and mental disorders such as anxiety, depression, posttraumatic stress, and attention deficit disorder with or without hyperactivity. They did not exhibit any sexually transmitted diseases. They lack hygiene and self-care behaviors, and information concerning how to take care of their health and body. They have suicidal thoughts,

http://dx.doi.org/10.5772/68000

make suicide attempts, and have self-mutilation behaviors. The most common drugs that are prescribed for their various mental disorders are sertraline, valproate, carbamazepine, fluoxetine, and risperidone.

- Most of them have not initiated sexual activity and have been victims of sexual abuse within their families. They have little information about contraception, their anatomy and physiology, and sexually transmitted diseases. They have not been pregnant and have not had any children.
- Engaging in risk behaviors such as drinking alcohol or smoking, ideation and suicide attempts, and self-mutilation behaviors.
- Having unstable, informal, and not lasting heterosexual couple relationships. They frequently exchange partners, and their relationship is seen more as a friendship. As they grow older, their couple relationships become more formal and durable. Their partners satisfy their affective needs, including support and security, and in their relationships they have not suffered intimate partner violence.
- Establishing at a younger age broad social networks of both genders, and at older ages the quality and size of these diminish since they have few friendships of their same gender. They usually associate with peers who rarely engage in risk behaviors and provide affection, support, security, and protection. However, their peer group can be a perpetrator of violence, especially at school.
- Coming from single-parent families, mainly single mothers who have elementary or middle school and get remarried. They typically live with the mother's family, which is why it is an extended family. Their family is also characterized by having a low socioeconomic status since they do not have the means necessary to survive, and low-paying jobs or underemployments, where there is physical, sexual, and psychological abuse, abandonment, and negligence because the mother works long hours and does not have time to take care of her children, and where there could be substance abuse, among others.
- Living in overcrowded communities without infrastructure and resources; occasionally, without access to basic services and with social problems such as substance use and abuse, alcoholism, and theft; and
- Entering a non-governmental or governmental organization due to domestic violence and lack of economic wealth. Over time, their institutional fellows become their new family. However, with age they become more reluctant to accept rules and regulations, and to follow orders. They choose to stay so they can complete their schooling, get job training and the chance of a better quality of life. It is worth mentioning that only because they remain in an organization they do not have to work for their own support and that of their families, they continue to study and do not engage in risk behaviors like substance abuse, sexual activity, and delinquent behavior.

Finally, Mexican sheltered, unsheltered or emergency sheltered girls are a social subgroup characterized by:

- Coming from either Mexico City (n<40%) or from different states of the Mexican Republic (n>50%).
- Lacking a stable housing, and having wandered through unsuitable housing spaces such as the street.
- Having an average age of ($M[SD]$=15.62[2.137]) years with an age range of 11–21 years, and a grade level of middle school. Many of them are illiterate, have severe problems with reading and writing, and their academic achievement and performance are low. In their family they do not receive any kind of support for their studies, and they are not self-sufficient because if they work it is for their own survival, hence they do not obtain the means necessary to continue studying. Also, they frequently do not have the papers required to enroll in the educational system such as birth certificates. Furthermore, only if they remain in an organization they are able to continue their schooling. Nonetheless, since it is a very transitory group that constantly moves from one place to another and from one institution to another, they often fail to complete their formal education.
- Participating in the formal economy system by having jobs with low salaries such as assistant cook and saleswoman, and in the informal economy system doing jobs like street vendor, housemaid, and begging. They can carry out criminal acts such as stealing, selling drugs, and prostitution, thus, exposing themselves to risks that diminish their bio-psycho-social well-being. While remaining in an organization, they do not work because they are provided with vocational training in skills such as tailoring, gastronomy, hotel industry, and stylist, among others, in order to contribute to their life project.
- Consuming drugs, alcohol, and tobacco. The most frequently used drugs are inhalants, for their low cost and easy access, and marijuana. Substances are obtained within their peer group and/or the street. In fact, one of the reasons for abandoning an organization is to use drugs since it is forbidden due to institutional regulations. Nevertheless, they can introduce these substances in the organization.
- Having a deplorable physical and mental health status because they usually present oral and visual health problems; diverse skin, respiratory, and gastrointestinal diseases; malnutrition; anxiety; depression; personality disorders; and, less frequently, psychosis; and sexually transmitted diseases, the most common being human papilloma virus, gonorrhea, vaginosis, genital herpes, and, occasionally, AIDS. The drugs that are typically prescribed for their treatment of mental disorders are sertraline, carbamazepine, clonazepam, valproic acid, haloperidol, and risperidone. Their hygiene habits and self-care behaviors are deficient, and they have little knowledge regarding how to take care of their body and health. They have suicidal thoughts, make suicide attempts, and have self-mutilation behaviors. If they are not in an organization, they rarely attend health centers to receive medical and psychological assistance, and they do not have sufficient resources to afford such care, such as the required papers (*e.g.*, birth certificate, immunization record, address, social insurance). Moreover, they suffer discrimination for their status and are denied these services, which is why only when they remain in an organization they have access to health care centers and receive proper care. In the absence of treatment, their health status deteriorates, thus shortening their life.

http://dx.doi.org/10.5772/68000

- Having an active sexual life that could have originated as a consequence of having suffered sexual abuse within the family. They usually have multiple sexual partners both formal and informal, have unprotected sex, and have sexual relationships under the influence of alcohol and/or drugs. Their relationships can be heterosexual, homosexual, or bisexual. As a result of their sexual behavior, they frequently get pregnant and abort. If they have children, they are abandoned or given for adoption. If they decide to keep them, their child-rearing practices are ambivalent being both violent and loving and caring. They can have children with several different partners, who stay for a while with them and then end the relationship or abandon them. They practically have very less information on contraception, sexually transmitted infections, how to take care of themselves, and their anatomy and physiology. Since they lack a stable housing, a well-paid job, and money, they commonly do not receive proper care when pregnant, when having an abortion or a sexually transmitted disease, thus leading to premature death.

- Engage in risk behaviors such as drug use and abuse, self-mutilation, suicide ideation and attempts, and delinquent activities (*e.g.*, theft, drug selling, and prostitution). One of the reasons they abandon the organization in which they remain is to continue these risk behaviors.

- Having several heterosexual couple relationships, and, to a lesser extent, homosexual and/or bisexual. Their couples provide security, support, affection, and protection, but they can also become a source of violence, substance abuse, delinquent activity, and they can influence them to abandon the organization where they reside. Furthermore, they can become dependent on their partner, finding it extremely difficult to end the relationship even if it is unhealthy and quite harmful.

- Establishing stable and long-standing friendships of both genders with their peer group. Their closer social networks typically comprise three to five persons of their age or one or two years older. Their peer group provides affection, support, security, and protection. In fact, it becomes their new family that they can abandon the organization just to follow their friends, and their means to satisfy their basic and emotional needs. Nonetheless, their peer group can induce them to risk behaviors and can turn into a source of violence.

- Developing in dysfunctional environments characterized by risk behaviors, physical, sexual, and psychological abuse, emotional and physical abandonment, educational, physical, and psychological neglect, lack of resources necessary to survive, drug use and abuse, extended and reconstituted families with lousy jobs or underemployment, where they are forced to work, to collaborate with domestic chores, and to take care of their siblings, and where they lack educational, recreational, and employment opportunities.

- Living in communities immersed in social problems such as poverty, overcrowded settlements, drug abuse, delinquency, insecurity, violence, gang activity, and lack of access to basic services (*e.g.*, water, paving, electricity, sewage, garbage collection, public transportation, etc.), among others.

- Entering a non-governmental or governmental organization due to domestic violence and/or abandonment to satisfy their basic needs, and to obtain better life conditions. Their permanence in an institution is temporal and brief of no more than about four years. Still, they form close ties with their companions in the organization that often endure over the years. They have problems with the institution's staff because they do not agree with the employees' attitudes and behaviors (*e.g.*, "There is preferential treatment", "Some are granted internet access and they do not have consequences if they do not do their chores while others, like me, are overloaded with housework and scolded for everything", "They do not let me explain myself, and they do not believe me", "I get yelled for everything"), with the institutional regulations, norms and rules, and, lastly, not all of the organization's activities please them and they are bothered that they are mandatory. Moreover, among their companions, there are several problems of coexistence, and physical and psychological violence. Under these circumstances, it is usual for the girl to abandon the organization for a transitory lifestyle without the possibility to have an independent and autonomous life which leads to social adaptation.

To summarize, the main differences among girls at risk of homelessness and those sheltered, unsheltered, or emergency sheltered reside in that the former preserve their family, school and community ties, the majority of them are not sexually active, do not use drugs but with increasing age drink alcohol and smoke, stay for a long period of time in an institution for homeless people, their romantic relationships are unstable and informal, have not been pregnant or had abortions, the institutional staff and their housemates constitute their new family, do not work, continue studying, are victims of sexual abuse, and physical and psychological violence to a lesser degree, and their social networks vary with age so that with an increase in age, their size and quality diminish, and are conformed mainly of the same gender, although they usually relate to boys, however the bonds established with other girls are closer and deeper. The latter have little or no contact whatsoever with their family and community of origin, they form school ties only if they attend school, they work, the majority do not continue studying unless they remain in an institution for homeless people, suffer at a higher rate of sexual, physical, and psychological abuse, use and abuse drugs and, less frequently, alcohol and tobacco, have an active sexual life, have been pregnant and had abortions, their romantic relationships are enduring and more formal, they can get involved in homosexual and/or bisexual relationships, engage in risk behaviors including delinquent activities, their peer group forms their new family, stay for a brief period of time at shelters, organizations or homeless facilities, and their social networks include risk groups and individuals of both genders.

As it can be noted, homeless girls grow up in dysfunctional family environments, among communities immersed in social problems without effective governmental and societal solutions, in educational systems contrary to their needs and interests, and in precarious economic areas [9], with null or few opportunities of obtaining a better quality of life. Also, they suffer from sexual and labor exploitation, physical, sexual, and psychological abuse, and discrimination at a higher rate than the rest of the population [10], and they represent the fastest growing population [11], because currently, no country is without the presence of homeless women [12]. As a result of living under such extreme vulnerable conditions, they have low self-concept and self-esteem, they lack self-efficacy and coping skills. They present anxiety and depression more frequently than men [13, 14]; they are impulsive, aggressive,

http://dx.doi.org/10.5772/68000

emotionally unstable, and attempt suicide three to four times more than men [15]. They have difficulties to establish and maintain healthy relationships and to be socially competent, and they exhibit emotional problems [16]. This deficiency can extend to adulthood [17]. However, they rarely receive proper care since the majority does not have access to health care centers or are even denied attention [18]. Subsequently, they have poor psychological adjustment [19], which affects their ability to functionally adapt and cope with environmental stressors [20], thus diminishing their well-being and quality of life [21]. Consequently, homeless girls grow up in environments that favor the occurrence of physical and socio-emotional damages and that make it difficult to live with an acceptable minimum of well-being and security, including the dissatisfaction of their essential needs, which entails a series of repercussions that negatively affect their well-being, specifically their health. Therefore, the phenomenon increases leading to an exponential growth of social problems like delinquency, substance use and abuse, unwanted pregnancies, homeless families, violence, insecurity, poverty, low educational level and unemployment, and diseases including sexually transmitted infections, among others. All of these, in turn, augment the socioeconomic, educational, labor, and cultural backwardness of a country, which results in indirect and direct costs to society in terms of resources and efforts [22]. Although it is an overwhelming reality, previous studies with similar groups have focused mainly on socioeconomic and cultural matters, as well as risk behaviors, dismissing the relevance of psychological aspects and mental health [23]. So, particularly in Mexico, information about the psychological functioning and its associated factors in homeless girls are partially and superficially known. These data are crucial to help reduce their vulnerability, and to provide them with skills that induce changes in the short, medium and long term [24], because a person's ability to be productive, proactive, and prosocial is a function of his or her health.

1.1. Well-being and its associated factors

Mental health is not just the absence of mental disorders. It is defined as a state of well-being by which the individual is aware of his or her own abilities, can cope with the normal stressors of life, work productively and fruitfully, and is able to make a contribution to his or her community. It includes subjective well-being, autonomy, competence, intergenerational dependence, and recognition of the ability to perform intellectually and emotionally [25]. For that reason, mental health is the foundation of individual well-being and effective functioning of the community. Thus, it constitutes the degree of psychological adjustment determined by, among other elements, the level of perceived subjective well-being. A higher subjective well-being leads to a better mental adjustment and a greater satisfaction with current living conditions [26].

Subjective well-being is the degree in which an individual generally judges the quality of his or her life as favorable and feels satisfied with it [27]. It has a cognitive and an emotional dimension. The cognitive dimension alludes to the judgment that is made about the satisfaction that one has with life. On the other hand, the emotional aspect refers to experiencing positive and negative emotions. A high level of well-being entails favorably assessing personal satisfaction and having positive feelings more frequently than negative emotions [28].

Consequently, a positive perception of subjective well-being associates with a series of benefits in various areas that go from mental health and longevity to labor performance and satisfaction, income, and the establishment and maintenance of healthy interpersonal relationships [29]. Likewise, having a high level of subjective well-being not only benefits people, but also society in general. Those who feel satisfied with their lives are more altruistic, engage in prosocial activities, participate in charity events and for community development, and are more tolerant of the government. As a result, promotion and preservation of the subjective well-being of human beings can contribute to the formation of a more stable, productive, and functional society [27]. In contrast, individuals dissatisfied with their current living conditions may present aggressive behavior, anxiety, depression, suicidal thoughts and ideation, sexual risk behavior, substance use and abuse including alcohol, eating disorders, and health problems. Furthermore, they tend to be more physically and psychologically victimized, have difficulty adapting to their surroundings, and have low academic performance and achievement. Due to this, their mental health status deteriorates [30]. So, subjective well-being is a factor affected by various components: sociodemographic issues [31] like socioeconomic status, schooling, unemployment, civil status, and psychosocial characteristics [32] such as anxiety, depression, and social and emotion regulation skills, among others. Learning and acquiring social skills such as assertiveness and an adequate regulation of emotions positively affect the psychological well-being of an individual; while the presence of mental disorders such as depression and anxiety affect it negatively. That is, as previous research has referred [33], anxiety and depression decrease the index of subjective well-being, while assertiveness and functional emotion regulation strategies increase it. For this reason, the Top-down Theory of Subjective Well-being [34] was used as the theoretical framework to guide this research since it explains well-being from a series of intrapersonal factors such as various psychosocial characteristics like the ones mentioned above. This theory states that personality, temperament, social comparison, goal-orientation, and social adaptation affect subjective well-being. That is, intrapersonal factors that include cognitive and emotional characteristics determine to a greater extent the degree of subjective well-being. Internalizing (*e.g.*, anxiety, depression, etc.) and externalizing (*e.g.*, aggression, impulsiveness, deficit attention, etc.) disorders, and psychosocial characteristics (*e.g.*, self-concept, self-esteem, self-efficacy, emotion regulation, assertiveness, coping skills, problem-solving abilities, etc.) form part of these cognitive and emotional aspects linked together as the intrapersonal factors associated to such well-being.

Anxiety and depression are among the most prevalent disorders worldwide [35]. These disorders occur two times more often in women than in men [36]. They are the first cause of disability in the female population [37], and they usually present together; this comorbidity is more frequent in girls [38], which leads to a greater bio-psychosocial deterioration [39]. Likewise, they inhibit and interfere with the acquisition of assertive behaviors since the ability to be assertive is not inherited and it is not something innate or immovable; it is learned through practice [40]. Consequently, it depends on personal, social, cultural, and emotional characteristics. For this reason, cognitive, emotional, and behavioral aspects of anxiety and depression associate with negative thoughts and disruptive actions concerning social aptitudes, thus increasing aggressive or passive behavior, and the inability to adequately defend one's right and to express personal thoughts and feelings appropriately; which is why life

http://dx.doi.org/10.5772/68000

satisfaction decreases [41]. As such, the lack of assertiveness makes the person feel socially incompetent and unable to adapt to a particular group and cope with stressful circumstances. Similarly, anxiety and depression are associated with low levels of emotion regulation [42], specifically with the use of dysfunctional emotion regulation strategies. These strategies, in turn, augment the possibility of exhibiting anxiety and depression [43]. Emotional regulation includes any strategy aimed at maintaining, increasing, or suppressing an ongoing affective state. As such, it refers to active attempts to influence anxiety and depression and constructively express emotions at the right time and place so that the person's resources are protected without self-confidence diminishing [44], which is why emotion dysregulation increases anxiety and depression.

Unassertiveness, emotion regulation deficiencies, anxiety, and depression, are factors that involve various negative consequences if not treated early and effectively, which is concerning, since these, as mentioned, can persist into adulthood. Some of them are [45]:

a. low self-concept and self-esteem, and decreased perceived self-efficacy,

b. lack of problem-solving and coping skills,

c. social incompetence and absence of significant affiliative relationships,

d. risk behaviors and affiliation to high risk groups,

e. behavioral, anxiety and emotional disorders,

f. isolation, withdrawal, rejection and social exclusion, antisocial behavior, physical and psychological victimization; and

g. low academic performance and school achievement, school dropout and unemployment.

The deterioration of the quality of life and interpersonal functioning, along with the need for assistance and health care, represents a burden to the individual and to society because of the costs generated directly and indirectly [46]. These consequences are devastating for people and their communities, since it leads to further worsening of their subjective well-being and thus, of their physical and mental health. This results eventually in an early death.

To sum up, subjective well-being is important for health preservation and maintenance, and for having a better quality of life. Its absence negatively affects the mental health of individuals, leading to mental disorders [47] and the loss of physical and psychological integrity. Nonetheless, information surrounding issues of subjective well-being among this particularly diverse population is scarce [48], and little is known about the health and well-being of people who live on the streets although their lifestyle involves health risks [49]. Likewise, worldwide there are few studies addressing these specific factors in vulnerable people such as homeless girls. In fact as far as it has been reviewed, in Mexico, information concerning the well-being and its associated factors in homeless girls is only partially and superficially known. What is more, homeless girls have been scarcely considered as an object of study—a common circumstance in societies with a predominantly male orientation that discriminate and exclude women [50]. Consequently, the phenomenon grows day by day surpassing societal response. Therefore, to remedy the lack of work and to encourage the design and implementation of effective policy actions based on the data collected, this research was carried out to develop and test a path model of subjective well-being in homeless girls taking into account the following intrapersonal factors: anxiety, depression, assertiveness, and emotion regulation. This would provide information essential to enhance the well-being of this collective and promote their physical and mental health, thus, contributing to diminish their vulnerable conditions.

Lastly, it should be noted that worldwide research on subjective well-being in homeless people has not taken into account anxiety, depression, emotion regulation, and assertiveness as its predictors. The extant research has focused on the effect of other variables such as: social support and expectations about the future in homeless young adults ages 18-23 years old in Texas, the United States [48]; the repercussions of weather in homeless people in Australia [51]; stigmatization, sexual involvement, and school enrollment in homeless children and orphans in Nigeria [52]; the cognitive development of homeless children and orphans in Cambodia, Ethiopia, India, Kenya, and Tanzania [53]; and sexual abuse, loneliness, and connectedness in homeless youth ages 16-23 years old in Texas, the United States [54]. Only a study carried out in Scotland in the year 2002 [55] has examined well-being specifically in the female homeless population of 18 years and older, although from a qualitative perspective. However, the research focused on how these women built their social identity, self-concept, and psychological well-being in relation to their experience of living in homeless facilities. This once again highlights the existence of little information about subjective well-being and its associated factors. In addition, the present work is one of the first studies to assess some psychosocial characteristics (anxiety, depression, emotion regulation, and assertiveness) as predictors of such well-being in homeless girls.

2. Research data

2.1. Method

2.1.1. Participants

A total of 240 homeless girls between the ages of 6–23 years ($M[SD]$=11.13[3.47]) were chosen with an intentional sampling method from various non-governmental organizations of Mexico City. Of these, 77% were born and lived in Mexico City, while 20% came from other states of the Mexican Republic. As many as 85% did not use or abuse any kind of substance, including alcohol and tobacco, 70% were in elementary school and 23% in middle school, 84% professed the catholic religion, 44% lived with their mothers during holidays, 16% remained in the organization all the time, and 65% entered an institution due to lack of resources, while 28% because of family abuse. Likewise, the girls had stayed in an institution for an average period of 42 months ($M[SD]$=42.30[37.346]). Lastly, at least 50% had suffered some form of abuse either within their families, group of peers, or home community (**Figure 1**).

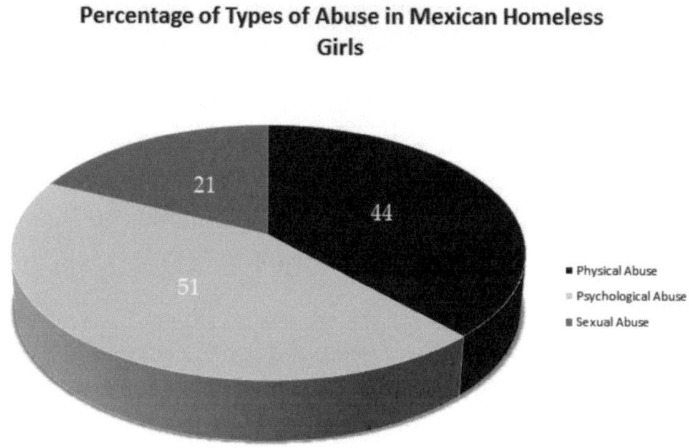

Figure 1. Percentage of the 240 homeless girls who have been victims of physical, psychological, or sexual abuse.

For being included in the current research, participants were required to reside in a non-governmental organization (*e.g.*, institutions, shelters, centers) which exclusively attends homeless children and youth. They were excluded in case they would not collaborate in the study and/or abandon their institution. Finally, it is important to mention that participation was voluntary with the acceptance of the legal guardians of the girls, when they were available, and/or the organization.

2.1.2. Measurement

Since commercial and standardized instruments are seldom valid for these vulnerable populations [56], all scales had to be developed during this study following the culturally relevant psychometric validation process proposed by [57] and, specifically for Mexican culture, by [58]. Items of each instrument are rated on a 6-point Likert scale from One (Never) to Six (Always) that also include equivalent percentage intervals (One: 0% of the time—Six: More than 80% of the time).

The self-report questionnaires measure anxiety symptoms, depression symptoms, assertiveness, emotion regulation, and subjective well-being:

- *Anxiety:* Self-report questionnaire that assesses anxiety symptoms with eight items grouped in two factors that together explain 56% of variance and both have a 0.59 Cronbach's Alpha.
- *Depression:* Self-report questionnaire that assesses depression symptoms with eight items grouped in two factors that together explain 50% of variance and both have a 0.75 Cronbach's Alpha.
- *Assertiveness*: Self-report questionnaire that assesses assertive behaviors with nine items grouped in two factors that together explain 52% of variance and both have a 0.64 Cronbach's Alpha.
- *Emotion regulation*: Self-report questionnaire that assesses emotion regulation strategies with seven items grouped in two factors that both explain 50% of variance and both have

a 0.60 Cronbach's Alpha. It also has an extra question to explore the type of strategies individuals use to manage and control their emotions when they experience anxiety and/or depression symptoms.

- *Subjective well-being:* Self-report questionnaire that assesses subjective well-being with 32 items grouped in five factors that altogether explain 61% of variance and have a Cronbach's Alpha ranging between 0.85 and 0.90. Furthermore, it evaluates the intensity and frequency of 10 emotions (five positive and five negative) using a 10-point scale from 0 (Never/None) to 10 (Very Often/A lot).

It is important to mention that these scales developed for the current study have not yet been used in other research with homeless people due to the fact that in Mexico it is a line of research in development with a long way for consolidation. Also, it is a group seldom taken into account for study.

2.1.3. Procedure

Taking into account that this social group is constantly moving from place to place and is difficult to approach [59], non-governmental organizations that assist homeless children and youth were approached in order to acquire information more rapidly and have easier access to this population. Work with this collective was possible thanks to the permissions granted by the institutions. Also, voluntary participation of the girls from the organizations to collaborate in the research was requested. The general objectives of the study were explained and it was emphasized that all data obtained would be kept confidential and used for study purposes only. Likewise, the doubts the girls had were clarified while paying attention not to bias their responses. There was no time limit for the applications, which were held in classrooms previously assigned by the organization's staff. Subsequently, the girls were thanked for their collaboration but as a means to guarantee their volunteer role, they did not receive any material reward for their participation. The questionnaires were applied individually in the form of interviews with an average time of 15–30 min approximately. The realization of this research required only the organization's approval and no individual informed consent, having previously noted that all the ethical procedures and guidelines specified in the Psychologist Code of Ethics [60] would be followed throughout the whole study. As participation was voluntary, the girls provided their written consent signing with their fingerprint. Their refusal to answer a question or all of them was respected at all times. Lastly, statistical analyses were conducted using SPSS version 22. These included descriptive and correlation analyses employing Pearson correlation. To test the path model, AMOS version 22 was also used.

2.2. Findings and conclusions

Among the main problems of homeless girls are anxiety and depression, since 82% presented anxiety symptoms and 47% depressive symptoms. Other features were the lack of assertiveness with 22% being unassertive, experiencing negative emotions more than 80% of the time and with great intensity (*i.e.*, on a scale from One [never] to Ten [always], great intensity corresponds to at least eight), and employing dysfunctional emotion regulation strategies to manage and control emotions with 67% using strategies like self-mutilating behaviors, displaying hostile and aggressive behaviors toward others (*e.g.*, revenge, retaliation, etc.), using and abusing alcohol and drugs, being impulsive, and isolating themselves socially.

Moreover, it was found (**Table 1**) that anxiety and depression symptoms are positively associated, so that when one of them increases, so does the other and vice versa. Depression symptoms decrease the possibility of behaving assertively and of being satisfied with current living conditions. Additionally, they inhibit the use of functional emotion regulation strategies and augment those that are dysfunctional, which in turn exacerbates anxiety and depression symptoms. Meanwhile, assertive behaviors decrease anxiety and depression symptoms, increase the level of subjective well-being and the probability of using functional emotion regulation strategies that enhance perceived life satisfaction and social skills, and lessen anxiety and depression indexes.

Also, according to the path model analysis carried out (**Figure 2**), the main predictors of subjective well-being for this group of homeless girls are depression symptoms and functional emotion regulation strategies, since they directly and significantly affect the level of satisfaction. As such, depression symptoms decrease the level of subjective well-being while functional emotion regulation strategies increase it. Meanwhile, anxiety symptoms and dysfunctional emotion regulation strategies, through their influence on depressive symptoms, negatively affect well-being, thus reducing it. Assertiveness, by diminishing depressive symptoms, augments the index of well-being. Moreover, functional emotion regulation strategies inhibit depression symptoms, hence increasing the level of subjective well-being. Also, assertiveness decreases anxiety and depression symptoms, thus leading to a greater perceived life satisfaction. Furthermore, functional emotion regulation strategies lessen anxiety and depression indexes and enhance the level of assertiveness. In contrast, dysfunctional emotion regulation

	Dysfunctional ER strategies	Anxiety symptoms	Depression symptoms	Assertiveness	Subjective well-being
Functional ER strategies	**.344****	**–.192****	**–.294****	**.357****	**.438****
Dysfunctional ER strategies	–	**.317****	**.252****	.006	–.020
Anxiety symptoms	–	–	**.541****	**–.311****	**–.269****
Depression symptoms	–	–	–	**–.401****	**–.544****
Assertiveness	–	–	–	–	**.307****
Subjective well-being	–	–	–	–	–

Note. ER, emotion regulation.
**$p \leq .05$.

Table 1. Correlation analysis of variables associated with subjective well-being in Mexican homeless girls (N=240).

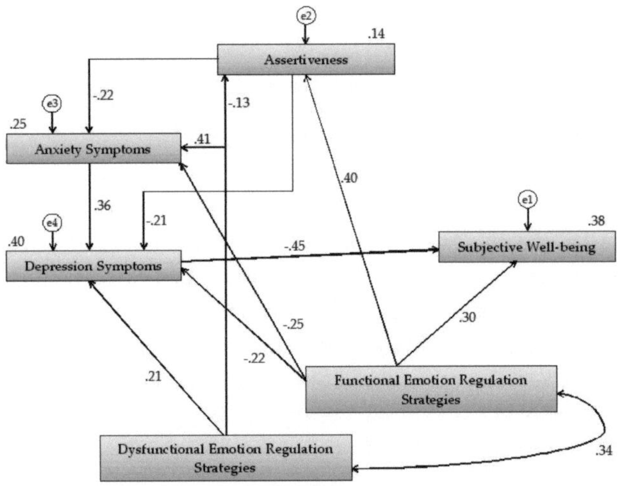

Figure 2. Final version of the Path Model Analysis of Subjective Well-Being in homeless girls. Standardized regression weights are shown. All regression weights were significant ($p \leq .05$) including the correlation between emotion regulation strategies.

strategies augment anxiety and depression symptoms and reduce social skills. As expected, both emotion regulation strategies are significantly correlated to each other.

To summarize, depression symptoms are a significant predictor of subjective well-being, and through these, anxiety symptoms, assertiveness and emotion regulation strategies affect either positively or negatively the index of well-being. Therefore, depression symptoms constitute a mediating variable of the effects that other variables have upon subjective well-being. Additionally, according to results obtained in this study, emotion regulation strategies are the most relevant factors of the model because not only do they affect all the other variables including depression, but also subjective well-being by either increasing it or reducing it.

Finally, fit indexes obtained, $x^2(3)=1.116$, $p=.773$; NFI=.997, RFI=.985, IFI=1.005, TLI=1.026, CFI=1.000, $p=.000$; RMSEA(90% IC)=.000(.000–.073), $p=.889$, reveal that the model reliably predicts subjective well-being in a group of homeless girls regarding various psychosocial characteristics such as assertiveness, emotion regulation strategies, anxiety and depression symptoms. Thus, this path model provides an excellent fit to the data collected and the studies reviewed. Consequently, the model can be used in further research to explain subjective well-being in similar population.

As it can be seen, findings revealed that one of the key problems afflicting homeless girls is the presence of anxiety and depression symptoms. This coincides with previous studies in other countries [49, 61, 62] that show that such groups suffer from these because of growing up among vulnerable conditions that constitute one of the predominant reasons for abandoning their homes and entering an organization. Also, it was observed that they tend to be unassertive because they exhibit greater difficulty in defending their rights and interests; they cannot express their ideas and feelings without being aggressive, and are hostile and unable to establish healthy relationships. With respect to emotion regulation skills, the strategies they use to manage and control their emotions are mainly dysfunctional since they exacerbate the unpleasant emotional experience, do not solve the problem, and do not cope effectively with stressful or adverse circumstances.

Furthermore, results indicated that higher rates of anxiety correlate with higher levels of depression and vice versa. This concurs with previous research done in other countries, which states that anxiety and depression are comorbid disorders [63], for they share common cognitive, emotional, behavioral, and physiological symptoms. Additionally, anxiety and depression symptoms are associated with lower levels of assertiveness and emotion regulation. These disorders inhibit and interfere with the acquisition of assertive behaviors leading to poor functioning and social adjustment [64], since they make the person feel socially incompetent and incapable of adapting to a specific group. Assertiveness, on the other hand, allows people to effectively cope with stressors by increasing the ability to cope with stress. This diminishes emotional and behavioral problems [65]. A lack of adequate emotion regulation skills can lead to depression [66]; while appropriate strategies reduce anxiety and depression indexes [67] and augment assertive behaviors. As for the level of subjective well-being, it decreases when presenting anxiety and depression symptoms [68], and increases with the use of effective emotion regulation strategies [69] and with the acquisition of assertive skills. This is because individuals with adequate social skills assess their daily experiences as less stressful and functionally adapt to society unlike those who lack such abilities [70].

Lastly, concerning the path model tested, it was found that depression symptoms and functional emotion regulation strategies are direct predictors for subjective well-being, while the rest of the factors, anxiety symptoms, assertiveness and dysfunctional emotion regulation strategies, act as indirect predictors. This is consistent with the Top-down Theory of Subjective Well-being [34] that states that cognitive and emotional characteristics of individuals determine well-being, where depression [71] is a primordial cognitive aspect.

Anxiety and depression reduce the index of well-being. This agrees with recent research that states that such symptoms are significantly associated with a low level of subjective well-being. Anxiety symptoms decrease well-being through their effect on depression by augmenting it, which in turn increases its effects. These results are consistent with evidence that demonstrates that anxiety precedes depression [72], which can be explained by the fact that anxiety can substantially increase stress and interfere with daily functioning, that in turn leads to depression [73]. On the other hand, appropriate management and control of emotions is positively correlated with life satisfaction [74] because it modifies unpleasant emotions and its associated negative thoughts that lead to disruptive behaviors. Hence, functional emotion regulation strategies enhance the index of subjective well-being, which enables the individual to move from a dysfunctional cycle to a more functional one in which he or she feels satisfied with him- or herself and with his or her surroundings. Additionally, when emotion management and control is functional, it reduces anxiety and depression symptoms. In contrast, when emotion regulation is dysfunctional, it increases such symptoms. These outcomes are consistent with previous work that shows that functional emotion regulation strategies positively correlate with resilience and negatively with depression and anxiety. These disorders have a high and positive correlation with dysfunctional strategies [75]. When negative emotions related to anxiety and depression are handled effectively, it is easier to modify and control associated dysfunctional cognitions and attitudes. Therefore, the person is able to effectively cope with anxiety- and depression-inducing events. For that reason, functional emotion regulation strategies constitute a protective factor against presenting anxiety and depression [76]. As a result, it is expected that such emotion regulation strategies increase subjective well-being directly and indirectly by diminishing depression symptoms, as shown in this study. To finish, functional and dysfunctional emotion regulation strategies affect the index of assertiveness. Dysfunctional emotion regulation strategies are associated to a greater social incompetence and maladjustment [77], and thus, to a lack of assertiveness. On the contrary, functional emotion regulation strategies augment the probability of being assertive since it is a prerequisite for the effective use of social skills [78]. This enhances the level of subjective well-being by diminishing anxiety and depression symptoms in a way that it indirectly affects the satisfaction of life. In other words, assertiveness is a protective factor against the development of disorders like anxiety and depression because it allows a person to cope with stressors effectively. Consequently, self-trust and logical expression of emotions and ideas increase, anxiety lessens, social relationships improve, respect for others' rights is fostered, and the ability to cope with stress gradually augments [79]. However, the fact that assertiveness did not directly predict subjective well-being may be because subjective well-being is more associated to the quality and quantity of affiliative relationships of an individual [80] than with assertive behaviors. That is, social support is significantly correlated with perceived subjective well-being [81]. For example, in Japan [82] it was found that lack of perceived emotional and instrumental social support is associated with mental well-being.

To recapitulate, these results provide preliminary evidence of factors that significantly predict subjective well-being in Mexican homeless girls by increasing or decreasing it. To the best of my knowledge this is one of the first models of subjective well-being with various psychosocial characteristics as its predictors in this collective. Therefore, the clinical usefulness of the present study resides in designing and implementing interventions that take into account such factors in a way that the possibility of reaching developmental and emotional milestones increases, and homeless girls learn healthy social and emotion regulation skills and how to cope with problems when they arise, especially in the context of Latin American cities. This would lead to a greater well-being during childhood and youth [83], which will positively influence their quality of life [84].

In conclusion, subjective well-being is an essential factor for health and longevity [29]. Vulnerable populations as homeless girls lack subjective well-being which deteriorates their mental health. The promotion of mental health is achieved through actions that create environments and living conditions that enable individuals to adopt and maintain healthy lifestyles. Further research has to replicate the current results in order to obtain more information concerning well-being and its associated factors to have a more comprehensive understanding of the phenomenon and to promote mental health. Also, more studies are required to see how these variables behave in homeless girls from various countries and different cultures. In this way, it is possible to identify needs, characteristics, and dysfunctional and functional behaviors related to well-being that are present or absent in homeless girls. This data leads to a better design of sustainable and efficacious interventions that induce long-term changes and increase the probability that this collective adequately adapts to society with opportunities of obtaining an improved quality of life.

http://dx.doi.org/10.5772/68000

Finally, homelessness is a complex multifactor phenomenon product of economic, political, and social circumstances interrelated. These circumstances aggravate as a result of the lack of government's response and effective lines of actions. Among homeless people, children, particularly girls are the most vulnerable groups. Homeless girls are a severely neglected and marginalized group with an impaired well-being, physical and mental health deficiencies, and restricted opportunities of obtaining a better quality of life. All of this is in spite of the fact that childhood and adolescent development is a crucial factor for the consolidation of intellectual, physical, and social aptitudes that define a healthy adulthood; and that during these periods the foundations of health in adulthood, are established. In order to offer practical solutions from a psychosocial perspective to a phenomenon that leads to the deterioration of a nation and to a lack of individual and social well-being, it is vital to continue carrying out studies in which vulnerable people such as this collective are approached. It is the only way to obtain valid information through which it is possible to design effective lines of action, for early intervention leads to an exponential growth in health status in adult life, which in addition to providing incalculable well-being, is an investment in physical and mental health. On the other hand, it is a means for avoiding future pathologies, minimizing possible sequelae, and reducing risk factors.

Author details

Susana Castaños-Cervantes

Address all correspondence to: susycc84@gmail.com

Iberoamerican University, Mexico City, Mexico

References

[1] World Health Organization. Vulnerable groups [Internet]. n.d. Available from: http://www.who.int/environmental_health_emergencies/vulnerable_groups/en [Accessed: 2016-06-14].

[2] Human and Social Development. Children. Mexico: State Development Plan of Chihuahua 2004-2010; 2010.

[3] Institute for the Attention and Prevention of Addictions in Mexico City. Vulnerable groups [Internet]. n.d. Available from: http://www.iapa.df.gob.mx/work/sites/iapad/resources/PDFContent/50/poblaciones_vulnerables.pdf [Accessed: 2016-04-21].

[4] Rice E, Milburn N, Rotheram-Borus M, Mallet S, Rosenthal D. The effects of peer group network properties on drug use among homeless youth. The American Behavioral Scientist. 2005; 48(8):1102-1123.

[5] Scanlon T, Tomkins A, Lynch M, Scanlon F. Street children in Latin America. British Medical Journal. 1998; 316(7144):1596 1600.

[6] Oakley P, Salazar M. Children and violence: the case of Latin America. Colombia: Save the Children Fund (United Kingdom) and Third World Editors; 1993.

[7] Richards R, Smith C. Environmental, parental and personal influences on food choice, access and overweight status among homeless children. Social Science & Medicine. 2007; 65(8):1572-1583.

[8] Castaños S. Brief cognitive-behavioral group therapy for homeless women [thesis]. Mexico City: National Autonomous University of Mexico; 2014.

[9] Guerrero P, Palma E. Representaciones sociales sobre educación de niños y niñas de calle de Santiago y Quito. Revista Latinoamericana de Ciencias Sociales, Niñez y Juventud. 2010; 8(2):1025-1038.

[10] Misganaw A, Worku Y. Assessment of sexual violence among street females in Bahir-Dar town, North West Ethiopia: a mixed method study. Public Health. 2013; 13(825):1-8. DOI: 10.1186/1471-2458-13-825

[11] Committee Opinion. Health care for homeless women. Obstetrics & Gynecology. 2013; 122(4):936-940.

[12] Jabeen Z, Azra. A comparative study on mental health of street children living with their families and runaways from families. Indian Streams Research Journal. 2013; 2(12):1-6.

[13] Raffaellia M, Koller S, Reppold C, Kuschick M, Krum F, Bandeira D, et al. Gender differences in Brazilian street youth's family circumstances and experiences on the street. Child Abuse & Neglect. 2000; 24(11):1431-1441.

[14] Tevendale H, Lightfoot M, Slocum S. Individual and environmental protective factors for risky sexual behavior among homeless youth: an exploration of gender. AIDS and Behavior. 2009. 13:154-164.

[15] Capuzzi D, Gross D. Youth at risk: a prevention resource for counselors, teachers and parents. USA: Pearson-Merrill Prentice Hall; 2008.

[16] Ganesh A, Campbell D, Hurley J, Patten S. High positive psychiatric screening rates in urban homeless population. The Canadian Journal of Psychiatry. 2013; 58(6):353-360.

[17] United Nations International Children's Emergency Fund. Progress for childhood. A balance on childhood protection. Number 8; 2009.

[18] Merscham C, van Leeuwen J, Mcguire M. Mental health and substance abuse indicators among homeless youth in Denver, Colorado. Child Welfare. 2009; 88(2):93-110.

[19] Votta E, Farrell S. Predictors of psychological adjustment among homeless and housed female youth. Journal of the Canadian Academy of Child and Adolescent Psychiatry. 2009; 18(2):126-132.

[20] Elegbeleye A. Predictors of the mental health of orphans and vulnerable children in Nigeria. Ife PsycholgIA. 2013; 21(2):170-180.

[21] Oppong K, Meyer-Weitz A, Petersen I. Correlates of psychological functioning of homeless youth in Accra, Ghana: a cross-sectional study. International Journal of Mental Health Systems. 2015; 9(1):http://www.ijmhs.com/content/9/1/1.

[22] Xie R, Sen B, Foster M. Vulnerable youth and transitions to adulthood. New Directions for Adult and Continuing Education. 2014; 143:29-34. DOI: 10.1002/ace.20102

[23] Saddicha S, Linden I, Krausz M. Physical and mental health issues among homeless youth in British Columbia, Canada. Are they different from older homeless adults? Journal of the Canadian Academy of Child and Adolescent Psychiatry. 2014; 23(3):200-206.

[24] Tagurum Y, Chirdan O, Bello D, Afolaranmi T, Hassan Z, Iyaji A, et al. Situational analysis of orphans and vulnerable children in urban and rural communities of Plateau State. Annals of African Medicine. 2015; 14(1):18-24. DOI: 10.4103/1596-3519.148714

[25] World Health Organization. What is mental health? Questions and answers online [Internet]. 2007. Available from: http://www.who.int/features/qa/62/es/index.html [Accessed 2014-03-15].

[26] Friedman H, Kern M, Reynolds C. Personality and health, subjective well-being and longevity. Journal of Personality. 2010; 78(1):179-215.

[27] Diener E, Ryan K. (2009). Subjective well-being: a general overview. South African Journal of Psychology. 2009; 39(4):391-406.

[28] Eid M, Larsen R. The science of subjective well-being. USA: The Guildford Press; 2008.

[29] Diener E, Chan M. Happy people live longer: subjective well-being contributes to health and longevity. Applied Psychology: Health and Well-Being. 2011; 3(1):1-43.

[30] Escribà-Agüir V, Ruiz-Pérez I, Montero-Piñar M, Vives-Cases C, Plazaola-Castaño J, Martín-Baena D. Partner violence and psychological well-being: buffer or indirect effect of social support. Psychosomatic Medicine. 2010; 72:383-389.

[31] Sarracino F. Determinants of subjective well-being in high and low income countries: do happiness equations differ across countries? Journal of Socio-Economics. 2013; 42:51-66.

[32] Malo S, Navarro D, Casas F. El uso de los medios audiovisuales en la adolescencia y su relación con el bienestar subjetivo: análisis cualitativo desde la perspectiva intergeneracional y de género. Athenea Digital. 2012; 12(3):27-49.

[33] Derdikman-Eiron R, Indredavik M, Bratberg G, Taraldsen G, Bakken I, Colton M. 2011. Personality and social sciences gender differences in subjective well-being, self-esteem and psychosocial functioning in adolescents with symptoms of anxiety and depression: findings from the Nord-Trondelag health study. Scandinavian Journal of Psychology. 2011; 52:261-267. DOI: 10.1111/j.1467-9450.2010.00859.x

[34] Diener E. Subjective well-being. Psychological Bulletin. 1984; 95(1-2):542-575.

[35] Abela J, Hankin B. Handbook of depression in children and adolescents. USA: The Guildford Press; 2008.

[36] Rafful C, Medina-Mora M, Borges G, Benjet C, Orozco R. (2012). Depression, gender, and the treatment gap in Mexico. Journal of Affective Disorders. 2012; 138(3):165-169.

[37] Noble R. Depression in women. Metabolism Clinical and Experimental. 2005; 54(1):49-52.

[38] Frank L, Matza L, Revicki D, Chung J. 2005. Depression and health-related quality of life for low-income African-American women in the U.S. Quality of Life Research. 2005; 14:2293-2301. DOI:10.1007/s11136-005-6541-1

[39] Appleton P. Children's Anxiety: A Contextual Approach. USA: Routledge; 2008.

[40] Flores M, Xool M, Molina Y. Self-concept and self-esteem and their relation with assertiveness in children. In: AMEPSO (Mexican Social Psychology Association), editors. Social Psychology in Mexico. Mexico: AMEPSO; 2010. p. 1065-1072.

[41] Rivera S, Flores M, Euán T, Castañeda P. Attachment and assertiveness during childhood. In: AMEPSO (Mexican Social Psychology Association), editors. Social Psychology in Mexico. Mexico: AMEPSO; 2010. p. 309-315.

[42] Mohapatra S, Agarwal V, Sitholey P, Arya A. 2014. A clinical study of anxiety disorders in children and adolescents from North Indian children and adolescents clinic. Asian Journal of Psychiatry. 2014; 8:84-88.

[43] Pe M, Raes F, Kuppens P. The cognitive building blocks of emotion regulation: ability to update working memory moderates the efficacy of rumination and reappraisal on emotion. PLos One. 2013; 8(7):e69071. DOI:10.1371/journal.pone.0069071

[44] Silva J. (2005). Emotion regulation and psychopathology: the model of vulnerability-resilience. Revista Chilena de Neuro-Psiquiatría. 2005; 43(3):201-209.

[45] Crawford M, Manassis K. Anxiety, social skills, friendship quality, and peer victimization: an integrated model. Journal of Anxiety Disorders. 2011; 25:924-931.

[46] Slesnick N, Guo X, Brakenhoff B, Feng X. Two-years predictors of runaway and homeless episodes following shelter services among substance abusing adolescents. Journal of Adolescence. 2013; 36:787-795.

[47] Leimkühler A, Keller J, Paulus N. Subjective well-being and male depression in male adolescents. Journal of Affective Disorders. 2007; 98:65-72.

[48] Barczyk A, Thompson S, Rew L. The impact of psychosocial factors on subjective well-being among homeless young adults. Health & Social Work. 2014; 39(3):172-180. DOI: 10.1093/hsw/hlu020.

[49] Seager J, Tamasane T. Health and well-being of the homeless in South African cities and towns. Development Southern Africa. 2010; 27(1):63-83. DOI: 10.1080/03768350903519358.

[50] Venumadhava G, H.Y P. 2013. Girl child abuse. Indian Streams Research Journal. 2013; 3(2):2-7. DOI: 10.9780/2230-7850/322013/2196

[51] Pendrey C, Carey M, Stanley J. Impacts of extreme weather on the health and well-being of people who are homeless. Australian Journal of Primary Health. 2014; 20:2-3. http://dx.doi.org/10.1071/PY13136.

[52] Adejuwon G, Oki S. Emotional well-being of orphans and vulnerable children in Ogun state orphanages Nigeria: predictors and implications for policy. Ife PsychologIA, 2011; 19(1): 1-18. DOI: 10.4314/IFEP.V19I1.64571.

[53] Escueta M, Whetten K, Ostermann J, O'Donnell K, The Positive Outcomes for Orphans Research Team. Adverse childhood experiences, psychosocial well-being and cognitive development among orphans and abandoned children in five low income countries. International Health and Human Rights, 2014; 14(6): 1-13. DOI: 10.1186/1472-698X-14-6.

[54] Rew L. Relationships of sexual abuse, connectedness, and loneliness to perceived well-being in homeless youth. JSPN. 2002; 7(2):51-63.

[55] Stephen D. Young women construct themselves: social identity, self-concept and psychosocial well-being in homeless facilities. Journal of Youth Studies. 2000; 3(4):445-460. DOI: 10.1080/13676260020006740.

[56] Hesse M, Thiesen H. The use of the ADHD Self-Rating Scale (ASRS-6) in the homeless psychometric properties, alcohol use and self-nurse agreement. Journal of Addictions Nursing. 2013; 24(2):108-115.

[57] Geisinger K. Cross-cultural normative assessment: translation and adaptation issues influencing the normative interpretation of assessment instruments. Psychological Assessment. 1994; 6(4):304-312.

[58] Reyes I, García L. 2008. Psychometric culturally relevant validation procedure: an example. In: AMEPSO (Mexican Social Psychology Association), editors. Social Psychology in Mexico. Mexico: AMEPSO; 2008. p. 625-630.

[59] Coker T, Elliot M, Kanouse D, Grunbaum J, Gilliland M, Tortolero S. et al. 2009. Prevalence, characteristics and associated health and health care of family homelessness among fifth-grade students. American Journal of Public Health. 2009; 99(8):1446-1452. DOI: 10.2105/AJPH.2008.147785

[60] Mexican Society of Psychology, Civil Association. Psychologist code of ethics. Mexico: Trillas; 2004.

[61] Strehlau V, Torchalla I, Kathy L, Schuetz C, Krausz M. 2012. Mental health, concurrent disorders and health care utilization in homeless women. Journal of Psychiatric Practice. 2012; 18(5):349-360.

[62] Sarajlija M, Jugovi A, Zivaljevi D, Merdovi B, Sarajlija A. Assessment of health status and quality of life of homeless persons in Belgrade, Serbia. Vojnosanit Pregl. 2014; 71(2):167-174. DOI: 10.2298/VSP1402167S.

[63] Shamsuddin K, Fadzil F, Ismail W, Shah S, Omar K, Muhammad N, et al. 2013. Correlates of depression, anxiety and stress among Malaysian university students. Asian Journal of Psychiatry. 2013; 6:318-323.

[64] Settipani C, Kendall P. Social functioning in youth with anxiety disorders: association with anxiety severity and outcomes from cognitive behavioral therapy. Child Psychiatry and Human Development. 2013; 44:1-18. DOI: 10.1007/s10578-012-0307-0

[65] Poulou M. 2014. How are trait emotional intelligence and social skills related to emotional and behavioural difficulties in adolescents? Educational Psychology. 2014; 34(3):354-366. DOI: 10.1080/01443410.2013.785062

[66] Berking M, Wirtz C, Svaldi J, Hofmann S. 2014. Emotion regulation predicts symptoms of depression over five years. Behavior Research and Therapy. 2014; 57C(1):13-20. DOI: 10.1016/j.brat.2014.03.003

[67] Gross J. Handbook of emotion regulation. USA: The Guildford Press; 2007.

[68] Sola J, López R, Padilla D, Sánchez M. 2013. Anxiety, psychological well-being and self-esteem in Spanish families with blind children: A change in psychological adjustment? Research in Developmental Disabilities. 2013; 34:1186-1890.

[69] Ho M, Cheung F, You J, Kam C, Zhang X, Kliewer W. The moderating role of emotional stability in the relationship between exposure to violence and anxiety and depression. Personality and Individual Differences. 2013; 55:634-639.

[70] Ozben, S. 2013. Social skills, life satisfaction, and loneliness in Turkish university students. Social Behavior and Personality. 2013; 41(2):203-214.

[71] Galinha I, Pais-Ribeiro J. Cognitive, affective and contextual predictors of subjective wellbeing. International Journal of Wellbeing. 2011; 2(1):34-53. DOI:10.5502/ijw.v2i1.3

[72] Frank-Briggs A, Alikor A. Anxiety disorder amongst secondary school children in an urban city in Nigeria. International Journal of Biomedical Science. 2010; 6(3):246-251.

[73] Kendall P. Child and adolescent therapy: cognitive-behavioral procedures. USA: The Guildford Press; 2012; 143-234.

[74] Gutiérrez M, Goncalves T. Assets for development, school adjustment and subjective well-being for adolescents. International Journal of Psychology and Psychological Therapy. 2013; 13(3):339-355.

[75] Min J, Yu J, Lee C, Chae J. Cognitive emotion regulation strategies contributing to resilience in patients with depression and/or anxiety disorders. Comprehensive Psychiatry. 2013; 54:1190-1197. DOI: 10.1016/j.comppsych.2013.05.008

[76] Vanderhasselt M, Koster E, Onraedt T, Bruyneel L, Goubert L, De Raedt R. 2014. Adaptive cognitive emotion regulation moderates the relationship between dysfunctional attitudes and depressive symptoms during a stressful life period: a prospective study. Journal of Behavior Therapy and Experimental Psychiatry. 2014; 45(2):291-296. DOI: 10.1016/j.jbtep.2014.01.003

[77] Mihalca A, Tarnavska Y. 2013. Cognitive emotion regulation strategies and social functioning in adolescents. Procedia-Social and Behavioral Sciences. 2013; 82:574-579.

[78] Fehlinger T, Stumpenhorst M, Stenzel N, Rief W. 2013. Emotion regulation is the essential skill for improving depressive symptoms. Journal of Affective Disorders. 2013; 144:116-122. DOI: 10.1016/j.jad.2012.06.015

[79] Vatankhah H, Daryabari D, Ghadami V, Naderifar N. The effectiveness of communication skills training on self-concept, self-esteem and assertiveness of female students in guidance school in Rasht. Social and Behavioral Sciences. 2013; 84:885-889. DOI: 10.1016/j.sbspro.2013.06.667

[80] Carlsson F, Lampi E, Li W, Martinsson P. Subjective well-being among preadolescents and their parents—Evidence of intergenerational transmission of well-being from urban China. Journal of Socio-Economics. 2014; 48:11-18.

[81] Li B, Ma H, Guo Y, Xu F, Yu F, Zhou Z. Positive psychological capital: a new approach to social support and subjective well-being. Social Behavior and Personality. 2014; 42(1): 135-144.

[82] Ito K, Morikawa S, Okamura T, Shimokado K, Awata S. Factors associated with mental well-being of homeless people in Japan. Psychiatry and Clinical Neurosciences. 2014; 68:145-153. DOI:10.1111/pcn.12108.

[83] Centers for Disease Control and Prevention. Children's mental health [Internet]. 2015. Available from: http://www.cdc.gov/ncbddd/spanish/childdevelopment/mentalhealth.html [Accessed: 2016/07/24]

[84] Palomba R. Quality of life: concepts and measures. In: CELADE/Population Division, CEPAL. Quality of life and support networks of seniors. Workshop. July 2002. Santiago, Chile: 2002.